Microscale Manipulations in Chemistry

CHEMICAL ANALYSIS

A SERIES OF MONOGRAPHS ON
ANALYTICAL CHEMISTRY AND ITS APPLICATIONS

Editors

P. J. ELVING · J. D. WINEFORDNER

Editor Emeritus: I. M. KOLTHOFF

VOLUME 44

A WILEY-INTERSCIENCE PUBLICATION

JOHN WILEY & SONS
New York/London/Sydney/Toronto

Microscale Manipulations in Chemistry

T. S. MA
Professor of Chemistry
City University of New York

V. HORAK
Professor of Chemistry
Georgetown University

A WILEY-INTERSCIENCE PUBLICATION

JOHN WILEY & SONS

New York / London / Sydney / Toronto

Library of Congress Cataloging in Publication Data:

Ma, Tsu Sheng, 1911–
  Microscale manipulations in chemistry.

  (Chemical analysis; v. 44)
  "A Wiley-Interscience publication."
  Includes bibliographical references and index.
  1. Microchemistry. I. Horák, Václav, chemist, joint author. II. Title. III. Series.
  QD79.M5M3   544'.83   75–20093
  ISBN 0–471–55799–4

Printed in the United States of America

10 9 8 7 6 5 4 3 2 1

PREFACE

Based on my experience in teaching, consulting, and research in the field of microchemistry for four decades, I have strived to prepare a monograph for general use. In this endeavor, it was my good fortune to have received the cooperation of Professor V. Horak who had taught microchemical techniques at Charles University, Prague, before moving to Georgetown University, Washington, D.C.

Microchemistry, I submit, is concerned with the principles and methods of using the minimum quantity of working material to obtain the desired chemical information. Hence microchemistry encompasses many areas of chemical experimentation; it is an attitude to perform laboratory work. The microchemical approach can be applied whenever an experimental datum is required, be it for the composition of matter (chemical analysis), preparation of a compound (chemical synthesis), or study of reaction mechanisms. Anyone with some knowledge of microchemistry realizes that the microchemical approach as a rule saves time and chemicals. For this reason microchemistry also can be utilized advantageously in chemical education where the main purpose is to demonstrate chemical principles.

Although microchemistry is generally practiced by analytical chemists, it has gradually penetrated into other areas of chemical investigations. For instance, 25 years ago at New York University I initiated a course on microtechniques of organic chemistry that included micromethods for the synthesis, separation, and purification of organic compounds. I presented a report at the International Microchemical Meeting in Vienna in 1955, and I proposed to use the milligram as the unit in describing experimental procedures. It is now accepted. Perusal of the current research papers, such as those on synthetic organic chemistry and physical organic chemistry, will attest to the fact that a large number of experiments have been carried out below the gram level. This trend is expected to continue as new instruments and techniques render it unnecessary to employ gram amounts of the working material to get the chemical information.

This book deals with the general methods of microchemical manipulation. It is intended to serve as a textbook and also as a comprehensive sourcebook. The first two chapters elucidate the background. Subsequent chapters are organized from the standpoint of the working material (solid mixture, solution, or impure liquid), and subjects are treated systematically

according to the purpose of the relevant experiment. Because of the nature
of this book, numerous illustrations are included. Some sections entail
description and discussion that might appear to be obvious. It should be
noted, however, that neglect of the obvious often is the cause of failure in
microscale manipulations.

I should like to express my appreciation to the many journals, publishers, and companies for their generous permission to reproduce the illustrations in this book. Special thanks are due to my fellow microchemists who
supplied me with original photographs of their apparatus that had not
been published elsewhere. Credit is due to my daughter Juliana Mei-Mei
for typing a large portion of the manuscript.

T. S. MA

New York, New York
May, 1975

CONTENTS

CHAPTER 3   FRACTIONATION OF SOLID MATERIALS   76

Microscale Manipulations in Chemistry

INTRODUCTION

I. GENERAL REMARKS

While microchemical techniques have been employed by scientific workers for over one hundred years, it was at the beginning of this century that Emich [1] made a systematic study of this subject and published the *Lehrbuch der Mikrochemie*, which dealt with methods of chemical experimentation using milligram (mg = 10^{-3} g) quantities of material. At about the same time, Pregl [2] developed methods to determine carbon, hydrogen and other elements in organic compounds with 3- to 5-mg samples, this being a nearly 100-fold reduction of the then prevailing required quantity, for which he was awarded the Nobel Prize in 1923. Subsequently, another approach to microchemistry was pursued by Feigl [3] who discovered extremely sensitive chemical reactions by which microgram (μg = 10^{-6} g) to nanogram (ng = 10^{-9} g) of a substance in very dilute solution can be detected. The works of these pioneers have been immortalized in a museum (Palais de la Découverte) in Paris [4].

In recent decades, there has been a gradual change of attitude among chemists toward the suitable sample size for research and analysis. As the abovementioned micromethods became generally accepted, the quantity of working material shifted from the decigram to the milligram scale. The rapid advance in chromatography and the analytical application of spectroscopy, radiochemistry, and enzymatic methods have pushed the sample size to the nanogram and picogram (pg = 10^{-12} g) regions. It is worthy of note also that the feasibility of using micro quantities of working material to perform chemical analysis render possible studies on pollution and the environment, which are expected to be of high priority in the 1970s.

Whereas the development of sensitive tests and refined instruments leads to micro methods for qualitative and quantitative analysis using extremely small amounts of working materials, it is incumbent upon the chemist to make the sample amenable to these methods. For instance, a high-resolution mass spectrometer coupled with a suitable computer is capable of pinpointing mass species from 1 μg of an organic compound [5], but it is very difficult to obtain a sample that is not contaminated with many times its weight of stopcock grease or the like. In the case of quantitative organic elemental analysis, when decigram samples are taken, a little surface moisture usually does not affect the determination. On the other hand, if the analysis is performed with 20- to 100-μg sample [6, 7], even the breath of the operator may significantly vitiate the results.

This monograph is concerned with the ways and means by which small amounts of material can be handled. The guiding principles to obtain a representative sample for microanalysis will be discussed. The techniques to prepare the analytical sample, prior to the detection or determination step, will be described.

II. SIZE OF THE MICRO SAMPLE FOR ANALYTICAL PURPOSES

A. SAMPLE SIZE GOVERNED BY THE CONTAINER, MEASURING DEVICE, AND METHOD

Microchemistry may be defined as the branch of chemical science which deals with the principles and methods for using the minimum quantity of working material to obtain the desired chemical information [8]. If the answer wanted is an analytical datum, either qualitative or quantitative, the size of the micro sample is controlled by the container (e.g., flask, capillary), the measuring device (e.g., balance, buret, spectrometer), and the method available. The capacity of the container is dependent on the operation. Thus Thornett [9] has described an apparatus for liquid-liquid extraction with a solvent volume of 1 ml. It consists of a capillary U tube connecting two vessels in which liquids may be mixed and separated by the operation of a plunger, and is particularly suitable for use with solutions that are dangerous to handle. Franklin and Keyzer [10] have described a steam-distillation device by means of which the essential oil content of a single leaf can be determined. Larrabee [11] has described a planchet-to-planchet still to distil a 0.1-ml sample and collect the distillate in 0.01 mole of a suitable reagent. Neuhoff [12] has designed an apparatus for centrifuging 10 μl to 1 ml of liquid up to 20,000 rev/min, while Bowers and Haschemeyer [13] have described an ultrafiltration cell to concentrate 0.25-ml samples. Natelson [14] has constructed an instrument for the automated handling, dilution, and dispensing of 25-μl samples of biological fluids. When the analysis requires reaction at high temperatures, microgram amounts of the sample can be heated in sealed capillaries [15, 16]. Gas holders for analyzing 2 ml down to 5 μl of gases have been reported [17–19], and a device capable of delivering 10^{12} to 10^{19} atoms of gas with about 1% error has been described [20]. Emission spectroscopy has been performed with 50-μg samples placed in 0.2 g of graphite pellet [21], whereas the apparatus for thermal analysis requires from 1 to 20 mg of material [22, 23]. For infrared spectrometry, Klein and Ulbert [24] have described a pelleting technique to hold 1 to 4 μg of sample.

Microbalances which handle samples in the microgram range are available from several manufacturers [25–27] or can be constructed according to specifications described in the literature [28, 29]. So far as manipulation is concerned, the problem is not so much with the precision of the measurement as with the technique for transporting extremely small amounts of material free from contamination or loss in transit. Measurement of liquid volume in the microliter region is accomplished by means of the syringe [30], pipet [31], or buret [32]. Pasternak [33] has reported that with good equipment, 5 μl of liquid can be measured with a precision of 2%. The limitations of other measurements vary according to the operations. Siggaard-Anderson and Oliver [34] have described a 130-μl spectrophotometric cell of 15-mm path, while Kuzmin and co-workers [35] use a 0.02-μl flowthrough cell. Alimarin and Petrikova [36] have measured the pH of a solution in the range 1.68 to 9.20 within ±0.04 units, using a cell on the microscope stage. Evans [37] has measured the current produced by the electrolysis of 10^{-9} equivalent of solution placed on the surface of the electrode.

The sensitivity of the analytical method sets a limit on the sample size. Probably the most sensitive methods are based on radioactivity, which can analyze samples in the picogram (10^{-12} g) region [38, 39]. Next come the electrochemical methods which deal with nanogram (10^{-9} g) amounts of sample [40]. Atomic absorption and atomic fluorescence spectrometry can perform analysis in the 10^{-9}-g region [41, 42], while arc-emission spectrometry requires decimilligram (10^{-4} g) samples, and other spectrochemical methods usually employ microgram quantities of material [43]. From microgram to milligram amounts of sample are used in the "wet chemical" methods based no gravimetry and titrimetry [8, 44]. The technique for determining the titrimetric endpoint sometimes controls the sample size. Thus colorimetric titration uses a microliter volume of 0.1 N solution [45], while cryoscopic titration uses 0.2 ml of 0.01 N solution [46]. In general, the analysis of inorganic materials [47] can be accomplished with samples much smaller than those for organic analysis, especially those for the structural elucidation of organic compounds [48, 49, 49a].

Chromatographic separations usually accompany analytical detection or determination, and the sample size is governed by the instrument as well as the detection technique. Lacourt [50] has shown that the limiting quantities for paper chromatography are between 0.01 and 20 μg; Kirk and co-workers [51] have reported the same range for electrophoresis. Tschesche and Frank [52] have determined 0.01 μmol of amino acids by ion-exchange chromatography. The advantage of microscale manipulation is often shown in the analysis of mixtures using chromatographic processes. It can easily be demonstrated on a thin-layer silica gel plate that 10 μg of azo dyes separate nicely, while larger samples tend to overlap [53]. In the investigation of natural rubber, Alishoev and co-workers [54] have reported that the pyrolysis of microgram samples in a gas chromatograph gives higher relative yields of volatile products than with milligram samples. The application of radiochemistry greatly enhances the sensitivity of gas-chromatographic detection. Thus it has been shown that the ^{63}Ni electron-capture technique is 20 times more sensitive to fluoroesters than is the flame ionization detector, and allows 10 ng of steroids to be determined [55] through their fluoro derivatives.

B. SAMPLE SIZE GOVERNED BY THE NATURE OF THE PROBLEM

In the application of microscale manipulation, the sample size is frequently dependent on the nature of the analytical problem. Occasionally the whole sample is in the milligram region, as in the case of a hair sample or bloodstain in forensic analysis [56]. Sometimes the material to be analyzed is large, but only negligible amounts can be taken for analysis: Investigations of the colors, used in old paintings and of the ink in oracle bones [57] are typical examples.

Many biochemical research projects necessitate techniques to work with small amounts of organic compounds. Recently, for the study of a simple tripeptide hormone, Guillamin, Folkers, and co-workers reported that 250,000 hypothalami of sheep brain were used to recover 1 mg of the peptide [58]. For the analysis of the blood of an infant [59], or the cerebral fluid of an animal, it is incumbent to employ a sample size of 0.1 ml or less. By contrast, it is not necessary to reduce urine analysis below the 0.5-ml scale, since a sufficient amount is usually available [60].

Varieties of analytical problems that require microscale manipulation techniques are concerned with trace constituents in the environment or in natural products, or with the study of trace by-products in chemical synthesis, especially in the pharmaceutical industry. This type of problem may be considered as dealing with "micro yields," since the objective is to isolate the analytical sample from an immensely large amount of foreign matter. Some examples are given in the following section.

III. TREATMENT OF "MICRO YIELDS"

In the investigation of the constituents of plants, it usually occurs that a component which is present in very small quantity merits study because of a special property such as pharmacological activity. The aim of the analyst is then to employ the minimum amount of the plant material to obtain the desired information in the most economical manner. For instance, Bournias [61] has described a method to determine pigments, sugars, and amino acids using a few milligrams of fresh leaves by means of thin-layer chromatography. For screening alkaloids, it has been demonstrated that paper chromatography requires 9 hr, whereas thin-layer chromatography requires about 2 hr [62]. Gas chromatography is extensively used in the analysis of plant aromas [63] and flavors [64]. Numerous chemical agents which serve as antagonistic agents [65] or attractants [66, 67] are produced in minute amounts by animals and plants, and their isolation and elucidation became possible only through the application of microscale manipulation. This is also true for the analysis of trace constituents of seawater [68] and terestial objects [69].

In many cases of organic synthesis such as those based on the Grignard [70] and Friedal-Crafts [71] reactions, there are minor side products. Investigation of the minor products present in the reaction mixture relies on microscale operations. Recently Fox and Windsor [72] have demonstrated the synthesis of seven α-amino acids in yields of 0.002–0.007% by heating formaldehyde with ammonia. Mislow and co-workers [72a] synthesized trimesitylmethane from bromomesitylene with a yield of 1.1%. Certain photochemical reactions form products in micro yields [73–75]. Trace amounts of elements are obtained in radiochemical reactions [76, 77], and the experimental work is invariably performed in the microgram region [78, 79]. Preparation of compounds labeled with radioactive isotopes such as ^{14}C and ^{35}S are usually carried out on the microscale.

Biochemical studies dealing with the fate of metabolites generally involve the analysis of micro amounts of isolated materials [80, 81]. This is particularly true in research on drug metabolism, since the drug determination in the nanogram region are not unusual.

Recent attention to pollution and environmental problems emphasizes the need for microscale manipulation related to trace analysis. In these problems, while the test material (e.g., air, water, soil) may be plentiful, the compounds to be identified or determined are in micro concentrations [84–86]. Tracking such compounds, separating and isolating the micro yields, and evaluating the validity of analytical results are important considerations. Thus DeQuasie and Grey [87] have proposed using stable isotope of carbon and sulfur for air-pollution studies. Lisk [88] has discussed the problems in the analysis of pesticide residues; Bergstroem-Nielsen [89] has published a mathematical model to show the relation between recovery and the actual amount of residue present. Rasmussen and Went [90] have described a procedure to detect nanogram amounts of organic matter in 5 cm^3 of air sample.

IV. CONSIDERATIONS IN HANDLING THE MATERIAL FOR ANALYSIS

A. PREVENTION OF SAMPLE CHANGES

In any analytical problem, it is apparent that the sample to be processed should not undergo any chemical change before the final step of detection or determination. This is of special importance when the amount of working material is in the micro region, because (a) the environment has proportionally very large influence on the material and (b) the change may easily be overlooked. For instance, a hygroscopic material stored in glass-stoppered 500-ml bottle can be analyzed without precaution by quickly weighing out a 0.5-g sample, whereas 1 mg of this material might be spoiled by being exposed to the breath of the analyst. Since the objective of chemical analysis is to assay the material as originally received, one should always keep in mind the changes that may occur while the material is waiting for, or during, treatment. Alteration of the whole material, or certain components thereof, due to the effect of heating during grinding or extraction, of light on storage, and of the atmospheric oxygen, carbon dioxide, ammonia, moisture, acid vapors, alkalinity of glass, or the like may have considerable significance on the accuracy and reliability of the analytical results. Every effort should be made to prevent or minimize such changes. Thus Maravalhas [91] has designed an apparatus for the steam distillation of essential oils in which the condensing oil does not come into contact with hot vapor during distillation, so that the possibility of decomposition is avoided. Long and O'Brien [92] have described a jointless apparatus for the preparation of saturated solutions from air-sensitive solvents and solutes. For the purpose of keeping the sample under the same atmospheric condition as the original material, Cross [93] has proposed a humidity control system in which the mixing ratio of wet and dry streams of air can be adjusted to give the required humidity in a thermostatically controlled chamber. In order to prevent change in blood samples, Laessig [94] has reported the use of Tergitol-NPX and sodium fluoride to stabilize the material for alcohol determination, while Michod and co-workers [95] have recommended treatment with heparin of polythene microtubes to retard the rate of coagulation.

B. PREVENTION OF SAMPLE LOSSES ON PRETREATMENT

Loss of the micro sample or some of its constituents may occur during any of the series of operations in the analytical procedure, such as filtration and concentration. In chromatography, a part of the analytical material may remain at the origin and not move. Hence serious mistakes may result if these factors are not taken into consideration. The filtration losses during analytical extractions have been studied by Marsh and Schill [96], who reported that filter paper gave the highest losses and glass-fiber paper [97] the lowest. Wilson and co-workers [98] have observed that during filtration of a raw sugar solution after clarification with basic lead acetate, the sucrose concentration has a positive bias in the initial portion of the filtrate owing to preferential absorption of water by the filter paper. To minimize the effect on subsequent polarimetry, the first 10 ml of filtrate is discarded. As an alternative to filtration, centrifugation may be used; the latter procedure, being free from preferential-absorption and evaporation effects, gives the better estimate of polarization values.

At very low concentrations the adsorption of materials on container walls may be significant. Differences due to this effect have been found when tubes made of glass, polyethylene, and nitrocellulose, respectively, were used for centrifugation. Shephard [98a] has studied the adsorption of moisture by some materials used in instrument construction.

The adsorption of various volatile compounds on silica gel, activated charcoal or a mixture of molecular sieves and their recovery by release into the head-space vapor or by solvent extraction have been investigated by Palamand and co-workers [99]. Silica gel and molecular sieves adsorb alcohols, aldehydes, ketones, esters, and terpenes more strongly than does activated charcoal. These compounds are not released by heating for 10 min at 110°C, but are released by heating after injection of 0.5 ml of water.

For the extraction of lipids, Mitchell [100] has shown that the lower limit to the amount of material required in the conventional method was set by the loss accompanying the transfer to and from the extraction thimble, and he has designed a microextractor that eliminates this transfer and gives accurate results with 5 to 50 mg of material containing 0.3 to 5 mg of lipid. For the determination of thiamine, Gupta and co-workers [101] have studied the effect of pH and the concentration of bromothymol blue on the extraction of the thiamine-dye complex by chloroform and found that the optimum pH falls in a narrow range 5.2 to 6.6. For the estimation of drugs in biological tissue, Feldman and Gibaldi [102] have observed that the aqueous extraction of phenolsulphonphthalein from homogenates of rat intestinal tissue is nearly quantitative, whereas significant losses occur after incubation of the material.

In quantitative organic elemental analysis, low results sometimes are caused by the premature volatilization of the sample in the combustion tube. A simple technique to prevent such losses is to place a piece of solid carbon dioxide above the section of the combustion tube containing the sample [103]. A thermoelectric cooling device to be clipped onto the combustion tube has also been designed for this purpose [104].

Losses of analytical sample in analysis for pesticide residues deserve special attention because of their low concentrations in the working material. Chiba and Morley [105] have observed that losses of DDT, γ-BHC, and dieldrin during filtration, partitioning, and washing are relatively small but cumulative. The most significant loss arises during evaporation to dryness. A micro column and collection tube has been recommended for the concentration of 10 ml of solution to a volume of 0.1 to 0.3 ml in a steam bath [106]. On the other hand, Moats and Kotula [107] have found that the elution rates for single-step chromatographic cleanup of pesticide residues from Florisil columns and from carbon-Celite columns can be increased to 250 and 100 ml per minute, respectively, without adversely affecting recoveries of pesticides.

C. SAMPLING

Obtaining representative samples is of utmost importance in microanalysis. The reader is referred to an excellent discussion of this subject by Benedetti-Pichler [108]. One of his personal experiences in the preparation of correct analytical samples is cited here for illustration [109]. He was called upon to referee a case of violation in which the Food and Drug Administration investigator charged that certain mercurial tablets contained insufficient amount of the drug, while the manufacturer insisted that quantities larger than the upper limit had been used in the preparation. Upon investigation, Benedetti-Pichler found that the lower value was obtained when the contents of the gelatin capsule were rinsed out, while the value claimed by the manufacturer was indeed correct when the whole capsule was dissolved. The discrepancy was now easily explained by the diffusion of the mercury compound into the gelatin. Since whole tablets were administered when the drug was prescribed, the sampling method specified by Food and Drug Administration was in error. The procedures for detection and determination of the sorption of drugs by plastics have been reviewed [109a].

For the preparation of analytical sample from bulk material, Rowland [110] has described a simple sample-divider constructed from a phonograph turntable and a plastic dish of six equal-sized and -shaped segments. The dish is placed on the turntable and allowed to rotate at 20 rev/min while the powder is fed by gravity into the rotating segments via a funnel firmly clamped above the turntable. It is apparent that this device does not retain all volatile components in the material. Gooden [111] has designed a turntable for the automatic sampling of granulated pesticides. A sieve receiver pan is fixed on a turntable, and a metal cup is placed within the pan. A funnel is so adjusted above the pan that its outlet just clears the cup, and the funnel and cup are set off center with respect to the pan. The turntable is rotated at 45 rev/min, and the stock sample is passed through the funnel; the portion for testing falls within the cup, and the remainder falls in the pan. The position of the cup within the pan determines the ratio of the weight of test sample to stock sample. Care should be taken when adapting these devices to microscale manipulation so that representative samples are obtained.

Extraction is a general method for obtaining samples of trace components. It should be noted, however, that samples from different extraction procedures may vary considerably. In the evaluation of extraction and cleanup methods for the analysis of DDT in plant materials, Ware and Dee [112] have observed that (a) hexane-ethanol (2:1) is the best solvent mixture for extraction, (b) mixing the material and allowing it to stand in the solvent for 16 hr is the most thorough extraction method, (c) elution of the compound from a Florisil column with 20% of dichloromethane in light petroleum increases the total residue recovery by 18% over that obtained by elution with ethyl ether–pentane (3:17), and (d) exhaustive Soxhlet extraction with methanol–chloroform after the initial extraction yields an additional 5% of residue. Recently, it has been reported that the use of chloroform–methanol (1:1) in a high-speed mixer provides the most efficient technique for extracting DDT and related pesticides [113].

It should be noted that techniques vary for sampling diverse materials— for example, ores and concentrates; metals and alloys; inorganic, organic, and metallo-organic substances; gases, liquids, solids, and slurries; raw materials, intermediates, and finished products; pure synthetic compounds; and so on. An incorrect sampling method may invalidate analytical results. Grant [113a] has studied the sampling and preparation errors in trace analysis.

D. PREVENTION OF CONTAMINATION

After satisfactorily representative samples are obtained, the next serious problem is the prevention of contamination during microscale manipulation and analysis. There are three culprits responsible for contamination, namely, containers and reaction vessels, chemical reagents, and the atmosphere. Disposable containers such as sample-weighing pans, platinum crucibles, and pycnometers have been recommended [114]; they are particularly helpful for routine repetitive analysis, since reused containers are apt to retain similar material from prior tests.

In order to minimize contamination due to reagents, chemicals employed for microchemical experiments should be purified and then stored in separate small vials or ampul [115] instead of in large bottles. Blank tests should always be carried out; otherwise the analytical results may be unreliable. Spitz [116] has reported that erroneous results in the evaluation of purity or identification of organic compounds after column or thin-layer chromatography may be caused by impurities in silica gel. Fritze and Gietz [117] have shown that reagent contamination leading to the formation of metal-protein complexes presents problems in the trace analysis of metals in blood. Employing an ion-exchange technique for urine analysis, Tompsett [118] has found that the eluates are not entirely suitable for spectrophotometric examination, because of interfering material derived from the resin. Plasticizers are easily extracted from plastic tubings and containers.

The preponderant influence of the environment on micro samples was discussed in a previous section (Section IV.A). It is important therefore, to prevent contamination due to the atmosphere and surrounding. Since oxygen and water vapor are the common contaminants, many devices have been proposed to eliminate these substances in the environment. Thus Eubanks and Abbott [119] have described a gas purification and pressure control system for inert atmosphere boxes. Water is removed by means of molecular sieves, and oxygen by means of copper powder on an inert carrier. This system permits not only in-line determinations of water and oxygen, but also regeneration of purifying agents without interrupting the operation of the box. It has been reported that the concentration of water and oxygen can be decreased to less than 25 and 50 ppm, respectively, after seven recirculations.

If the problem of contamination has not been effectively solved, the analytical results on micro samples are understandably open to question. The recent debate on "polywater" is a typical example. In this case microgram amounts of material were collected in fine capillaries for the measurement of physical constants and subjected to spectroscopic analysis. Since the purity of the sample could not be verified, conflicting data were reported. One preparation was reported to be a solution of sodium lactate, the primary constituent of human sweat, indicating that the sample was a product of biological contamination [120]. Another sample was shown by electron spectroscopy to contain sodium, potassium, sulfate, chloride, nitrate, borate, silicate, and carbon-oxygen compounds as well as traces of other elements, apparently derived from the reaction vessel and capillary container [121]. It is interesting to note that numerous theoretical calculations have been published dealing with the molecular structure of "polywater" without first ascertaining the validity of the experimental results [122]. The importance of microscale manipulation in order to get a reliable analytical sample cannot be overemphasized.

V. CONSIDERATIONS IN HANDLING MICRO APPARATUS

A. CLEANLINESS AND ORDERLINESS

Cleanliness and orderliness are two essential requirements in microscale manipulations. As mentioned by Benedetti-Pichler [123], not only all apparatus containing the material under investigation, but also everything that gets into contact with it, including the hands, must be clean. Bench top and shelves should be free from dust at all times. This prevents the micro apparatus and hands from being contaminated.

The need of orderliness in keeping micro apparatus and in arranging experiments is obvious. Dirty apparatus should not be permitted to accumulate; it should be cleaned immediately. The success of cleaning is also better assured when it is still known what the vessel contained, so that the proper solvent can be chosen for the removal of the substances adhering to the walls. Utensils which have been attacked by the reagents should be put aside.

B. CLEANING OF REUSABLE GLASSWARE

The recent trend toward using disposable glassware with guaranteed cleanness is commendable. It saves time and eliminates the error resulting from contaminated glassware. However, if the disposable glassware must be cleaned for the particular experiment, the main advantage of disposable glassware is no longer realized.

The cleaning of reusable glassware is one of the important aspects in successful microscale experimentation. This factor should be considered when the apparatus is selected for the experiment. Thus, vessel shapes that may cause the accumulation of materials and dirt in relatively inaccessible parts should be avoided, particularly when the washing is to be done in a machine. An apparatus which can be disassembled into several parts [124] is preferable to a complicated one-piece device. Ground-glass joints with rough surface can be a source of trouble due to their tendency to adsorb certain materials. Hence the "clear seal joints" [125] (precision-made smooth surface joints, used without lubricant) are advantageous; this is also true for fritted discs. In the case of experiments which require a high degree of purity of all components, a blank run to test the quality of the glassware should be carried out.

Different methods are employed in cleaning glassware, with respect to both reagents and processes. The following cleaning fluids are most common: water, dilute or concentrated acids (e.g., nitric acid, sulfuric acid–dichromate), dilute ammonia solution, sodium or potassium hydroxide solution, organic solvents, detergents. The techniques include soaking, washing with a brush, washing in an automatic laboratory dishwasher, and washing by ultrasonic devices. In many cases, the cleaning requires several steps in a certain sequence. For example, organic solvents are used prior to soaking in concentrated nitric acid. Spots resulting from organic chemicals resistant to washing may be removed by baking the glassware at 600°C in an oven for annealing glass. Knorr [126] has described a method for ultracleaning glassware as follows. Used steroid-contaminated glassware is first cleaned manually with methanol and water, then washed in a laboratory washing machine without the addition of wetting agents, the last rinse being with water. After drying at 120°C, flasks and tubes are closed with aluminum foil and heated in an oven to 500°C during 2 hr, kept at this temperature for 1 hr, and allowed to cool slowly overnight in the closed oven. This process destroys traces of proteins, steroids, and organic compounds which often remain after the usual washing, and does not endanger the glassware through breakage.

VI. OBJECTIVES AND ORGANIZATION OF THIS MONOGRAPH

This monograph has three main objectives. (1) It is intended to be a guidebook for the practicing chemists who deal with small samples, particularly for those who are faced with the situation that the amount of working material is at a level below the capability of the available equipment. Suitable microscale apparatus and techniques can be chosen from the various methods discussed. (2) It serves as a laboratory manual for advanced students who wish to improve their experimental skill and broaden their knowledge. (3) It is a comprehensive treatise on microchemical manipulations, with extensive references, for microchemists to keep abreast of the recent developments in the field.

For the sake of clarity and convenience in utilization, this monograph is organized from the standpoint of the working material (e.g., solid, liquid) as well as the purpose of the experimental operation (e.g., to separate all components of a mixture, to process a single compound for organic elemental analysis, etc.). Since the same apparatus or technique may be employed in several different operations, cross references are given in the appropriate sections. Pertinent literature is cited in each section. Within the limits of space, as many methods as possible are described in detail. As a rule, the applications and limitations of the different apparatus or techniques are briefly mentioned.

REFERENCES

1. F. Emich, *Lehrbuch der Mikrochemie.* Bergmann, München, 1911.
2. F. Pregl, *Die quantitative organische Mikroanalyse.* Springer, Berlin, 1916.
3. F. Feigl, *Tröpfelreaktionen.* Akademische Verlag, Leipzig, 1931.
4. M. Harmelin, *Chim. Anal.*, **51**, 603 (1969).
5. F. W. McLafferty, private communication.
6. G. Tölg, *Ultramicro Elemental Analysis.* Wiley, New York, 1970.
7. R. Belcher, *Submicro Methods of Organic Analysis.* Elsevier, Amsterdam, 1966.
8. T. S. Ma, in F. J. Welcher (Ed.), *Standard Methods of Chemical Analysis*, 6th ed., Vol. II, p. 357. Van Nostrand, Princeton, 1963.
9. W. H. Thornett, U.K.A.E.A. Report, RCC-M, 173 (1964).
10. W. J. Franklin and H. Keyzer, *Anal. Chem.*, **34**, 1650 (1962).
11. M. G. Larrabee, *Anal. Biochem.*, **1**, 151 (1960).
12. V. Neuhoff, *Arzneim.-Forsch.*, **18**, 629 (1968).
13. W. F. Bowers and R. H. Haschemeyer, *Anal. Biochem.*, **25**, 549 (1968).
14. S. Natelson, *Microchem, J.*, **13**, 433 (1968).
15. B. J. Marcus and T. S. Ma, *Mikrochim. Acta*, **1969**, 816.
16. N. Nicolaides and H. C. Fu, *Lipids*, **4**, 83 (1969).
17. G. Hoffmann, *Dtsch. Lebensm. Rundsch.*, **61**, 177 (1965).
18. H. K. Pratt and C. W. Greiner, *Anal. Chem.*, **29**, 862 (1957).
19. M. J. Marshall and G. Constabaris, *Can. J. Chem.*, **31**, 842 (1953).
20. A. Weinstein and H. C. Friedman, *Rev. Sci. Instrum.*, **35**, 1083 (1964).
21. D. M. Ellen, *J. Forensic Sci Soc.*, **5**, 196 (1965).
22. J. J. Durez, P. Tissot, and R. Monnier, *Helv. Chim. Acta*, **50**, 822 (1967).
23. S. Patai, Y. Halpern, L. Esterman, and M. Weinstein, *Isr. J. Chem.*, **6**, 445 (1968).
24. W. J. de Klein and K. Ulbert, *Anal. Chem.*, **41**, 682 (1969).
25. Mettler Instrument Corp., Princeton, N.J.
26. Cahn Division, Ventron Instrument Corp., Paramount, Calif.
27. Oertling Ltd., London.
28. J. van Lier, *Rev. Sci. Instrum.*, **39**, 1841 (1968).
29. D. H. Fine and L. Glasser, *J. Sci. Instrum.*, **42**, 109 (1965).
30. Hamilton Co., Whittier, Calif.
31. R. H. Laessig, *Anal. Chem.*, **40**, 2205 (1968).
32. A. Haack and E. Rudy, *Mikrochim. Acta*, **1968**, 23.
33. A. Pasternak, *Chem. Anal. (Pol.)*, **13**, 593 (1968).
34. O. Siggaard-Anderson and D. Oliver, *Scand. J. Clin. Lab Invest.*, **21**, 92 (1968).

35. S. V. Kuzmin, V. V. Matveev, E. K. Pressman, and L. S. Sandakhchiev, *Biokhimiya*, **34**, 706 (1969).
36. I. P. Alimarin and M. N. Petrikova, *Zh. Anil. Khim.*, **23**, 1042 (1968).
37. D. H. Evans, *Anal. Chem.*, **37**, 1520 (1965).
38. M. T. Kelley, "Modern Trends in Radiochemical Analytical Methods." 6th International Symposium on Microtechniques, Graz, 1970.
39. S. Mlinko, T. Szarvas, I. Gacs, and K. Payer, "Mikromethoden in der organischen Isotopen-Gasanalyse." 6th International Symposium on Microtechniques, Graz, 1970.
40. I. P. Alimarin, M. N. Perrikova, and T. S. Kokina, "Electrochemical Ultramicromethods of Analysis." 6th International Symposium on Microtechniques, Graz, 1970.
41. R. Woodriff and R. A. Stone, *Apl. Opt.*, **7**, 1337 (1968).
42. T. S. West, "Determination of Very Small Amounts of Materials by Atomic Absorption and Atomic Fluorescence Spectrometry." 6th International Symposium on Microtechniques, Graz, 1970.
42a. V. Neuhoff and M. Weise, *Arzneim.-Forsch.*, **20**, 368 (1970).
43. V. Svoboda and I. Kleimann, *Anal. Chem.*, **40**, 1534 (1968)
44. N. D. Cheronis and T. S. Ma, *Organic Functional Group Analysis*. Wiley, New York, 1964.
45. A. A. Benedetti-Pichler, *Microtechniques of Inorganic Analysis*. Wiley, New York, 1942.
46. S. Ebel, *Z. Anal. Chem.*, **244**, 23 (1969).
47. H. Flaschka, "Inorganic Microanalysis." 6th International Symposum on Microtechniques, Graz, 1970.
48. W. Simon, "Moderne Methoden zur Structuraufklarung organischer Verbindungen." 6th International Symposium on Microtechnique, Graz, 1970.
49. D. R. Lids, Jr., in C. N. Reilly (Ed.), *Advances in Analytical Chemistry and Instrumentation*, Vol. 5, p. 235. Wiley, New York, 1966.
49a. W. Steck and B. K. Bailey, *Can. J. Chem.*, **47**, 3577 (1969).
50. A. Lacourt, *Mikrochim. Acta*, **1956**, 700.
51. A. Karler, C. L. Brown, and P. L. Kirk, *Mikrochim. Acta*, **1956**, 1585.
52. H. Tschesche and C. Frank J. Chromatogr., **40**, 296 (1969).
53. A. Fono, A. Sapse, and T. S. Ma, *Mikrochim. Acta*, **1965**, 1103.
54. V. R. Alishoev, V. G. Berezkin, I. B. Nemerovskaya, B. M. Kovarskaya, E. I. Zalalaev, Z. P. Markovich, E. A. Porkrovskaya, and O. S. Fratkin, *Vysokomol. Soedin.*, **A 11**, 247 (1969).
55. P. W. Wilson, D. E. M. Lawson, and E. Kodicek, *J. Chromatogr.*, **39**, 75 (1969).
56. N. C. Jain and P. L. Kirk, *Microchem. J.*, **12**, 229 (1967).
57. A. A. Benedetti-Pichler, *Anal. Chem.*, **9**, 149 (1938).
58. *Chem. Eng. News*, Dec. 14, 1970, p. 39.
59. S. Natelson, *Microtechniques of Clinical Chemistry*. Thomas, Springfield, Ill., 1957.
60. R. F. Coward and P. Smith, *J. Chromatogr.*, **39**, 496 (1969).

61. M. Bounias, *Chim. Anal.*, **51**, 76 (1969).
62. P. E. Haywood and M. S. Moss, *Analyst*, **93**, 737 (1968).
63. F. Drawert, W. Heiman, R. Emberger, and R. Tressl, *Chromatographia*, **2**, 57 (1969).
64. R. P. W. Scott, *Chemy Ind.*, **1969**, 797.
65. R. H. Whittaker and P. P. Feeny, *Science*, **171**, 758 (1971).
66. M. Jacobson and M. Beroza, *Science*, **140**, 1367 (1963).
67. U. E. Brady, J. H. Tumlinson III, R. G. Brownlee, and M. S. Silverstein, *Science*, **171**, 802 (1971).
68. A. E. Werner and M. Waldichuk, *Anal. Chem.*, **34**, 1674 (1962).
69. P. G. Simmonds, G. P. Shulman, and C. H. Stembridge, *J. Chromatogr. Sci.*, **7**, 36 (1969).
70. M. S. Kharasch and O. Reinmuth, *Grignard Reactions of Non-metallic Substances.* Prentice-Hall, New York, 1954.
71. G. E. Olah (Ed.), *Freidel-Cratfs and Related Reactions.* Wiley-Interscience, New York, 1963.
72. S. W. Fox and C. R. Windsor, *Science*, **170**, 984 (1970).
72a. P. Finocchiaro, D. Gust, and K. Mislow, *J. Am. Chem. Soc.*, **96**, 2165 (1974).
73. J. G. Calvert, *Photochemistry.* Wiley, New York, 1966.
74. D. C. Neckers, *Mechanistic Organic Photochemistry.* Rheinhold, New York, 1967.
75. P. Ausloos (Ed.), *Fundamental Process in Radiation Chemistry.* Wiley, New York, 1968.
76. G. Friedlander and J. W. Kennedy, *Introduction to Radiochemistry.* Wiley, New York, 1949.
77. G. R. Choppin, *Experimental Nuclear Chemistry.* Prentice-Hall, Englewood, 1961.
78. M. Cefola, *Mikrochim. Acta*, **1967**, 732.
79. W. Helbig, *Z. Anal. Chem.*, **182**, 15(1961).
80. V. H. Cheldelin, *Metabolic Pathways in Microorganism.* Wiley, New York, 1961.
81. K. Hoffmann, *Fatty Acid Metabolism in Microorganism.* Wiley, New York, 1962.
82. R. I. Dorfman and F. Ungar, *Metabolism of Steroid Hormones.* Academic, New York, 1965.
83. R. T. Williams, *Detoxication Mechanisms.* Wiley, New York, 1959.
84. M. B. Jacobs, *Chemical Analysis of Air Pollutants.* Wiley-Interscience, New York, 1960.
85. F. A. Patty (Ed.), *Industrial Hygiene and Toxicology,* Wiley, New York, 1963.
86. P. Drinker and T. Hatch, *Industrial Dust*, 2nd ed. McGraw-Hill, New York, 1954.
87. H. L. DeQuasie and D. C. Grey, *Am. Lab.*, December 1970, p. 19.
88. D. J. Lisk, *Science*, **170**, 589 (1970).
89. M. Bergstroem-Nielsen, *Dtsch. Lebensm. Rundsch.*, **65**, 1631 (1969).

90. R. Rasmussen and F. W. Went, *Science*, **144**, 566 (1964).
91. N. Maravalhas, *Chem. Anal.*, **53**, 23 (1964).
92. A. M. Long and O. O'Brien, *Chemy Ind.*, **1968**, 1764.
93. N. L. Cross, *J. Sci. Instrum.*, Ser. 2, **1**, 65 (1968).
94. R. H. Laessig, *Microchem. J.*, **13**, 561 (1968).
95. J. Michod, C. Platsoukas, and J. Frei, *Z. Klin. Chem. Klin. Biochem.*, **7**, 455 (1969).
96. M. Marsh and G. Schill, *Svensk. Farm. Tidskr.*, **64**, 921 (1960).
97. T. S. Ma and A. A. Benedetti-Pichler, *Anal. Chem.*, **25**, 999 (1953).
98. R. A. Wilson, C. G. Smith, R. H. James, and R. R. Wallace, *Analyst*, **93**, 773 (1968).
98a. W. Shephard, *Rev. Sci. Instrum.*, **44**, 234 (1973).
99. S. R. Palamand, K. S. Markl, and W. A. Hardurick, *Proc. Am. Soc. Brew. Chem.*, **1968**, 75.
100. P. Mitchell, *Helv. Chim. Acta*, **27**, 961 (1944).
101. V. D. Gupta, D. E. Cadwallader, H. B. Herman, and I. L. Honigberg, *J. Pharm. Sci.*, **57**, 1199 (1968).
102. S. Feldman and M. Gibaldi, *J. Pharm. Sci.*, **57**, 1234 (1968).
103. T. S. Ma, *Chin. Chem. Soc.*, **15**, 112 (1947).
104. H. Trutnovsky, *Mikrochim. Acta*, **1968**, 371.
105. M. Chiba and H. V. Morley, *J. Assoc. Off. Anal. Chem.*, **51**, 55 (1968).
106. J. A. Burke, P. A. Mills, and D. C. Bostwick, *J. Assoc. Off. Anal. Chem.*, **49**, 999 (1966).
107. W. A. Moats and A. W. Kotula, *J. Assoc. Off. Anal. Chem.*, **49**, 973 (1966).
108. A. A. Benedetti-Pichler, in W. G. Berl (Ed.), *Physical Methods in Chemical Analysis,* Vol. 3, p. 184. Academic, New York, 1960.
109. A. A. Benedetti-Pichler, personal communication; see also F. L. Schneider, Mikrochim. Acta, **1967**, 743.
109a. B. Skaletzki and H. Wollmann, *Pharmazie,* **28**, 1 (1973)
110. E. O. Rowland, *Miner. Mag.*, **33**, 524 (1963).
111. E. L. Gooden, *J. Agric. Food Chem.*, **10**, 397 (1962).
112. G. W. Ware and M. K. Dee, *Bull. Environ. Contam. Toxicol.*, **3**, 375 (1968).
113. F. M. Whiting, J. W. Stull, W. H. Brown, M. Milbrath, and G. W. Ware, *J. Dairy Sci.*, **51**, 1039 (1968).
113a. C. L. Grant, personal communication; ASTM Special Technical Publication No. 540 (1973).
114. L. Cahn and W. J. Cadman, *Anal. Chem.*, **30**, 1580 (1958).
115. T. S. Ma and R. F. Sweeney, *Mikrochim. Acta*, **1956**, 191.
116. H. Spitz, *J. Chromatogr.*, **42**, 384 (1969).
117. K. Fritze and R. J. Gietz. *J. Radioanal. Chem.*, **1**, 265 (1968).
118. S. L. Tompsett, *Analyst*, **93**, 740 (1968).
119. I. D. Eubanks and E. J. Tbbott, *Anal. Chem.*, **41**, 1708 (1969).
120. D. L. Rousseau, *Science*, **171**, 170 (1971).

121. R. E. Davis, D. L. Rousseau, and R. D. Board, *Science*, **171**, 167 (1971).
122. *Chem. Eng. News*, June 9, 1970, p. 7; July 13, 1970, p. 19.
123. A. A. Benedetti-Pichler, *Identification of Materials,* Springer, New York, 1964, p. 68.
124. V. Brezina, *Chem. Prum.*, **21**, 245 (1971).
125. Available from Wheaton Scientific Co., Millville, N.J.
126. D. Knorr, *Z. Klin. Chem. Klin. Biochem.*, **9**, 175 (1971).

THE LABORATORY FOR MICROSCALE MANIPULATIONS: GENERAL CONSIDERATIONS

I. ORGANIZATION OF THE MICROCHEMICAL LABORATORY

A laboratory dealing with chemical experimentation using materials below the gram scale may be considered a microchemical laboratory. The experimental operations may involve the preparation of new or known chemical species, separation, isolation, detection, identification, estimation, and/or precise quantitative analysis. Apparatus and equipment used may vary from the simple test tube and spot plate to sophisticated combustion furnaces, chromatographs, optical instruments, radioactivity counters, mass spectrometers, etc. Understandably, very few establishments possess all the equipment amenable to microchemical work, and the organization of the laboratory depends on the need of the experimental processes. Three basic principles may be mentioned, however, for organizing the laboratory and selecting techniques for microchemical experimentation: (1) The various apparatus and equipment should be placed in close proximity. (2) Techniques should be chosen which guarantee that the working specimen will be protected from contamination and losses. (3) Transfer of the material from vessel to vessel should be limited. In this monograph we are primarily concerned with the processing of the working material so as to obtain microanalytical samples suitable for chemical tests or physical measurements. Some general considerations are discussed in the following sections.

A. LOCATION FOR MICROSCALE OPERATIONS

Since microscale manipulation is usually associated with some experimental work preceeding it or analytical procedure following it, the ideal location for micro operations is in the same laboratory where the other experiments are conducted, or an adjoining one. When they are in the same room, it is advisable to put up a convenient partition to confine the area in order to eliminate air drafts and to minimize the danger of contaminating the micro samples with the dust and fumes generated in the laboratory. Many microscale experiments are preferably carried out while the operator sits down comfortably; hence the benches should be of suitable height and provided with leg room. Small equipment should be kept in trays or drawers with partitions. It is best if large apparatus is partly assembled and stored in shelves and cabinets. Needless to say, the space for microscale manipulation should be meticulously clean and neat. All operations should be conducted in such manner that, in case some sample escapes from the container, it will fall on the bench top instead of going down to the floor.

Some microscale operations are self-contained (e.g., ultramicro techniques in biochemistry [1] and clinical analysis [2, 2a], microgram transformation of inorganic compounds [3], chemical microscopy [4], etc), and occupy very little space in the laboratory. For working with materials which are sensitive to air and moisture, a controlled-atmosphere chamber should be conveniently located. It may be a simple plastic bell jar [5] or a specially designed steel box [6, 7]. When working with dangerous materials such as radioactive chemicals, it is prudent to do all manipulations in trays lined with towels.

B. GENERAL SUPPLIES FOR MICROSCALE MANIPULATIONS

While the equipment used for microscale manipulation varies from laboratory to laboratory, there are general items which can be purchased from the supply houses [8] and assembled when the microchemical laboratory is organized. Thus the common apparatus will be on hand when needed, and delay of experimental operation because of waiting for equipment is avoided. These items are discussed in the succeeding sections in this chapter. Apparatus that is required for specific processes or can be easily made when needed is treated in other chapters in connection with the particular experiments.

An assortment of standard tapered or "clear seal" [8a] joints and tubing of suitable sizes should be bought, washed clean, and stored in a dust-free place. It is advantageous to use clean glass parts to make the micro apparatus instead of doing the cleaning after the apparatus has been fabricated. Intelligent planning of the laboratory calls for a collection of "universal apparatus" which can be used interchangeably for several operations. An example [9, 10] is shown in Fig. 2.1. Reaction vessels with capacities from 5 to 25 ml are made from ␓ 14/20 outer joints. These vessels can serve for reflux under a water condenser (a, b) or air condenser (c). Micro-Soxhlet extraction is illustrated in d, while sublimation is carried out with the attachment shown in e. When a liquid mixture is to be fractionated, the operation is shown in f. Removal of the bulk of solvent from a reaction mixture is depicted in g. A universal assembly

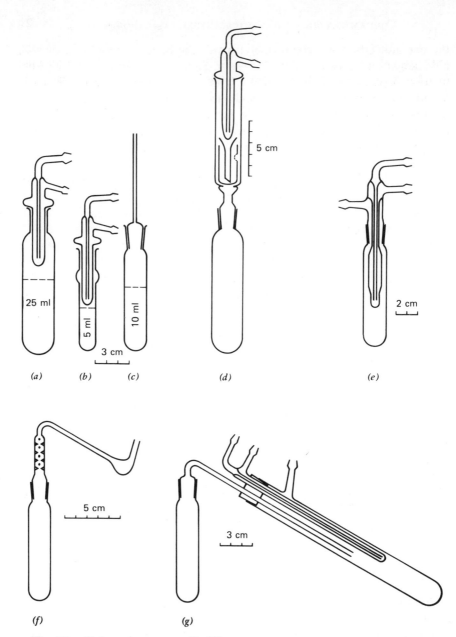

Fig. 2.1. Universal apparatus [9, 10].

recommended for biochemical operations has been described by Dubbs [11]; it can be used for homogenization, filtration, extraction, reflux, distillation, and drying. Marhoul and Ruzicka [12] have proposed glass and polyethylene apparatus for filtration, separation of crystals, vacuum evaporation, and transfer of liquid. Mares [13] has described a microapparatus for lipid methanolysis in which the whole operation can be carried out without transporting the solution and with a minimum of loss. Graves and Vincent [14] have constructed an apparatus that permits product synthesis, solvent removal, and sublimation to be carried out in one reaction flask fitted with interchangeable heads.

II. CONTAINERS FOR HANDLING SMALL SAMPLES

A wide variety of containers are used in microscale manipulations. Some examples are shown in Fig. 2.2. These vessels have capacities between 1 to 25 ml. In order to standardize microchemical equipment, the British Standards Institute [15] and the Specifications Committee of the Division of Analytical Chemistry, American Chemical Society [16] have published guidelines and specified the dimensions of a number of the containers used in microchemistry. It can be seen in Fig. 2.2 that a majority of the vessels have tapered, instead of flat or round, bottoms. Thus a reasonable ratio between the diameter of the liquid surface and its depth is maintained for a wide range of volumes. This facilitates the transfer of sample with minimal losses. The same principle is utilized in the mini-vials [17] (Fig. 2.3) which became commercially available recently.

The filterbeaker shown in Fig. 2.4 is a useful container for operations that involve the formation of a precipitate that need not be transferred to another vessel in the subsequent step. The sample and reagents are introduced through the top opening. The precipitate is collected on the fritted disc by tilting the filterbeaker with its left arm connected to the aspirator. The filterbeaker also can be used to prepare solutions in the following manner: The sample is introduced through the top opening. Solvent is added, and the beaker is heated. The insoluble material is retained by the fritted disc when the solution is filtered.

In designing containers for microscale manipulations, versatility and minimization of dead space are important considerations. The modified Warburg manometer flask [18] (Fig. 2.5) is an example. By sealing a glass partition to divide the body of the flask, it serves as a double-chamber vessel, and the dead space above the solutions is greatly reduced. Such a vessel may be used when two solutions are to be introduced into the same apparatus without being mixed until it is desired to do so.

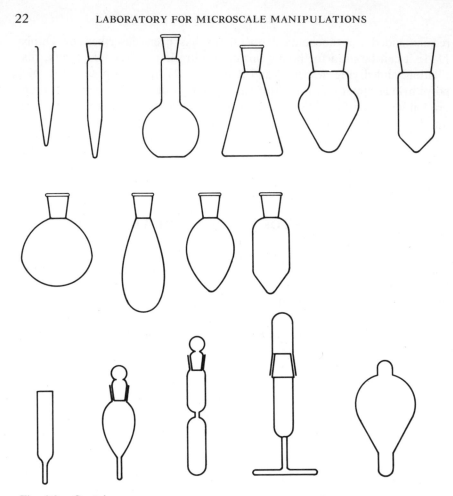

Fig. 2.2.   Containers.

Knobloch and Mudrova [19] have described a device for measuring and mixing of small volumes of solution by incorporating a bulb in the capillary tube, as illustrated in Fig. 2.6. Eckfeldt and Shaffer [20] have designed a semiautomatic precision pipet to deliver identical sample volumes of approximately 1 ml. Holmes [21] has constructed a graduated pipet for rapid and accurate delivery of 0.2-ml portions of viscous liquids.

In handling minute quantities of material, contamination from the walls of the container and from lubricant at the joints, and possible leaks in the system, are sources of concern. Therefore, in the analysis of trace elements, the use of plastic vessels has been recommended so as to eliminate the presence of all foreign cations. Smith [22] has described a separatory

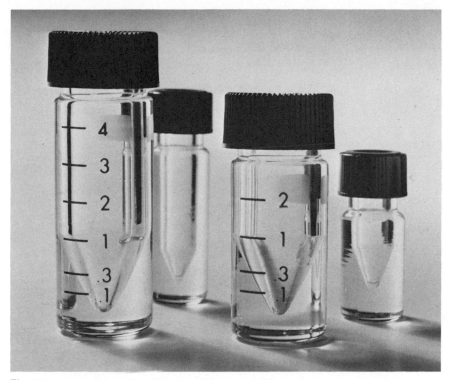

Fig. 2.3.   Mini-vials [17]. (Courtesy Alltech Associates, Inc.)

Fig. 2.4.   Filter beaker. (Courtesy A. H. Thomas Co.)

funnel made from polyethylene tubing of 25-mm outside diameter and 20-mm bore. As depicted in Fig. 2.7, the stopcock is replaced by a screw clamp, thus avoiding the use of lubricating grease. PTFE (e.g., Teflon) stopcocks are always used without grease.

In the attempt to prevent leaks in ground-glass equipment when working with corrosive chemicals such as fuming nitric acid and thionyl chloride, Quin and Greenlee [23] have found that a joint sealed with molten vinylidene chloride polymer is resistant to leaking for periods up to 10 days. Teflon sleeves matching glass joints are commercially available.

Capillary tubes or capillaries are common containers for holding vol-

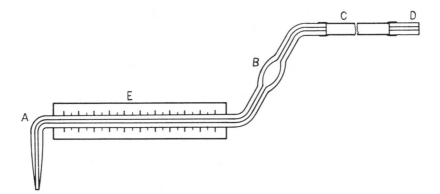

Fig. 2.5. Reaction vessel with glass partition [18]. (Courtesy *Anal. Chem.*)

Fig. 2.6. Equipment for measuring and mixing small volumes of solutions [19]: *A*, horizontal microburet; *B*, bulb; *C*, rubber tubing; *D*, closing capillary; *E*, supporting plate with a millimeter scale.

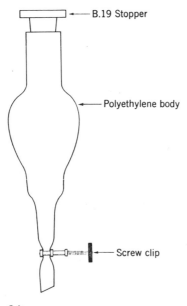

Fig. 2.7. Polyethylene separatory funnel. (Courtesy *Lab Pract.*)

umes below the milliliter region. Glass tubing of 3- to 4-mm bore is the most suitable. Capillaries of smaller diameter are made by drawing a tube of convenient size. Ordinary 8-mm tubing is used to prepare thick-walled capillaries by collecting glass in the flame, while thin-walled capillaries are preferably drawn from 12-mm tubing. (Commercial melting-point tubes can be utilized as microcontainers after sealing; their capacity is about 100 μl.) With few exceptions, soft glass is employed to prepare capillary tubes, since it can be worked over an ordinary gas burner. Hard glass requires an oxygen flame. Borosilicate (Pyrex) capillaries also tend to be brittle. The reader who needs instruction in making capillaries is referred to the detailed directions given by Benedetti-Pichler [24] and Schneider [25].

Disposable containers are recommended wherever feasible. Capillary tubes belong to this category. Not only is a fresh capillary free from contamination of the previous sample; it is much easier to prepare a new capillary than to clean an old one for reuse. Swift [26] has described a technique to prepare microbeakers for use in microscopy. Cahn and Cadman [27] have advocated the use of disposable platinum "crucibles" prepared as follows. A 12-mm square platinum foil, weighing about 12 mg, is folded as shown in Fig. 2.8. It is suspended in a wire hanger and weighed relative to a tare. The sample is added, and the foil is weighed again. The hanger is removed from the balance, and the square is folded as shown. The sample is completely enclosed in metal. If it is to be weighed after ashing, it is again inserted in the hanger.

Containers that are designed for holding micro samples for specific measurements or analytical procedures such as calorimetry, spectrometry, and thermal decomposition will be discussed in Chapter 9.

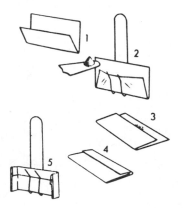

Fig. 2.8. Sequence of folding foil to make disposable platinum "crucible" [27]. (Courtesy *Anal. Chem.*)

III. TOOLS FOR MICROSCALE WORK

Some tools which are commonly used in microchemical experiments are briefly discussed below:

For Measurement and Delivery of Liquids. In Fig. 2.9 are shown varieties of pipets and syringes that can be used to deliver a volume of liquid at and below the milliliter region. As depicted in Fig. 2.10, the micropipet can be filled with the sample in four different ways, namely, (*a*) through oral suction, (*b*) by means of a rubber nipple, (*c*) by connection to the buret, and (*d*) by joining to a plunger. If a constant volume is to be delivered, the automatic-filling micropipets shown in Fig. 2.11 are recommended. The pipet is filled spontaneously by capillary action, or a slight excess of the sample is drawn into the pipet so that it overflows at the top of the capillary; on expelling, only the liquid inside the capillary will come out. For the measurement of fractions of a column, as in microtitrimetry using milliliters to microliters of solution, the microburet is employed. A review of the development of microburets has been given by Cheronis [28].

For Weighing. If the Mettler microbalance and Cahn electrobalance are available in the laboratory, naturally they serve well for weighing small samples. On the other hand, many microscale experiments can be

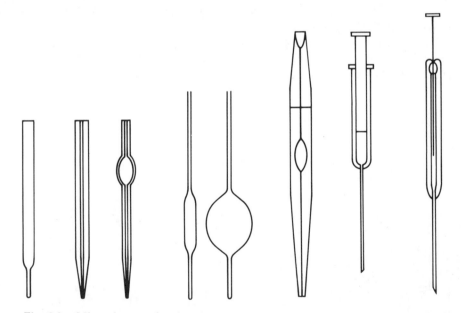

Fig. 2.9. Micropipets and syringes.

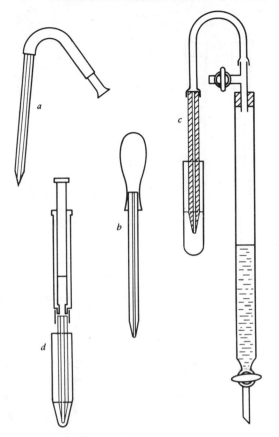

Fig. 2.10. Devices for filling micropipets.

carried out without the benefit of such precision equipment [28a]. An ordinary analytical balance or a simple two-pan scale may be adequate for working in the milligram region. For micromanipulation in the microgram region, a modified Salvioni balance [29] can be constructed by fixing one end of a quartz fiber or steel spring [30] in a clamp placed at one edge of a 15-×25-cm board, as shown in Fig. 2.12, with the other end carrying the weighing pan made of aluminum foil.

For Magnifying. Since microscale manipulations deal with either very small volumes or minute particles, it is often necessary to magnify the working material in order to compensate for the inadequacy of the human eye. The hand lens, desk-top enlarger, and stereoscopic dissecting microscope are convenient apparatus for 4- to 10-fold magnification of the specimen. The magnifier should have a wide field, so that experiments can

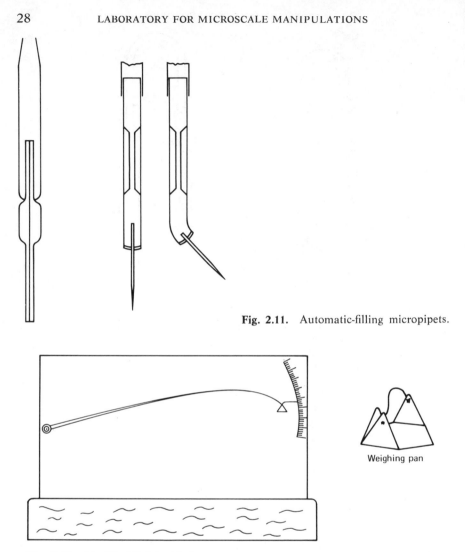

Fig. 2.11. Automatic-filling micropipets.

Weighing pan

Fig. 2.12. Modified Salvioni balance.

be performed on the whole specimen under continuous observation. A compound microscope is usually needed for working in the microgram or microliter region. The microscope is preferably fitted with a rotating stage that can be connected to the mechanical micromanipulator. The optical parts should consist of bright-field condenser that can be focused on a plane 10 mm above the stage, objectives with magnifications of 5, 10, and 20 diameters, and micrometer eyepiece with magnification 5×. Microchemical experiments as a rule do not require magnification higher than

100-fold. On the other hand, since reagents (sometimes corrosive chemicals) and micro tools are to be applied to the specimen, there should be a working distance of at least 5 mm between the specimen and the microscope objective. Ancillary equipment related to magnifying includes microscope slides, cover glasses, and spot plate.

For Transferring Solid Particles. Microspatulas for transferring milligram amounts of solid and semisolid material are conveniently prepared from Monel metal rod of 2-mm diameter by flattening its two ends; one end can be rectangular and the other end triangular, as shown in Fig. 2.13*a*. Commercial microspatulas made of aluminum have thick blades and are not suitable for picking up small crystals or powder; they should be modified by filing or grinding. A microspatula with a spoon shape is difficult to clean and tends to retain traces of previous sample. A glass needle (Fig. 2.13*b*) is used to move particles on the microscope slide or spot plate. It is made by drawing out a glass rod; a platinum wire sealed to a glass rod serves the same purpose. A needle with bent tip (Fig. 2.13*c*) is used when the working distance between the specimen and the magnifier is restricted. The needle can be brought under the microscope objective in a horizontal position and then turned on its axis to bring the point into contact with the specimen. Solid particles are picked up by means of forceps. Platinum-tipped forceps (Fig. 2.14*a*) and forceps to pick up tubular objects (Fig. 2.14*b*) are commercially available. An electrostatic tool for handling very small crystals has been described by Nicoll [31]. A fused-silica rod is coated with a thin conducting layer of tin oxide; one end of the rod is then ground flat and is given an electrostatic charge by pressing it onto a clean piece of silicone rubber. Alternatively, the tool can be made from a plastic rod coated with silver paste; one end is then

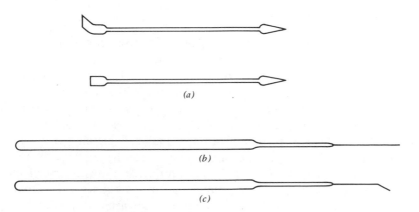

(a)

(b)

(c)

Fig. 2.13. (*a*) Microspatula; (*b*) glass needle; (*c*) needle with bent tip.

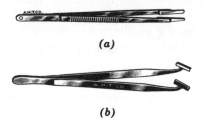

(a)

(b)

Fig. 2.14. (a) Platinum-tipped forceps; (b) forceps for handling capillaries. (Courtesy A. H. Thomas Co.)

lapped flat and charged by rubbing on wool. The charged end is used to pick up and handle small crystals. The specimen can be released readily if it adheres sufficiently well to a surface or if the electrostatic charge is allowed to decay. Bainbridge and Carter [31a] have designed a "mini-tong" for handling small and fragile radioactive specimens; objects with diameter less than 1 mm (e.g., a human hair) can be handled.

Miscellaneous. Sieves used in microanalytical work have been discussed by Brickey and co-workers [32]. A sieve handle is recommended which helps in maintaining the proper angle of the sieve. For holding fine wire, an inexpensive hand vise is commercially available [33]. The wire passes through the hollow handle and can be adjusted to any length. Horwitz and Wood [34] have constructed a simple holder that can be attached to a microscope stage or slide in order to manipulate the microneedle or micropipet. Bonting and Walters [35] have described a device for marking test tubes to contain volumes of 20 to 500 μl.

IV. DEVICES FOR HEATING, COOLING, AND TEMPERATURE CONTROL

A. MICROBURNERS

Various models of microburners are available from supply houses [8]. Depending on the source and kind of gas, however, they do not always work properly. A simple microburner can be easily constructed from a glass stopcock, as shown in Fig. 2.15. One arm of the stopcock is bent at a right angle, and an opening is made at position A, below which the tubing is slightly constricted. The opening A serves as an inlet for air. By sealing it with a piece of plastic tape and then piercing the tape by means of a needle, the amount of air coming in can be adjusted to furnish a suitable flame [36]. Charlett [37] has fabricated a microburner from a hypodermic needle cut off and ground to form a square end. Gunders [38], also using the hypodermic needle, has constructed a microburner

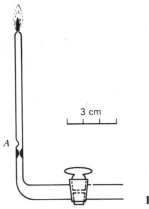

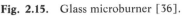

Fig. 2.15. Glass microburner [36].

for sealing capillary tubes with a minimum of risk to the contents. It operates on the Bunsen principle with methane-butane (9:1) as fuel. The flame does not come to a sharp point as do flames using compressed air or oxygen; nevertheless, the tip of the flame is small enough for the intended applications.

B. HEATING BLOCKS AND HEATING BATHS

While oil and sand baths can be used for heating microchemical reaction vessels, the metal block is recommended wherever feasible. The danger of contaminating the sample with the bath fluid must be considered, aside from the tedium of removing the fluid from the surface of small containers such as capillary tubes and microns. A simple metal block is shown in Fig. 2.16. It is machined from a bar of aluminum of 6- to 8-cm diameter. Holes of various diameters and depths are drilled on one end (*a*) to fit capillary tubes, microbeakers, microflasks, etc. Two shallow concentric wells are cut on the other end (*b*), on which two sizes of microscope slide covers can be placed, and a long hole is drilled perpendicular to the bar for the insertion of a thermometer. The end (*b*) is used for observation of fusion phenomena or for evaporation on a microscope slide. The aluminum block is heated either with a microburner or electrically by winding resistance wire or insulated heating tape around its body [36]. A sophisticated electronically controlled heating stage for microscale manipulation was described by Ma and Schenck [39], and has been improved recently by incorporating solid-state electronics and stainless-steel-sheathed cartridge heaters. Fig. 2.17 shows the apparatus; this

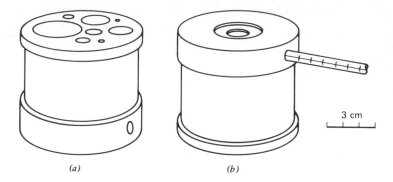

Fig. 2.16. Simple aluminum heating block. (*a*), top; (*b*) bottom.

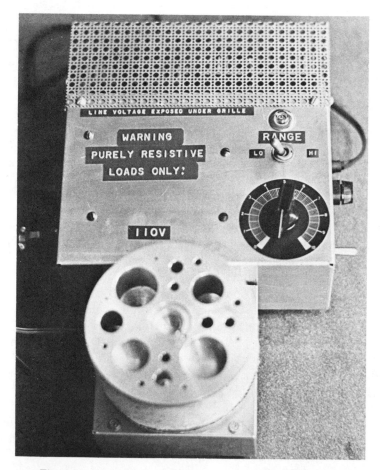

Fig. 2.17. Automatic temperature-controlled heating stage.

device [40] permits automatic control of temperature between 20 and 250°C to within ±0.5°C.

In Fig. 2.18 are shown two devices for heating a small volume of liquid at constant temperature. Depending on the boiling temperature of the liquid in the reservoir, the sample is exposed to constant temperature up to 200°C. The apparatus is heated to keep the liquid in the reservoir boiling under reflux. Drey and co-workers [40a] have described an electrically heated water bath for performing small-scale operations simultaneously.

Morgan and Welland [41] have constructed a constant-temperature heating block that can be used for rapidly attaining a temperature of 180°C and maintaining this to within ±0.5°C. Kuemmel [41a] has reported a simple arrangement for heating samples at constant rate from −196° to +200°C on a copper stage. Weir and Rutledge [42] have described a fluidized-solid constant-temperature bath. The heat-transfer medium consists of a bed of silica-alumina microspheres fluidized at 200 to 350°C by a current of air. A vacuum hotplate has been described by Nicholls [43]. It consists of an electrically heated stage which supports a stainless-steel plate; the latter makes direct contact with a glass funnel connected to a vacuum pump.

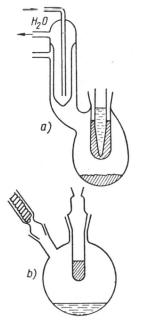

Fig. 2.18. Heating small containers and small volumes by means of vapors.

C. MISCELLANEOUS HEATING AND TEMPERATURE-CONTROL DEVICES

The electrical heating mantle [44], flexible heating tape, and the immersion heater are applicable to microscale vessels of capacity 25 ml or larger. For smaller containers, the heating wire may be sealed through the glass [45]. Heating under an infrared lamp is also a convenient device for micromanipulations.

A microfurnace for use in vacuum and inert atmosphere in which temperatures over 2000°C can be reached in a few seconds has been constructed by Presland and White [46]. Simple heating units to be used on the microscope stage has been described by Swift [47]. Devices for temperature control of hotplates, muffle and tube furnaces, and humidity ovens have been reviewed by Galloway [48]. For measuring the temperature of the contents of a micro reaction vessel whose opening is too narrow for the insertion of a thermometer, a thermocouple should be used instead. Lubb [49] has described a technique to connect the thermocouple wires to a closed rotating flask, as shown in Fig. 2.19.

D. COOLING DEVICES

Ice-salt and solid carbon dioxide–acetone (or –alcohol) are well-known cooling mixtures for operations that do not require low-temperature control within a narrow range. Gorbach [50] has constructed a micro cooling apparatus using the thermoelectric cooling element; the reaction mixture for microanalysis or micropreparation can be maintained between 12 and −22°C by means of this device. Recently aluminum cooling blocks which fit containers of 10- to 20-mm diameter have become commercially avail-

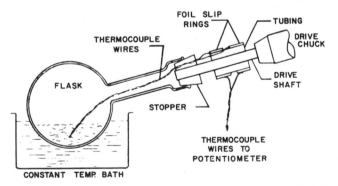

Fig. 2.19. Device for measuring temperature in a closed rotating flask [49]. (Courtesy *J. Chem. Educ.*)

able [51]; the temperature can be controlled from 25 to $-20°C$. Either air or tap water may be used for the heat exchanger. Virkola [52] has described a liquid bath using a 20% ethanol-water mixture as the circulating coolant, which can keep a stable temperature between 25 and 0°C to within ±0.25°. The bath consists of a steel bowl insulated with polystyrene and fitted with a thermostat with a heating element and thermal regulator. It is supported over a copper coil connected to a compressor unit.

Cooling samples to $-50°C$ or lower temperatures usually requires liquid nitrogen in a Dewar flask. Some recent modifications of the Dewars are mentioned below. While they were not designed for microscale operations, the apparatus can be adapted for such purposes. Foner [53] has described the construction of single-walled, space-saving Dewars. As shown in Fig. 2.20, it consists of a cylindrical reservoir B machined from polyurethane, which is attached to the nylon adaptor A, a nylon tail section T, and nylon bottom cap C. The pieces are joined with Adhesive 80 cement which seals all the joined surfaces and makes a smooth film to the base of B. The polyurethane is not soluble in this cement and is of sufficient density to withstand mechanical clamping and shock. The cement yields a slightly flexible seal at 77°K for years. Simons [54] has made an inexpensive Dewar vessel that allows absorption and emission spectroscopy in the ultraviolet and visible regions, at controlled temperature down to 77°K. Alon [55] has described a simple instrument for two-level control of liquid nitrogen in Dewars. Tyagi [56] has constructed a double-walled liquid-nitrogen container in nickel-silver with a flanged brass base that can be fitted on top of a brass box (with three windows and which can be evacuated). This cryostat can be used to study the optical and electrical properties of solids at a constant temperature in the range from 600 down to 77°K. The temperature can also be varied at the rate of 5°/sec. A com-

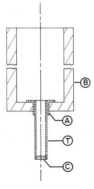

Fig. 2.20. Single-walled liquid N_2 "Dewar" [53]. B, polyurethane body; A, adaptor; T, tail section; C, end cap. (Courtesy *Rev. Sci. Instrum.*)

mercial cryocooling module [56a] is available that allows the sample temperature to vary from 150 down to 20°K in fluorescence, infrared, or laser Raman measurements.

Nassler [57] has reviewed low-temperature baths. A convenient way to maintain a temperature between -20 and $-70°C$ is to employ Dry Ice (solid carbon dioxide) and suitable organic liquids. Depending on the solvent used, reproducible temperatures are obtained (e.g., with carbon tetrachloride, $-23°C$; 3-heptanone, $-38°C$; cyclohexanone, $-46°C$; chloroform, $-61°C$). By mixing o- and m-xylene, a nearly linear temperature dependence can be achieved [57a].

V. DEVICES FOR EXTRACTION

A. EXTRACTORS FOR WORKING WITH SOLID MATERIALS

The popular apparatus for extraction of solid materials is the Soxhlet extractor, in which the sample is placed in a thimble and the solution overflows through a siphon tube into the solvent reservoir. Extensive study has been made of this type of extractor, and the British Standards Institution has published specifications together with recommendations on the size of thimble, flask, and condenser for use with extractors of 40-, 60-, 100- and 200-ml capacities, respectively [58]. While Soxhlet apparatus of 10-ml capacity is commercially available [58a], it is important that the reservoir in the flask should not be depleted, in order to prevent the compounds that have been extracted from depositing on the walls and being destroyed if they are heat sensitive. For handling minute quantities of solid material using small volumes of solvent, the assembly shown in Fig. 2.1d (see Section I.B) is the preferred device, even though it does not operate exactly on the Soxhlet principle. The glass funnel guides the solvent condensate to the bottom of the cup containing the sample; the resulting solution gradually overflows through the orifices near the rim and runs down to the reservoir. It is advisable to plug the orifices with filter paper to keep fine particles from going into the reservoir. Wasitsky [59] and Colegrave [60] have described microextractors in which the siphon tube is sealed to the bottom of the cup. It should be noted that, in order to allow an easy flow of the extractant, the siphon tube cannot have inside diameter less than 4 mm. The funnel and cup shown in Fig. 2.1d may be replaced by a cup with a sintered glass bottom, like the apparatus of Blount [61], Browning [62], and Hetterich [63]; such a device is useful when the compound to be extracted dissolves readily in the solvent.

Modifications of the Soxhlet type extractor with a view to improving

its performance have been advanced by many investigators. Since commercial Soxhlet thimbles are of relatively large pore size, Erdos [64] has
reported that they need to be impregnated with epoxy or melamine resin
before use for the retention of solid particles of size 10 to 100 nm.
Josephson [65] has modified the Soxhlet extractor so that it can be used
for extracting a series of samples without dismantling. The modification
consists of adding two side arms, one in the extraction chamber, and the
other in the flask containing the solvent. As illustrated in Fig. 2.21, the
former tube permits the introduction and removal of the extraction thimble, and the latter permits the solvent to be siphoned off after extraction.
O'Keefe [66] has described an apparatus that permits extraction at the
actual boiling point of the solvent. In order to minimize the exposure of
the extract to heat, Schirm [67] has incorporated an evaporator, which
consists of a three-bulb condenser heated by circulating through its jacket
a suitable vapor generated in a separate flask, between the extracting and
the unheated reservoir. Ensslin [68] has modified the Soxhlet apparatus
so that the extraction is carried out under reduced pressure. Kiger [69]
has published a description of an assembly that can be used for either
hot or cold extraction of the sample. After studying several types of extraction chamber for handling vegetable drugs, Kaminski [70] has reported a
design that gave superior results for most solvents. Hussain and Spanner

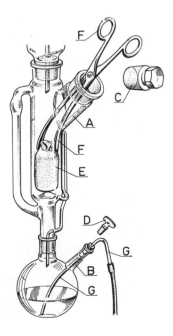

Fig. 2.21. Modified Soxhlet apparatus of Josephson [65]. (Courtesy *Clin. Chem. Acta.*)

[71] have described a Soxhlet microextractor suitable for use with 30 mg of fresh tissue. Franks [72] has constructed a semimicro apparatus for the determination of fatty matter in cotton fibers. Apparatus for the extraction of milligram quantities of lipids has been proposed by Mitchell [73] and by Scoggin and Tauber [74]. Antoszewski [75] has made a microextractor for pollen, seeds, and other biological materials. Bhatnagar and co-workers [76] have fabricated an apparatus for extracting fresh leaves, powdered leaves, and dried fruits.

B. EXTRACTORS FOR WORKING WITH LIQUIDS

The process of extracting from solutions is known as liquid-liquid extraction. For microscale operations, when the total volume of the solution and extracting solvent is less than 5 ml, test tubes and microns serve well. After stoppering the vessel and mixing the contents by agitation, the two layers are allowed to separate. Either the upper layer or the lower one can be withdrawn by means of a medicine dropper, as shown in Fig. 2.22a. By using the microne and the capillary pipet with a long fine tip (Fig. 2.22b), precise separation of the two solutions can be easily accomplished. If it is desirable to perform extraction at elevated temperatures, the vessel can be sealed, or closed with fluorocarbon (Teflon) tape and affixed to a shaking device. After equilibrium has been established, the vessel is opened and the two layers are separated as described above. An extraction pipet has been constructed by Gorbach [77], and a T tube for extraction by König and co-workers [78]; these devices are not as convenient as the microne. The separatory funnels shown in Fig. 2.23 are recommended for volumes from 5 to 20 ml; the capillary drain tubes permit sharp demarcation of the two liquid phases.

The various types of apparatus for extracting substances from solutions up to 1962 have been reviewed by Markov and Korinfskaya [79]. Sims and Adams [80] have described a device which is suitable for the extraction of a liquid at a series of increasing temperatures. It is provided with a water jacket, a magnetic stirrer, and baffles to prevent channeling. Heftmann and Johnson [81] have designed an extractor that operates under reduced pressure so that high-boiling solvents can be used for extracting heat-labile substances from aqueous solutions. Branica [82] has described an automatic microextraction apparatus for the extraction of solutes from solutions by means of immiscible solvents of densities different from those of the solutions. Hubbard and Green [83] have constructed an apparatus for multiple solvent extractions, which consists of four conical flasks mounted on a metal rack that can be attached to a shaking machine, and also an adjustable frame on which the rack can be hung for the separation

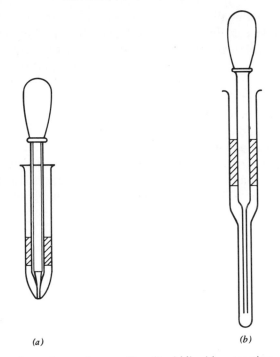

(a)   (b)

Fig. 2.22. Separation of two layers after liquid-liquid extraction: (a) in a test tube; (b) in a microne (tip of capillary pipet exaggerated).

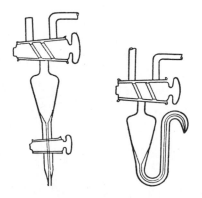

Fig. 2.23. Separatory funnel with capillary drain tube.

of the phases. By the use of two sets of four screw-cap flasks, each set mounted on a rack, eight samples can be submitted to solvent extraction in about the same time required for two samples by the manual shaking procedure.

Mieszkis [84] has improved the Clasper liquid-liquid extraction apparatus by incorporating the Soxhlet principle. Neves and Tavares [85] have described two devices that can be substituted for the thimble in a Soxhlet apparatus to enable extraction of liquids heavier or lighter than the solvent to be carried out without special extractors. Spikner and co-workers [86] have constructed a small-volume extractor from standard glassware, which can be used for the extraction of aqueous solution with organic solvents lighter than water.

Exhaustive liquid-liquid extraction can be carried out in several types of apparatus. The apparatus shown in Fig. 2.24 are easily fabricated and permit extraction from 2 ml or more of the sample. It may be mentioned that the universal-apparatus assembly shown in Fig. 2.1d (Section I.B) can be used for liquid-liquid extraction when the sample placed in the cup has a density greater than that of the solvent in the reservoir vessel. A fun-

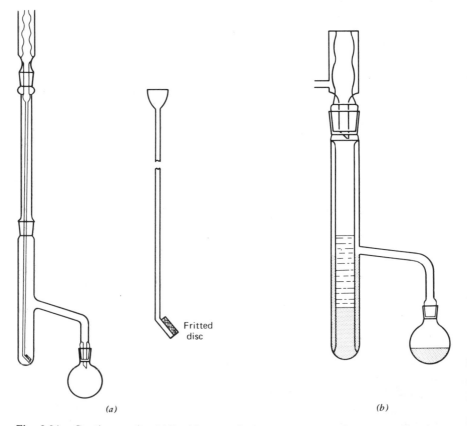

Fritted
disc

(a)   (b)

Fig. 2.24. Continuous liquid-liquid extraction apparatus: (a) for use with solvent lighter than sample; (b) for use with solvent heavier than sample.

nel with a bend at the bottom is used in place of the straight funnel, and the bottom opening is plugged with a tiny roll of filter paper. Some solvent is put into the funnel before inserting it into the cup in order to prevent the sample from entering the stem of the funnel. On the other hand, if the sample is lighter than the solvent, the funnel is replaced by a straight tube with a flange on top, similar to that depicted in Fig. 2.24b. The apparatus described by Wright and co-workers [87, 88], shown in Fig. 2.25, have wave-shaped extraction chambers. This device delays the solvent in its passage through the solution, thus prolonging the time of contact. The volume of solution required, however, is also increased. The extractor for solvent lighter than water requires 8 ml of sample, and that for solvent heavier than water 25 ml of sample. Liquid-liquid extraction apparatus for small-scale operation are commercially available (Fig. 2.26); they require 10 ml or more sample. The apparatus shown in Fig. 2.27 are suitable for operations in which the course of extraction needs to be monitered without dismantling the extraction assembly. When the solvent is

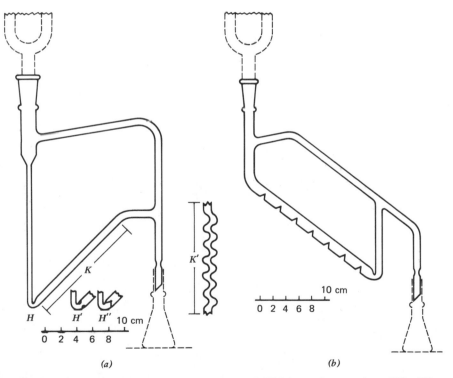

Fig. 2.25. Liquid-liquid extraction apparatus of Wright and co-workers [87, 88]: (a) for use with solvent lighter than sample solution; (b) for use with solvent heavier than sample solution. (Courtesy *Anal. Chem.*)

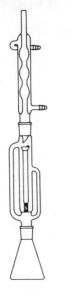

(a)   (b)

Fig. 2.26. Liquid-liquid extraction apparatus: (a) for solvent lighter than water; (b) for solvent heavier than water. (Courtesy Chatas Glass Co.)

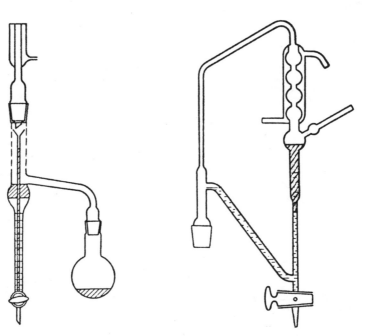

(a)   (b)

Fig. 2.27. Liquid-liquid extraction assembly with attachment for withdrawal of solution: (a) for use with solvent lighter than sample; (b) for use with solvent heavier than sample.

lighter than the sample, a portion of the sample solution can be withdrawn through the tap from time to time. When the solvent is heavier than the sample, the extract can be withdrawn for testing and then returned to the apparatus through the side arm.

Gleit [89] has fabricated a multipass countercurrent extractor from commercially available glass components. The apparatus was designed for stripping metal complexes from an aqueous phase into a lighter organic solvent, but has other potential uses. Ingamells [90] has described two devices for the extraction of metals from aqueous solution into organic phases that are heavier than water. Iffland [90a] has modified the Wehrli extractor with a solvent return for continuous extraction using solvents that are denser than the sample.

In view of the possible hazards involved, extractors for working with radioactive materials have been studied by several groups of investigators. Zaborenko and Alian [91] have described an apparatus for the continuous extraction of radioactive isotopes, with automatic recording of activity. Kienberger [92] has designed an automatic extractor for the purification of uranyl nitrate solution from various impure sources, by an extraction process of 15 operations. Jensen and Bane [93] have described a continuous ether extractor, which minimizes manipulations involving hazards and can be used to separate uranium from large amounts of fission-product activity or from various nonradioactive components. The device, shown in Fig. 2.28, enables the solution from which the uranium has been extracted to be transferred directly and safely into a shielded active-waste container without dismantling the apparatus. It also permits an aliquot of this solution to be segregated for confirmation that all the uranium has been extracted, or for further analysis. Wall [94] and Kemp and Ponting [95] have described micro mixer-settlers for continuous countercurrent solvent extraction.

Clarke and Kalayci [96] have constructed a simple rotary extractor suitable for isolating drugs from biological fluids without forming unbreakable emulsions. Hadd and Perloff [97] have designed an automatic extractor, working at room temperature, that can be used for extracting urine, blood, and tissue homogenates with solvents heavier or lighter than water. Milwidsky [98] has described an apparatus for the automated extraction, with subsequent washing, of unsaponified or unsulfonated compounds from soap or detergent. The reader is referred to the work of Craig [99] on the elaborate apparatus for countercurrent distribution.

C. EXTRACTION BY CODISTILLATION

Steam distillation of a few milliters of liquid or centigram quantities of solid material, such as leaves and fruits, can be conveniently carried out

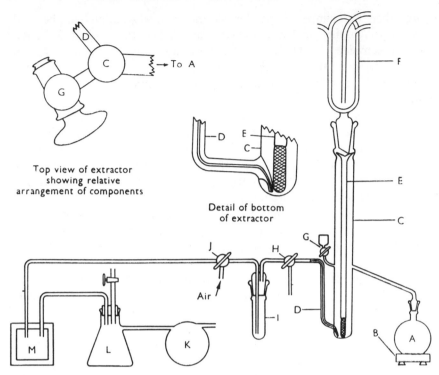

Fig. 2.28. Continuous ether extractor for radioactive materials. (Courtesy *Analyst*.)

by putting the sample in a test tube with side arm. Steam enters through a delivery tube from the top while the side-arm of the test tube is connected to the condenser. Alternatively, the universal reaction vessel (see Section I.B) can be used; the assembly is shown in Fig. 2.29. The receiver can collect up to 1 ml of the organic condensate. Before the distillation starts, water is added to the receiver until it reaches the upper bend. In this manner, organic liquid heavier than water will settle down in the reservoir, as illustrated in Fig. 2.29a, and can be withdrawn by means of a capillary pipet. On the other hand, if the organic distillate is lighter than water, it will rest on the water surface and can be recovered by disconnecting the receiver, as shown in *b*.

The steam-distillation apparatus shown in Fig. 2.30 uses the steam generator as heating jacket. Bohm [100] has described a similar design that incorporates a valve and leveling tube in the outer flask to control the pressure of the steam automatically. Antonacopoulos [101] has described an improved apparatus for the quantitative distillation of steam-volatile substances. Steam is generated in a wide-mouthed round flask and, during

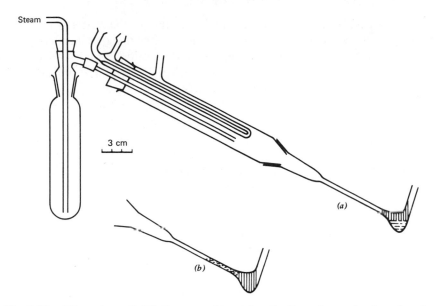

Fig. 2.29. Micro steam-distillation assembly: (*a*) collection of organic phase heavier than water; (*b*) collection of organic phase lighter than water.

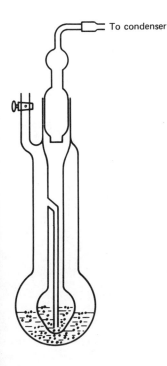

Fig. 2.30. Steam-distillation apparatus with jacket.

45

the main operation, passes through a tube leading to the bottom of a tube-shaped vessel containing the sample, which vessel fits into the mouth of the flask so that its contents are heated by the boiling water, and is connected with a still head and condenser. During the preliminary heating, a tap-funnel situated near the top of the flask is kept open to prevent steam condensing in the sample tube. According to the author, this apparatus gives, in comparison with the conventional arrangement, increased yields of volatile matter in a smaller volume of distillate. Riley [102] has employed a small-scale multipurpose distillation unit for steam distillation. Seehofer and Borowski [103] have fabricated a steam-distillation apparatus that is applicable to small samples and has automatic cleaning arrangements.

When the volume of organic condensate is more than 1 ml, the extraction apparatus of Wasicky and Akisue [104] may be used. As shown in Fig. 2.31, the sample and water are placed in a round flask. The mixed vapors pass through the condenser, and the condensate separates into two phases in a vertical receiving tube. The organic phase, which is heavier

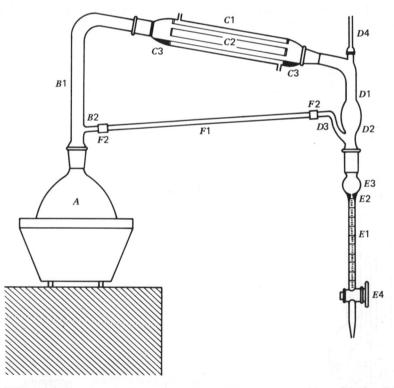

Fig. 2.31. Volatile-oil extraction apparatus of Wasicky and Akisue [104]. (Courtesy *Rev. Fac. Farm. Bioquim. São Paulo.*)

than water, is collected in the buret, while water runs back to the round flask through the side tube. If the organic phase to be recovered is lighter than water, it is collected above the water after the receiving tube has been previously filled with water up to the side tube. At the end of the experiment, water in the receiving tube is allowed to drain off, so that the organic phase runs into the buret to be measured.

D. MISCELLANEOUS SPECIAL EXTRACTION DEVICES

Grussendorf [105] has described two devices for solvent extraction. Solvent from a reservoir flows, under gravity, through the base of a tube containing the sample; from the top of the tube it is siphoned into a distillation vessel, the vapor being condensed and returned to the reservoir.

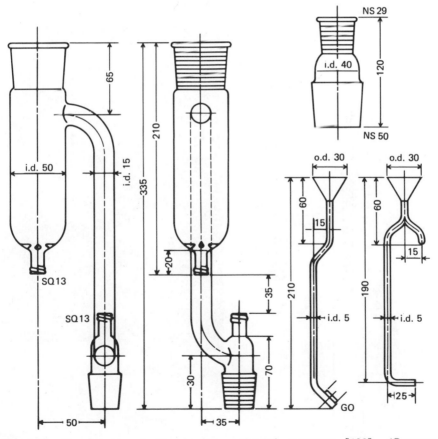

Fig. 2.32. Components of Trutnovsky's universal extractor [106]. (Courtesy *G.I.T. Fachz. Lab.*)

The apparatus is self-adjusting and, according to the author, more efficient than a Soxhlet apparatus. It can be modified to perform liquid-liquid extraction. Substances containing large amounts of water, such as fresh greenplants, cannot be extracted directly, and a centrifuge beaker is used for separation, following mixing with solvent, as a preliminary step to the extraction process. Trutnovsky [106] has described a simple universal extractor fabricated from several integral parts shown in Fig. 2.32. These components can be assembled into a run-through extractor, a Soxhlet extractor, or liquid-liquid extractors for solvents heavier or lighter than water.

Levy and Lifshitz [107] have constructed an apparatus for extracting fats that consists of three parts connected by ground-glass joints. A solid sample, immersed in the solvent throughout the extraction, can be extracted as in a Soxhlet extractor, or an aqueous sample can be extracted with a solvent lighter than water. Erdos [108] has described a micro-

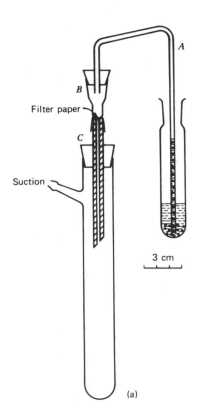

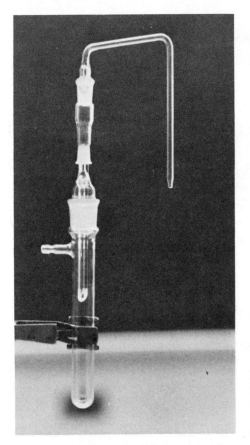

apparatus for liquid-solid extraction and for the subsequent removal of the solvent. Sutton and Vallis [109] have developed a phase-separating glass centrifuge applicable to solvent extraction procedures. It may be used for rapidly separating either immiscible liquid-liquid or liquid-solid phase mixtures.

VI. DEVICES FOR FILTRATION

A. GRAVITY FILTRATION APPARATUS

Filtration of 1 to 25 ml of solution and collection of 1 to 100 mg of precipitate are conveniently carried out with the apparatus shown in Fig. 2.33. Through the siphon A, solid and supernatant liquid can be trans-

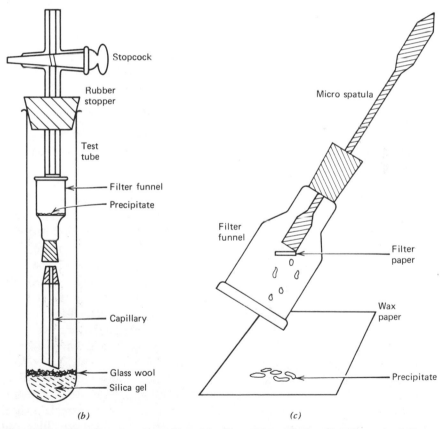

Fig. 2.33. Microfiltration assembly. (*a*) Illustration of the filtration: *A*, delivery tube; *B*, filter funnel; *C*, capillary. The photograph shows an all-glass apparatus and flexible clamp. (*b*) Drying the precipitate. (*c*) Removing the dried precipitate.

ferred quantitatively into the filter funnel *B*. The latter has a standard tapered joint for connection with the capillary tube *C*, which has a bore of 1.5 mm and is flattened at the top. These three components are assembled as follows: The capillary tube *C* is connected to the receiver (e.g., the test tube with side arm in Fig. 2.33, or the bell jar in Fig. 2.36), and a filter-paper disc of about 4-mm diameter is placed on the flat top, followed by one drop of the solvent to be employed for washing the precipitate. If necessary, slight suction may be applied to hold the filter-paper disc in place. Now the filter funnel *B* is joined to the capillary tube *C*. After affixing the siphon *A* to the filter funnel *B*, the reaction vessel containing the sample is brought into position. First, most of the supernatant liquid is withdrawn by suction into *B*, filtered through the filter paper disc and collected in the receiver. The washing solvent is then added to the reaction vessel. Finally, the precipitate together with the washing is transferred into the filter funnel *B* as illustrated in Fig. 2.33*a* (note that the solid particles should always be mixed with liquid while in the delivery tube). After the solid has been collected in the filter funnel *B*, the latter is separated from the siphon *A* and capillary tube *C*. The filter funnel *B* with its contents is placed in a drying device to remove the last traces of solvent; then the solid is recovered by pushing a rod through the narrow end of the filter funnel, as illustrated in Fig. 2.33*c*. This simple and inexpensive micro filtration apparatus is commercially available [40]. Since the filter funnels have standard tapered joints, multiple filtration experiments can be made without disconnecting the capillary tube *C* from the receiver if the filtrates need not be collected separately.

The filtration devices designed by Schwinger [110], Yagoda [111], and Donau [112] all employ capillary tubes to support the filter to collect milligram amounts of precipitate; these apparatus are not as convenient as the one described above. Stock and Fill [113] have modified the Schwinger filter and also constructed a filter which can handle more than 30 mg of product. Other workers have used the glass funnel for microscale filtration. Thus Soltys [114] has recommended a device consisting of a glass nail, known as the Willstätter button, inserted into the stem of a 45-degree funnel. Erdey and Buzas [115] have suggested using a glass or porcelain disc with perforations. Porcelain perforated discs are commercially available; they should be about 5 mm thick and beveled to fit the funnel (see Fig. 2.34*a*). Other commercial filters suitable for collecting small amounts of solid are the Hirsch funnel, micro Gooch crucible, porcelain crucible with porous disc, and sintered-glass filter. It is apparent that filtration by means of a glass nail, perforated glass or porcelain disc, and Hirsch funnel requires filter paper. While micro Gooch crucibles and sintered glass filters are provided with filter beds, it is often necessary to

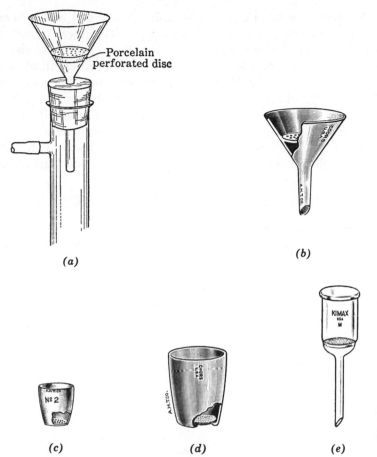

Fig. 2.34. Small-scale filters: (*a*) Porcelain disc on glass funnel; (*b*) Hirsch funnel; (*c*) Micro Gooch crucible; (*d*) Porcelain crucible with porous disc; (*e*) Sintered-glass filter. (Courtesy A. H. Thomas Co.)

add a layer of filter paper to prevent fine particles from clogging up the pores. Specifications for sintered glass funnels and other filtration apparatus have been published [116], but not all filters on the market are expected to conform to the standards. Fletcher [117] has described a filtration device for organic gravimetric analysis in which an Alundum thimble is used as the filtering crucible. It has considerably more filtering surface than noporous-walled filters of the same volume capacity.

The collection of the filtrate deserves attention. If the filtrate is to be further processed, it is advantageous to collect it in a container that need not be transferred to another vessel. A simple device is illustrated in

Fig. 2.35, which consists of a glass tube with side arm and two open ends, each being fitted with a one-hole rubber stopper. The lower rubber stopper carries a glass rod flattened at the top. Receivers of different sizes can be used by adjusting the position of the glass rod. After filtration, the upper stopper carrying the filter is disconnected, and the glass rod is pushed up so that the receiver containing the filtrate can be taken out easily. Another device, shown in Fig. 2.36, makes use of a bell jar which is kept airtight by resting it on a ground-glass plate or, preferably, a rigid rubber mat [118]. This arrangement allows direct delivery of the filtrate into wide-mouth containers such as evaporating dishes and watch glasses.

B. APPARATUS FOR INVERTED FILTRATION

The process of withdrawing the filtrate and leaving the precipitate in the original vessel is known as inverted filtration. Fig. 2.37 shows a

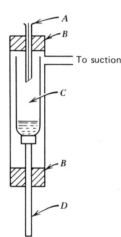

Fig. 2.35. Device for collecting filtrate: *A*, capillary tube or stem or filter funnel; *B*, one-hole rubber stoppers; *C*, receiver; *D*, glass rod with flat top.

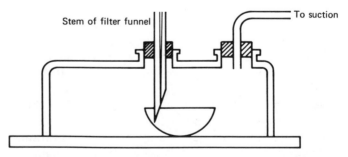

Fig. 2.36. Collection of filtrate in a dish.

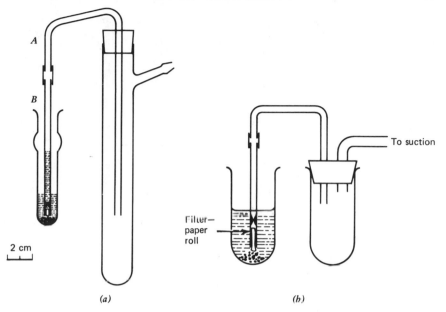

Fig. 2.37. Inverted filtration: (*a*) from universal reaction vessel; (*b*) from micro-beaker.

simple device for performing microscale inverted filtration. The filterstick *B* is prepared from a 4-mm glass tube (internal diameter, 2 mm) by making a capillary constriction about 10 mm from the bottom. The bottom of the filterstick is fitted with a tiny roll of filter paper, the end of which protrudes 1 mm beyond the rim. The filter stick is connected to the receiver by means of the bent tube *A*. The supernatant liquid is withdrawn by applying suction. It is apparent that the solid cannot be dried by suction as in gravity filtration. When the weight of the precipitate is to be determined, the filterstick should be weighed with the container before and after the experiment.

Various types of filter sticks have been proposed (see Fig. 2.38). The commercial models have sintered glass, porous porcelain, or platinum sponge as the filter mat. When the filtration is performed for the purpose of recovering the precipitate, filter sticks in which the filter mat does not flash with the rim should not be used; otherwise some of the precipitate will stay inside the filter stick and cannot be removed easily. Cooper [119] has described a filter stick having a glass bead within the beveled opening of a glass tube. Marhoul and Ruzicka [120] have designed a micro-funnel for inverted filtration.

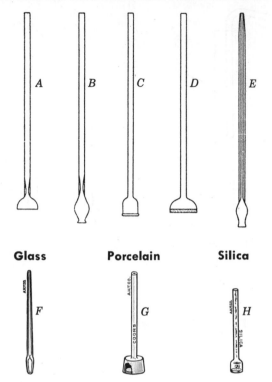

Fig. 2.38. Various types of filter sticks. (Courtesy A. H. Thomas Co.)

C. OTHER MICROFILTRATION DEVICES

The filtration devices discussed in the previous sections are not suitable for manipulating less than 1 ml of solution or for collecting precipitates in the microgram region. When the reaction vessel is an open container such as a microne, the fine capillary pipet is the equipment of choice for separating the liquid from the solid phase. The capillary should have 0.5- to 1-mm bore with one end drawn out to a tip of 10- to 20-mm length and not less than 0.2-mm bore. While the microne is being held in nearly horizontal position, as shown in Fig. 2.39a, the capillary pipet is inserted slowly so that its tip touches the upper portion of the solution, whereupon liquid enters the pipet by capillary action. The rate at which the liquid is withdrawn can be regulated by inclining the microne. It should be noted that the pipet should not rest along the walls of the microne (see Fig. 2.39 b); otherwise liquid will creep out along the outside of the pipet towards the rim of the microne. A syringe may be employed in place of the capillary pipet. Since the needle is opaque, however, movement of the

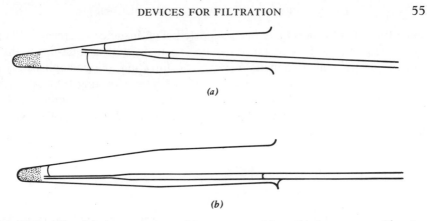

Fig. 2.39. Filtration in a microne: (*a*) correct position, (*b*) incorrect position for capillary pipet. (Courtesy Springer-Verlag.)

solution will not be seen, and fine particles entering the syringe cannot be detected immediately.

The collection of filtrate from a drop of solution on the microscope or spot plate is carried out by means of the capillary pipet, as illustrated in Fig. 2.40*a*. If the filtrate is not needed for further experiment, a simple way is to absorb the supernatant liquid onto a piece of blotting paper (see Fig. 2.40*b*).

Sealed capillary tubes are generally used as containers when microliter volumes of liquid or microgram amounts of solid are to be manipulated in closed vessels. Filtration in a closed capillary is carried out by centrifugation. The operation is described in detail in a later section on recrystallization (see Chapter 3, Section II.C).

Since flexible plastic containers became popular, devices using plastic

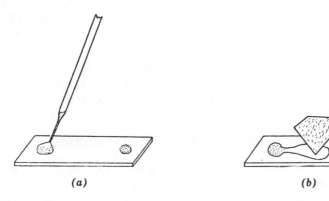

Fig. 2.40. Filtration on a slide: (*a*) with capillary pipet, (*b*) with blotting paper.

apparatus for filtration have been reported. For instance, Gerarde [121] has described a perforated polyethylene cap to fit with a filter-paper disc onto a plastic vial (Unopette reservoir); filtration is accomplished by squeezing the vial, as shown in Fig. 2.41, or by perforating the vial with a syringe. Bosewall and MacKay [122] have fabricated a semimicro filtration apparatus by incorporating a Gooch crucible with polyethylene addition tube and couplers.

D. DEVICES FOR FILTRATION UNDER CONTROLLED ATMOSPHERE

A general technique for performing microscale manipulations under controlled atmosphere involves the use of a "dry box," which is discussed in a later chapter (see Chapter 9, Section III.B). Simple devices, however, can be employed for microfiltration. For instance, two filter beakers can be connected as illustrated in Fig. 2.42, and a current of dry air or any suitable gas can be conducted through either terminal. Precipitation or crystallization is caused to take place in one filter beaker as it sits on the bench, while the other filter beaker remains empty. Then the positions of the two filter beakers are changed to those shown in the figure, and the solution is filtered from F_1 into F_2, while the solid is retained in F_1. This device is not recommended for filtering solutions that contain fine particles that may clog up the sintered glass disc. Sant and Prasad [123] have described an apparatus for inverted filtration under controlled atmosphere that is shown in Fig. 2.43; it can be adapted to small volumes. Kowala [124] has constructed an apparatus for filtration in an inert atmosphere that is suitable for use in a refrigerated centrifuge.

Shiba [125] has described a simple all-glass apparatus for hot filtration whereby flammable solvents can be used without fire hazard. The solvent-containing flask is connected to a specially constructed filter funnel, the constricted upper rim of which fits into the belled-out lower end of the condenser. There is no danger of loss of solvent or deposition of crystals on the filter paper. Horak [126] has designed a hot microfiltration apparatus in which the filter stick, connected by means of a ground-glass joint to an external jacket, is heated by the vapors of the solvent.

VII. DEVICES AND TECHNIQUES FOR DRYING

In chemical experimentation, the term "drying" may mean one of the following operations:

(1) Removal of the residue solvent (water or organic liquid) in a solid phase obtained in crystallization or extraction experiments.

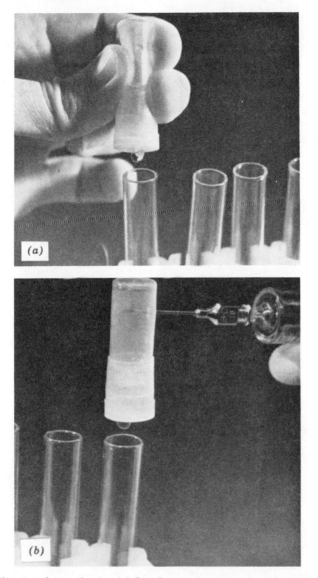

Fig. 2.41. Filtration from plastic vial [121]: (*a*) by squeezing Unopetto reservoir, (*b*) using syringe. (Courtesy *Microchem. J.*)

(2) Removal of volatile impurities collected on the surface of the solid sample during storage.

(3) Removal of water that is an original constituent of the sample (e.g., fresh leaves).

(4) Removal of water dissolved in an organic liquid or solution.

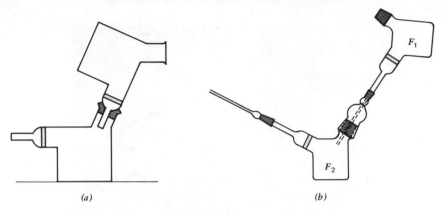

Fig. 2.42. Connecting two filter beakers for filtration: (*a*) directly, (*b*) through adapters. (Courtesy Springer-Verlag.)

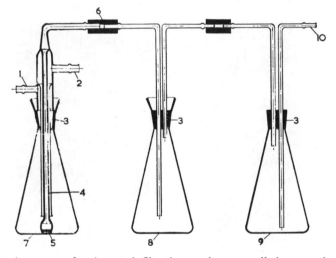

Fig. 2.43. Apparatus for inverted filtration under controlled atmosphere [123]: 1, gas inlet; 2, gas outlet; 3, B-24 joint; 4, capillary tube; 5, porosity-3 disc; 6, connector tubing; 7, 8, Conical flasks; 9, trap; 10, to filter pump. (Courtesy *Lab. Pract.*)

No special apparatus is needed for operation (4) in microscale manipulations. Drying is carried out in the same container holding the wet sample. Solid dehydrant is added and the mixture is agitated. It is important to use a dehydrant that does not disintegrate easily and is of sufficiently small dimensions that the crystals can move freely in the narrow capillary tubes or microne tip. Indicating silica gel, if applicable, is a useful drying agent, since it can be prepared in various sizes and provides direct information on when water has been completely removed from the sample.

The literature on drying devices for operations (1), (2), and (3) is legion. Numerous apparatus have been proposed for a variety of microchemical procedures. In this section, discussion will be confined to devices that serve general purposes. It is worthy of note that the drying of small quantities of material usually can be accomplished quickly and efficiently by means of simple equipment. Since the sample is small, the volatile matter to be removed, whether on the surface or imbedded in the solid phase, can be easily exposed and then expelled.

A versatile apparatus for drying at elevated temperatures in open air is shown in Fig. 2.44. It is fabricated from a 500-W tubular infrared lamp and has an aluminum reflector [127]. Having an area of 34×28 cm, it can accommodate several reaction vessels simultaneously. The temperature to which the vessels and their contents are exposed is dependent on the distance between each vessel and the infrared lamp. This apparatus can be used in a wide range of experiments for microanalysis and microsynthesis [128].

A common technique for the removal of water or organic solvent from small amounts of solid is to place the sample in a closed vessel containing a reagent which has great affinity for the vapors of the substance to be

Fig. 2.44. Apparatus for drying in open air [127]. (Courtesy *Mikrochim. Acta.*)

removed. Thus silica gel and phosphorus pentoxide may be used for the removal of water, paraffin for the removal of benzene, and sulfuric acid for the removal of pyridine. In order to facilitate the transfer of the vapors from the sample to the absorbent, the vessel is evacuated for a short period of time, either at room temperature or at elevated temperatuers. The vessel is then kept closed until the sample is "dried." Any device that permits the abovementioned steps to be performed will serve the purpose. A convenient and compact apparatus is shown in Fig. 2.45. It is an improved model of the automatic temperature-controlled drying apparatus

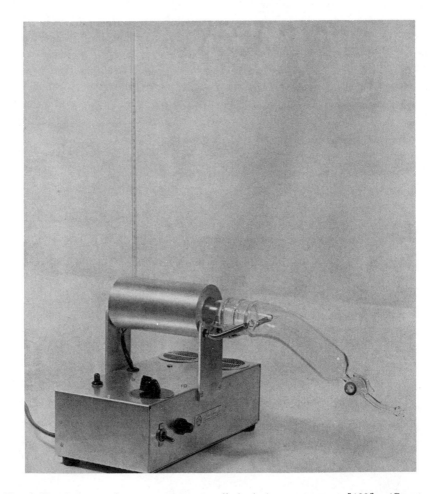

Fig. 2.45. Automated temperature-controlled drying apparatus [129]. (Courtesy Chatas Glass Co.)

described by Schenck and Ma [129], and is commercially available [40]. The closed vessel consists of a glass pistol connected to a stopcock for evacuation. The material to be dried is placed in one end of the pistol, which fits into the electrically heated chamber, while the reagent (absorbent) is placed in the cold, outer part of the pistol. The temperature can be set on the dial in the range from 25 to 195°C and maintained within ±1°. The thermometer ascertains that drying is carried out at the preset temperature. This apparatus has distinct advantages over the conventional Abderhalden-type vacuum drying apparatus, which requires a boiling liquid to heat the closed vessel. It indicates the exact temperature at which the sample is being dried, so that identical conditions can be reproduced for comparative analysis. Besides eliminating the tedium of changing the boiling liquids to attain different temperatures, the electric drying chamber does not release water and organic vapors into the laboratory. It occupies very little space and is less fragile than the Abderhalden apparatus.

Certain precautions should be taken in microscale drying that may be overlooked in macro experiments without causing serious errors. Many organic solids have significant vapor pressure at elevated temperatures. When the amount of sample is in the milligram region or less, an appreciable portion of the pure compound can be lost on evacuating the drying apparatus. With complex materials, more than one volatile component may be expelled. Hence, for analytical purposes, the drying condition should be controlled so that the procedure can be duplicated. A precipitate may contain solvent of crystallization, some of which may be liberated while the surface solvent is removed. For microanalytical work, it is desirable to raise the drying temperature in stages to see if it is possible to expel the solvent completely without affecting the substance under investigation.

Hecht [130] has undertaken an extensive review of the drying devices that have been proposed for use in microgravimetric analysis. Many apparatus described therein are designed for specific analytical containers such as the microcrucible, microboat, and filter beaker. The British Standards Institution has published the specifications for vacuum drying ovens for microchemical purposes [131]. Smith [132] has constructed a tubular desiccator that comprises a jacketed horizontal tube, fitted at each end with stopcocks. Heating to a selected temperature is achieved by boiling suitable liquids and allowing the vapors to circulate through the jacket. Mehr [133] has described a vacuum drying apparatus suitable for operation in the temperature range of approximately −50° to 90°C, by circulating water kept at the specified temperature, alcohol–carbon dioxide mixture, or a similar cooling fluid through a jacketed tube closed with a two-hole rubber stopper. Skinner [134] has advocated a set of simple micro vacuum desiccators for drying samples at ambient temperature. The apparatus is illustrated in Fig. 2.46. The vessel is designed to hold sintered glass crucibles or sample bottles. The desiccant is placed in the lower section, below the indented tips, and separated from the upper section by a pad of glass wool. A tightly fitted rubber stopper makes an adequate seal for most purposes. A rack of five desiccators makes available five separate reagents for immediate use. Where compounds must be stored under refrigeration in an anhydrous state, these small desiccators take up relatively little room in a refrigerator or deep freeze.

Another technique for expelling volatile residues in microscale manipulations involves passing a current of appropriate gas over the sample. For instance, the sintered glass filter tube containing the microsample can be dried by aspirating air through the vessel, as shown in Fig. 2.44, while being heated in the drying apparatus. The removal of solvents from thin-layer chromatographic plates is conveniently accomplished by blowing warm air onto the plate. When the microsample is in the capillary tube, the drying gas is conducted through a fine capillary, as illustrated in Fig. 2.47. The incoming gas should be purified by passing it through a wad of cotton, in order to prevent fine particles in the gas line from contaminating the microsample.

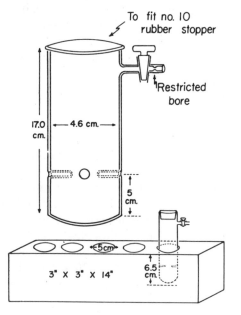

Fig. 2.46. Micro vacuum desiccator of Skinner [134]. (Courtesy *Anal. Chem.*)

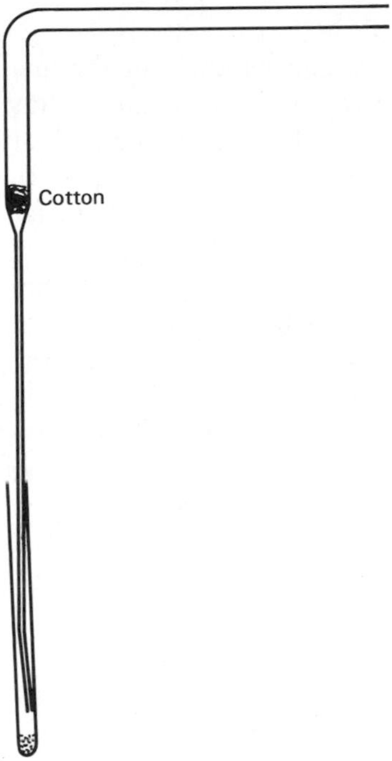

Cotton

Fig. 2.47. Drying crystals in a capillary tube.

VIII. DEVICES FOR DISINTEGRATION, HOMOGENIZATION, AND AGITATION

A. PURPOSE

In microscale experiments, the respective operations involving disintegration, homogenization, and agitation are performed for the purpose of obtaining a representative sample or attaining uniform reaction in the system. Thus, when a heterogeneous solid material is received for analysis, the first step is to test whether or not all components can be dissolved in a suitable solvent. If so, the preferable procedure will be to prepare the homogeneous solution and take aliquots for analysis. In case such an approach is not feasible, the material should be pulverized, and the particles thoroughly mixed. Contrastingly, a pure compound need not be pulverized before analysis. It should be remembered that grinding, mixing, and agitation may result in serious contamination and loss of sample when working with micro quantities of material. Beaulieu [134a] has reported on dust-free grinding.

B. GRINDING AND SIEVING DEVICES

The best grinding device depends on the size and hardness of the material submitted for analysis. In the milligram region, most organic solids can be reduced to small particles by pressing them between two pieces of microscope slide or glazed paper. Inorganic materials usually require a mortar and pestle. The smallest agate mortar (Fig. 2.48a) has a diameter of 25 mm. The porcelain mortar designed by Alber [135] has 20-mm inside diameter at the top, tapering to 7 mm at the bottom; it is 15 mm deep and is glazed all around except the outside bottom surface. Both ends of the pestle are glazed, the larger end being suitable for crushing the material prior to powdering it with the smaller end. After being finely ground, the powder is sifted with a sieve, which can be conveniently prepared as follows [136]. A piece of silk fabric or 100-mesh copper wire screen is spread over a 5-ml beaker, and the edges are then fastened by means of a rubber band or metal wire (Fig. 2.49). The powdered material

(a)   (b)

Fig. 2.48. Mortar and pestle: (a) agate mortar; (b) porcelain mortar of Alber [135]. (Courtesy A. H. Thomas Co.)

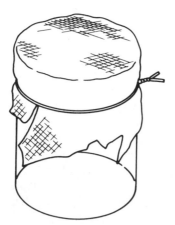

Fig. 2.49. Microsieve. (Courtesy Van Nostrand Co.)

is transferred onto the sieve and pushed with the microspatula. Particles that do not pass the sieve are returned to the mortar and reground. The powder in the beaker is thoroughly mixed, and a representative sample is obtained by quartering. A commercial microsieve set [137] consisting of four sieves of about 75-mm diameter and 25-mm depth is suitable for separating particles from 125 to 1000 μm.

Dangoumau [138] has described a grinding apparatus for 1 to 5 g of material that depends on vigorous vertical reciprocating motion of the sample with stainless steel balls and is fitted with a jacket to control the grinding temperature. Schlesinger, Nazaruk, and Reggel [139] have constructed a small mill that can grind as little as 0.5 g of coal or other solids. As shown in Fig. 2.50, the mill is a rotating thin cylinder in which smaller cylinders roll. The rolling cylinders are in contact, and a differential grinding action takes place between them, as well as crushing between them and the periphery of the mill. When the cylinder halves are bolted together, a cavity having a volume of less than 100 ml is formed. A small gear-reduction motor drives a pulley set, which rotates the mill arbor. The micromill that is commercially available [137] has a grinding chamber of approximately 50 ml and requires minimum sample size of 20 ml. Vorweck [139a] has described a disc mill that can reduce the sample in 10 seconds to particle size precisely adjusted to within 10 μm.

Bannister and Simler [140] have designed a self-feeding micromincer for animal tissues. Barton [141] has developed a grinding device capable of reducing 0.1 to 1 g of leaves and roots to fine powder in 3 min. A mixer mill is used with a tubular steel plunger reciprocating within a polyethylene weighing bottle. Stewart and Tipton [142] have described an inexpensive way to reduce granular samples of ash to a fine powder of particle size 75 to 150 μm. A piece of 20-mil tungsten wire, formed into a spiral in the shape of a cocoon around a $\frac{3}{8}$-in. tungsten carbide ball, is placed with the sample in a plastic vial with a polyethylene cap and agitated in a mixer mill for several periods of 15 sec. The spiral should slide freely inside the vial and should be at least half as long as the vial.

The disintegration of microgram amounts of solid is usually performed in the capillary tube or spot plate by means of a glass or metal rod. There is no special device for separating such small quantities on a sieve.

C. HOMOGENIZERS

Most homogenizers described in the literature are concerned with biological materials. Went [143] has constructed a high-speed microhomogenizer for use with 2.5 to 7 ml of sample. Vecerek and co-workers [143a] have described a homogenizer for processing 1 to 10 mg of material—for

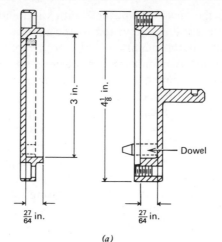

(a)

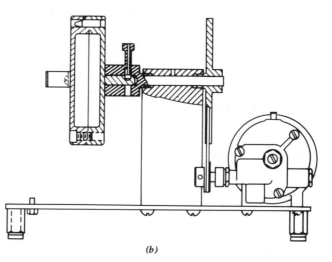

(b)

Fig. 2.50. Laboratory cylinder mill of Schlesinger, Nazaruk, and Reggel [139]: (a) cylinder halves; (b) mill assembly. (Courtesy *J. Chem. Educ.*)

example, liver. Alexander [144] has designed a laboratory homogenizer suitable for the preparation of suspensions of wet fibrous materials such as silage, wool, and biscuit meal. The rotatory apparatus induces continuous circulation of the suspension under preparation through a precision shear. The standard deviation of samples prepared in this way is ±2.8 mg in 480 mg, in respect to organic-matter content. Hikichi and Miltimore

[145] have described a device for homogenizing plant tissue ground in liquid nitrogen. It is designed on the principle of the common cement mixer and fabricated to facilitate mixing of the frozen powdered material while preventing operator contact with liquid nitrogen. This apparatus should be applicable to mixing any noncorrosive material in a granulated, powdered, or liquid form. The automated homogenization of tablets and capsules containing pharmaceutical mixtures has been studied by Ahuja, Spitzer, and Brofazi [146]. Cowgill [146a] has developed a method to homogenize 200 mg of powder material with 50 mg of graphite, using acetone and a sonicator, for optical-emission analysis.

While commercial blenders are common equipment in the chemical and biochemical laboratory, apparatus is not yet available for shearing and mixing in volumes of 100 ml or less. McCarter [147] has constructed a jacketed blender-reactor with which temperatures can be controlled over a wide range by proper selection of the cooling or heating medium circulating in the jacket.

D. AGITATION AND SHAKING DEVICES

Conventional stirrers are seldom used in microscale operations. If the reaction vessel is large enough for the insertion of stirrer blades, the latter are preferably fabricated from plastic material [148–150] instead of glass. Magnetic stirrers are suitable for agitation in small vessels. The stirring bar can be made by sealing a nail in capillary tubing, and the magnet can be affixed to a toy motor [151]. Smith [152] has described a device that consists of a magnet sealed in a plastic toothed wheel; the latter in turn is sealed in a plastic case, fitted with tubes so that the wheel can be rotated by means of running water, compressed air, or vacuum. This apparatus is commercially available [153]. Miwa [154] and Bhati and Anthony [155] have described a device made from two rod magnets encased in glass tubes joined midway to a glass collar in order to stir liquids in a round-bottomed flask without scratching or the stirrer being thrown off field.

Corliss [156] has proposed using a Teflon shield on top of the magnetic stirring unit to dissipate the heat evolved by the motor. Gugger and Mozersky [157] have described an efficient micro mixing device that has a mixing chamber that consists of a Teflon-coated bar magnet 2 mm in diameter and 7 mm long enclosed in a glass tube of 4-mm i.d. When the mixing chamber is placed on a magnetic stirrer, the primary motion obtained is rotation of the bar about its cylindrical axis. In addition, the bar wobbles, and there is some movement to and fro in the direction of the axis of the glass tubing, both of which enhance the stirring action.

Cerveny and Weitkamp [158] have constructed a magnetically stirred, high-pressure microreactor for use at temperatures up to 343°C.

Another magnetic stirring device proposed by Linderstrøm-Lang and co-workers [159] uses tiny beads known as "fleas" that consist of iron powder sealed in a thin shell of glass with diameter of about 2 mm. It is employed in ultramicro titrations in enclosed vessels [160]. A small electromagnet like the type for buzzers is mounted in such a manner that its poles are at the level of the surface of the liquid being stirred. The assembly includes a flasher that will alternately close and open the circuit to the magnet, thus causing the bead to rise and then to drop back.

When open microcontainers are used, passing a current of air or inert gas through the solution offers a simple and convenient way to stir the material. This operation is illustrated in Fig. 2.51. Walsby [161] has described a stirring method for microtitration to circumvent interference by voluminous precipitation during the experiment.

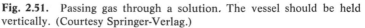

Fig. 2.51. Passing gas through a solution. The vessel should be held vertically. (Courtesy Springer-Verlag.)

Simple devices for shaking in microscale manipulations can be easily fabricated [162–165]. One apparatus is shown in Fig. 2.52; it consists of a platform supported by four wire springs. Fig. 2.53 illustrates a device that produces intense vibrations. The reaction vessel is affixed to the side of a small motor, which is equipped with an eccentric piece and suspended at four points by rubber tubings.

Shaking and vibrating apparatus for small volumes have appeared on the market recently. The tube rocker [166] shown in Fig. 2.54 can hold tubes of 6- to 35-mm diameter. A commercial ultrasonic device [167] has an adapter to fit a test tube containing as little as 2 ml of solution. A precision shaker [168] bath is claimed to permit precise control of temperature from −10° to 100°C, and shaking speeds of 20 to 200 oscillations per minute.

Robins [169] has constructed an apparatus for shaking volumes of liquids at constant temperature by immersing the apparatus in a water bath. Aggett [170] has compared piston-motion, rotating-disc, and radial-

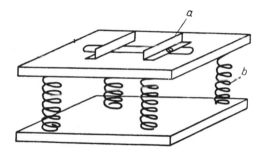

Fig. 2.52. Vibrating platform [163]: *a*, adjustable clamp, *b*, wire spring.

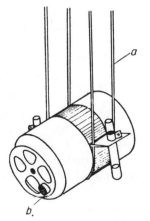

Fig. 2.53. Vibrating device using a small motor: *a*, rubber tubing; *b*, eccentric piece.

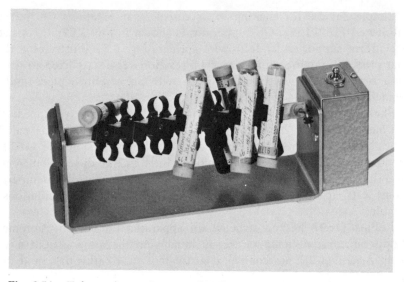

Fig. 2.54. Tube rocker and rotator [166]. (Courtesy Labindustries.)

motion shakers for the simultaneous shaking of 20 small samples at constant temperature, and reported that the radial-motion apparatus was the most efficient and reliable and that it required little maintenance.

REFERENCES

1. P. L. Kirk, *Quantitative Ultramicroanalysis.* Wiley, New York, 1950.
2. P. R. van Haga, *Clin. Chim. Acta,* **8**, 327 (1963).
2a. S. Natelson and D. J. Pochopian, *Microchem. J.,* **18**, 457 (1973).
3. A. A. Benedetti-Pichler and M. Cefola, *Anal. Chem.,* **14**, 813 (1942).
4. E. M. Chamot and C. W. Mason, Handbook of Chemical Microscopy, 3rd ed. Wiley, New York, 1958.
5. C. L. Gordon and R. B. Johannesen, *J. Res. Nat. Bur. Stand., A,* **67**, 269 (1963).
6. T. S. Ma and R. F. Sweeney, *Mikrochim. Acta,* **1956**, 198.
7. T. R. P. Gibb, Jr., *Anal. Chem.,* **29**, 584 (1957).
8. Some supply houses for microchemical equipment are: A. H. Thomas., Co., Philadelphia, Penn.; Fisher Scientific Co., Pittsburgh, Penn., and Fair Lawn, N.J.; Microchemical Specialties Co., Berkeley, Calif. See also *Analytical Chemistry's Laboratory Guide* and *Science's Guide to Scientific Instruments,* both appearing annually. See also Appendix B.
8a. Available from Wheaton Scientific, Millville, N.J.
9. T. S. Ma and J. M. Tien, *Microchem. J.,* **2**, 253 (1958).
10. T. S. Ma, *J. Chin. Chem. Soc., Ser. II,* **9**, 168 (1962).

11. C. A. Dubbs, *Anal. Chem.*, **21**, 1273 (1949).
12. A. Marhoul and V. Ruzicka, *Chem. Listy*, **60**, 1400 (1966).
13. P. Mares, *J. Chromatogr.*, **32**, 745 (1968).
14. B. B. Graves and B. F. Vincent, Jr., *Chem. Ind.*, **1962**, 2137.
15. For example, British Standards Institute, Micro-beakers, B.S. 1428: Part E2 (1954); micro-centrifuge accessories, B.S. 1428: Part E3 (1953), Part E2 (1954).
16. Committee on Microchemical Apparatus, Division of Analytical Chemistry, ACS, *Anal. Chem.*, **26**, 1186 (1954).
17. Alltech Associates, Inc., Arlington Heights, Ill.
18. H. K. Pratt and C. W. Greiner, *Anal. Chem.*, **29**, 862 (1957).
19. V. Knobloch and B. Mudrova, *Mikrochim. Acta*, **1970**, 235.
20. E. L. Eckfeldt and E. W. Shaffer, Jr., *Anal. Chem.*, **37**, 1624 (1965).
21. F. E. Holmes, *Chem.-Anal.*, **52**, 86 (1963).
22. J. D. Smith, *Lab. Pract.*, **19**, 186 (1970).
23. L. D. Quin and R. B. Greenlee, *J. Chem. Educ.*, **31**, 518 (1954).
24. A. A. Benedetti-Pichler, *Identification of Materials*, p. 99. Springer-Verlag, New York, 1964.
25. F. L. Schneider, *Qualitative Organic Microanalysis*, p. 4. Wiley, New York, 1946.
26. F. R. Swift, *Microchem. J.*, **11**, 221 (1966).
27. L. Cahn and W. J. Cadman, *Anal. Chem.*, **30**, 1580 (1958).
28. N. D. Cheronis, A. R. Ronzio, and T. S. Ma, *Micro and Semimicro Methods* [Vol. VI of A. Weissberger (Ed.), *Technique of Organic Chemistry*], p. 118. Wiley-Interscience, New York, 1954.
28a. N. D. Cheronis and T. S. Ma. *Organic Functional Group Analysis*, p. 73. Wiley, New York, 1964.
29. E. Salvioni, *Misura di Masse fra* g10^{-1} *e* g10^{-6}. Messina, 1901.
30. H. K. Alber, *Ind. Eng. Chem., Anal. Ed.*, **13**, 656 (1941).
30a. Pocket microscope, Walter C. McCrone Associates, Inc., Chicago, Ill.
31. F. H. Nicoll, *Rev. Sci. Instrum.*, **40**, 1108 (1969).
31a. J. E. Bainbridge and L. W. Carter, *Lab. Pract.*, **22**, 423 (1973).
32. P. M. Brickey, Jr., J. S. Green, J. J. Thrasher, and W. V. Eisenberg, *J. Assoc. Off. Anal. Chem.*, **51**, 873 (1968).
33. Pierce Previews, Pierce Chemical Co., Rockford, Ill., July, 1970.
34. C. Horwitz and S. Wood, *J. Sci. Instrum.*, **41**, 518 (1964).
35. S. L. Bonting and G. R. Walters, *Anal. Chem.*, **28**, 2035 (1956).
36. T. S. Ma, *J. Chin. Chem. Soc.*, Ser. II, **9**, 176 (1962).
37. S. M. Charlett, *Lab. Pract.*, **3**, 162 (1954).
38. E. Gunders, *Rev. Sci. Instrum.*, **38**, 433 (1967).
39. T. S. Ma and R. T. E. Schenck, *Mikrochimie*, **40**, 245 (1953).
40. Available from Chatas Glass Co., Vineland, N.J.
40a. C. N. C. Drey, F. J. Forrel, and R. Fuller, *Lab. Pract.*, **22**, 43 (1973).
41. D. M. Morgan and M. J. Welland, *Lab. Pract.*, **13**, 981 (1964).
41a. U. Kuemmel, *G-I-T Fachz. Lab.*, **15**, 142 (1971).
42. H. E. Weir and T. F. Rutledge, *J. Chem. Educ.*, **40**, 425 (1963).

43. S. F. Nicholls, *Lab. Pract.*, **12**, 152 (1963).
44. Ace Glass, Inc., Vineland, N.J.
45. A. A. Benedetti-Pichler, *Microtechniques of Inorganic Analysis*, p. 146. Wiley, New York, 1942.
46. A. E. B. Presland and J. R. White, *J. Sci. Instrum.*, Ser. 2, **2**, 67 (1967).
47. F. R. Swift, *Microchem, J.*, **7**, 437 (1963).
48. K. A. Galloway, *Lab. Pract.*, **4**, 334 (1954).
49. E. K. Lubbs, *J. Chem. Educ.*, **40**, 200 (1963).
50. G. Gorbach, *Mikrochim. Acta*, **1962**, 1035.
51. P. Virkola, *Lab. Pract.*, **18**, 852 (1969).
53. S. Foner, *Rev. Sci. Instrum.*, **40**, 1362 (1969).
54. J. P. Simons, *J. Sci. Instrum.*, Ser. 2, **1**, 872 (1968).
55. Y. Alon, *Rev. Sci. Instrum.*, **40**, 20 (1969).
56. R. C. Tyagi, *J. Sci. Instrum.*, Ser. 2, **2**, 995 (1969)
56a. Cryogenic Technology, Inc., Waltham, Mass., Bulletin TDS 5107T (1971).
57. J. Nassler in J. J. Lagowiski (Ed.), *The Chemistry of Non-aqueous Solvents*, Vol. 1, p. 213. Academic, New York, 1966.
57a. A. M. Philips, *J. Chem. Educ.*, **45**, 664 (1968).
58. British Standards Institution, B.S. 2071 (1954); B.S. 1428 Part L 1 (1963); B.S. 2071 (1964).
58a. Corning Glass Works, Corning, New York.
59. A. Wasitzky, Mikrochemie, **11**, 1 (1932).
60. E. B. Colegrave, Analyst, **60**, 90 (1935).
61. B. Blount, *Mikrochemie*, **19**, 162 (1936).
62. B. L. Browning, *Mikrochemie*, **26**, 54 (1939).
63. H. Hetterich, *Mikrochemie*, **10**, 379 (1932).
64. G. Erdos, *Magy. Kem. Lap.*, **24**, 513 (1969).
65. B. Josephson, *Clin. Chim. Acta*, **10**, 290 (1964).
66. J. C. O'Keeffe, *Lab. Pract.*, **7**, 97 (1958).
67. M. Schirm, *Arch. Pharm. Ber.*, **287**, 46 (1954).
68. W. Ensslin, *Z. Anal. Chem.*, **146**, 321 (1955).
69. A. Kiger, *Ann. Falsif.*, **50**, 162 (1957).
70. B. Kaminski, *Acta Polon. Pharm.*, **21**, 93 (1964).
71. A. Hussain and D. C. Spanner, *J. Sci. Instrum.*, **41**, 185 (1964).
72. F. Franks, *J. Tex. Inst.*, **47**, T 369 (1956).
73. P. Mitchell, *Nature*, **172**, 124 (1953).
74. J. K. Scoggin and O. E. Tauber, *Iowa State Coll. J. Sci.*, **28**, 165 (1953).
75. R. Antozewski, *Naturwissenschaften*, **45**, 42 (1958).
76. J. K. Bhatnagar, K. C. Loria, and C. K. Atal, *Indian J. Pharm.*, **27**, 10 (1965).
77. G. Gorbach, *Mikrochemisches Praktikum*, p. 42. Springer, Berlin, 1956.
78. O. König, W. R. Crowell, and A. A. Benedetti-Pichler, *Mikrochemie*, **39**, 281 (1949).
79. V. K. Markov and M. K. Korinfskaya, *Zavod. Lab.*, **28**, 1376 (1963).
80. R. P. A. Sims and G. A. Adams, *J. Am. Oil Chem. Soc.*, **35**, 139 (1958).

81. E. Heftmann and D. F. Johnson, *Anal. Chem.*, **25**, 1578 (1953).
82. M. Branica, *Arh. Kem. Zagreb*, **26**, 119 (1954).
83. G. L. Hubbard and T. E. Green, *Chem.-Anal.*, **53**, 119 (1964).
84. K. W. Mieszkis, *Analyst*, **79**, 109 (1954).
85. D. P. Neves and Y. Tavares, *Science*, **121**, 313 (1955).
86. J. E. Spikner, V. F. Ward, and J. C. Towne, *Chem.-Anal.*, **52**, 50 (1963).
87. M. Wayman and C. F. Wright, *Ind. Eng. Chem.*, *Anal. Ed.*, **17**, 5 (1945).
88. W. J. Chute and C. F. Wright, *Anal. Chem.*, **21**, 193 (1949).
89. C. E. Gleit, *Chem.-Anal.*, **51**, 20 (1962).
90. C. O. Ingamells, *Chem.-Anal.*, **53**, 55 (1953).
90a. D. C. Iffland, *Anal. Chem.*, **25**, 1577 (1953).
91. K. B. Zaborenko and A. Alian, *Zavod Lab.*, **28**, 1380 (1962).
92. C. A. Kienberger, *Anal. Chem.*, **29**, 1721 (1957).
93. K. J. Jensen and R. W. Bane, *Analyst*, **82**, 67 (1957).
94. G. P. Wall, A.E.R.E. Rep. CE/R 1730 (1955).
95. W. P. Kemp and K. W. Ponting, *Chem. Ind.*, **1957**, 1504.
96. E. F. C. Clarke and S. Kalayci, *Lab. Pract.*, **12**, 1095 (1963).
97. H. E. Hadd and W. H. Perloff, *J. Clin. Endocrinol Metab.*, **17**, 673 (1957).
98. B. M. Milwidsky, *Chem. Ind.*, **1969**, 411; personal communication, 1975.
99. L. C. Craig and D. Craig, "Laboratory Extraction and Countercurrent Distribution," in A. Weissberger (Ed.), *Technique of Organic Chemistry*, Vol. 3. Wiley-Interscience, New York, 1956.
100. E. Bohm, *Dtsch. Lebensm. Rundsch.*, **59**, 132 (1963).
101. N. Antonacopoulos, *Z. Lebensm. Unters.*, **113**, 113 (1960).
102. G. S. Riley, *Pharm. J.*, **194**, 320 (1965).
103. F. Seehofer and H. Borowski, *Beitr. Tabakforsch.*, **2**, 37 (1963).
104. R. Wasicky and G. Akisue, *Rev. Fac. Farm. Bioquim. Sao Paulo*, **7**, 399 (1969).
105. O. W. Grussendorf, *Chemy. Ind.*, **1966**, 52.
106. H. Trutnovsky, *G-I-T*, **11**, 580 (1967).
107. J. H. N. Levy and A. Lifshitz, *J. Assoc. Off. Anal. Chem.*, **50**, 1340 (1967).
108. J. Erdos, *Mikrochim. Acta*, **1961**, 515.
109. D. W. Sutton and D. G. Vallis, *J. Radioanal. Chem.*, **2**, 377 (1969).
110. F. Pregl, *Die quantitative organische Mikroanalyse*, 3rd ed., p. 244. Springer, Berlin, 1930.
111. H. Yagoda, *Mikrochemie*, **18**, 299 (1935).
112. J. Donau, *Mikrochemie*, **27**, 189 (1939).
113. J. T. Stock and M. A. Fill, *Lab. Pract.*, **6**, 38 (1957).
114. A. Soltys, *Mikrochemie*, Molisch Festschrift, 393 (1936).
115. L. Erdey and L. Buzos, *Magy. Kem. Foly.*, **61**, 443 (1955).
116. British Standards Institution, B.S. 1428: Part F 1 (1957); B.S. 1752 (1963); B.S. 1739 (1969).
117. A. N. Fletcher, *Anal. Chem.*, **9**, 1387 (1957).
118. J. R. Brown, *J. Chem. Educ.*, **34**, 165 (1957).

119. P. D. Cooper, *J. Chem. Educ.*, **41**, 85 (1964).
120. A. Marhoul and V. Ruzicka, *Chem. Listy*, **60**, 1401 (1966).
121. H. W. Gerarde, *Microchem, J.*, **7**, 321 (1963).
122. R. L. Boswall and D. C. Mackay, *Chem.-Anal.*, **52**, 54 (1963).
123. B. R. Sant and T. P. Prasad, *Lab. Pract.*, **19**, 76 (1970).
124. C. Kowala, *Chemy Ind.*, **1966**, 1029.
125. H. Shiba, *Anal. Chem.*, **26**, 943 (1954).
126. V. Horak, *Chem. Listy*, **48**, 616 (1954).
127. R. K. Maurmeyer and T. S. Ma, *Mikrochim. Acta*, **1957**, 563.
128. B. L. Marcus and T. S. Ma, *Mikrochim. Acta*, **1969**, 817.
129. R. T. E. Schenck and T. S. Ma, *Mikrochemie*, **40**, 236 (1953).
130. F. Hecht, "Geräte zur anorganischen Mikro-gewichtsanalyse," in F. Hecht and M. K. Zacherl (Eds.), *Handbuch der mikrochemischen Methoden*, Band I, Teil 2, pp. 181–188, 248–257. Springer, Wien, 1959.
131. British Standards Institution, B.S. 1428: Part G 2 (1957).
132. G. F. Smith, *Anal. Chim. Acta*, **17**, 192 (1957).
133. F. S. F. Mehr, *Lab. Pract.*, **15**, 671 (1966).
134. C. G. Skinner, *Anal. Chem.*, **28**, 924 (1956).
134a. P. L. Beaulieu, *Anal. Chem.*, **43**, 798 (1971).
135. H. K. Alber, *Ind. Eng. Chem., Anal. Ed.*, **13**, 656 (1941); A. H. Thomas Co., Philadelphia, Penn., Catalog No. 7329K.
136. T. S. Ma, "Quantitative Microchemical Analysis," in F. J. Welcher (Ed.), *Standard Methods of Chemical Analysis*, 6th edition, Vol. 2, p. 362. Van Nostrand, Princeton, 1963.
137. Available from Chemical Rubber Co., Cleveland, Ohio.
138. A. Dangoumau, *Chim. Anal.*, **36**, 179 (1954).
139. M. D. Schlesinger, S. Nazaruk, and L. Reggel, *J. Chem. Educ.*, **40**, 546 (1963).
139a. K. Vorwerck, *Muehle*, **107**, 257 (1970).
140. W. H. Bannister and J. Simler, *Lab. Pract.*, **16**, 874 (1967).
141. G. E. Barton, *Lab. Pract.*, **15**, 1255 (1966).
142. P. L. Stewart and I. H. Tipton, *Appl. Spectrosc.*, **22**, 58 (1968).
143. H. A. Went, *Biochem. Biophys. Acta*, **27**, 165 (1958).
143a. B. Vecerek, J. Krasny, J. Stepan, and L. Jilek, *Chem. Listy*, **67**, 975 (1973).
144. R. H. Alexander, *Lab. Pract.*, **18**, 63 (1969).
145. M. Hikichi and J. E. Miltimore, *Lab. Pract.*, **19**, 383 (1970).
146. S. Ahuja, C. Spitzer, and F. R. Brofazi, *J. Pharm. Sci.*, **57**, 1979 (1968).
146a. U. M. Cowgill, *Appl. Spectros.*, **28**, 455 (1974).
147. R. J. McCarter, *Rev. Scient. Instrum.*, **39**, 264 (1968).
148. G. F. Atkinson, *Chem.-Anal.*, **55**, 116 (1966).
149. C. Masson, *Lab. Pract.*, **18**, 851 (1969)
150. J. Bitz, *Rev. Sci. Instrum.*, **40**, 1344 (1969).
151. N. D. Cheronis and T. S. Ma, *Organic Functional Group Analysis*, p. 566. Wiley, New York, 1964.
152. G. F. Smith and A. H. Smith, *Talanta*, **11**, 1380 (1964).

153. G. Frederick Smith Chemical Co., Columbus, Ohio.
154. T. K. Miwa, *Chem.-Anal.*, **54**, 121 (1965).
155. A. Bhati and N. E. Anthony, *Lab. Pract.*, **16**, 328 (1967).
156. J. M. Corliss, *Chem.-Anal.*, **52**, 120 (1963).
157. R. E. Gugger and S. M. Mozersky, *Anal. Chem.*, **45**, 1575 (1973).
158. W. J. Cerveny and A. W. Weitkamp, *Rev. Sci. Instrum.*, **43**, 929 (1972).
159. K. Linderstrom-Lang and H. Holter, *Z. Physiol. Chem.*, **201**, 9 (1931).
160. D. Glick, *Technique of Cytochemistry*, p. 179. Wiley-Interscience, New York, 1949.
161. J. R. Walsby, *Anal. Chem.*, **45**, 2445 (1973).
162. E. J. Conway, *Biochem. J.*, **64**, 47 (1956).
163. M. Pecar, *Microchem. J.*, **3**, 557 (1959).
164. L. J. Burger, *J. Chem. Educ.*, **46**, 39 1969).
165. M. C. Elphick, *Lab. Pract.*, **19**, 1139 (1970).
166. Available from Labindustries, Berkeley, California.
167. Available from Heat Systems–Ultrasonics, Inc., Plainview, N.Y.
168. Available from GCA Precision Scientific, Chicago, Ill.
169. D. C. Robins, *Lab. Pract.*, **15**, 192 (1966).
170. J. Aggett, *Lab. Pract.*, **18**, 305 (1969).

FRACTIONATION OF SOLID MATERIALS

I. INTRODUCTORY REMARKS ON FRACTIONATION

A. NEED FOR SEPARATION PROCESSES

Among all laboratory experimental operations in chemistry, fractionation of the working material is probably the most essential and also most frequently carried out in the laboratory. Separation processes are often required in chemical analysis and synthesis, as well as in physical-constant measurements. Practically all naturally occurring substances subjected to chemical investigations are mixtures, while most man-made samples, such as commercial and industrial products, contain two or more chemical species. In chemical synthesis, since the aim is to produce a pure new compound, it is necessary to separate the desired compound from starting materials, by-products and impurities in order to obtain a specimen for elemental analysis or spectral inspection for the confirmation of the expected molecular formula. It is not surprising, therefore, that there are journals [1, 2], books [3, 4], and reviews [5] that deal with separation science exclusively.

Separation techniques warrant particular attention in microscale manipulations for three reasons. First, when the amount of working material is small, the chances of contamination and of loss of some portion of the original sample are magnified. Secondly, the experimental specimen may be irreplaceable because it is all the working material available. Finally, not all separation methods are suitable for application to small samples.

In this chapter and in the succeeding three chapters, we discuss the methods that are suitable for the separation of micro samples. It should be noted that the sample size varies with the separation process to be employed. Some procedures (e.g., thin-layer chromatography) are inherently micro methods, since they deal with working materials below the milligram region. Other procedures are adapted from gram-scale operations. In the latter case, it is prudent to know the lower limit of quantity with which the fractionation process can be achieved.

For convenience, we gather together the methods for the fractionation of solid materials in Chapter 3, and those for the fractionation of liquid materials in Chapter 5. It should be emphasized, however, that the differentiation between solid and liquid samples is arbitrary. For instance, a sample that consists of several liquids at room temperature may be treated as a solid mixture by solidifying the working material; conversely, solids can be fractionated in the molten state. Furthermore, when the analytical sample is received in the form of a homogeneous liquid, it may contain solutes that are solid compounds and can be obtained as such after the removal of the solvent.

In view of the fact that chromatography is applicable to solid, liquid, and gaseous substances, and because of the wide range of chromatographic fractionation techniques, we devote a chapter to chromatographic methods of separation (Chapter 4). Generally speaking, the aim of chromatography is to separate, and recover if feasible, all individual components in the working material.

In contrast with chromatography, a number of separation processes are concerned with only a portion of the substances in a complex mixture. Thus the particular portion that is separated usually still contains more than one compound. This kind of experiment is carried out either because it fulfills the need of the chemical investigation, or because one is restricted by the apparatus. For example, steam distillation is employed to determine the essential oils in plant materials; the distillate will comprise all steam-volatile compounds in the original sample. In the study of the oxidation of sterols in certain biochemical system, several ketosteroids may be produced, but the investigator may be satisfied with the extent of the oxidation reaction and not necessarily interested in the quantity of each ketosteroid obtained. This type of separation process are discussed in Chapter 6.

B. DISSOLUTION PROBLEMS

Except for a few methods that are based on particle-size differentiation (see Section VI, and Chapter 6, Section III.C), the first step in the separation experiment involves the dissolution of the working material, unless the sample is received originally as a homogeneous solution in the same solvent system to be employed in the separation procedure. While the selection of solvent and the conditions for dissolving the mixture are a routine matter treated in the conventional laboratory manuals, there are situations that require special techniques in order to optimize the capability of the fractionation process. Some examples are described below.

The determination of the constituents of lead alloys is of great importance to the electronics industry. Small samples are frequently encountered; therefore, accurate and sensitive techniques are necessary. Because the main constituents are tin and lead, sample dissolution is a problem. Thus, hydrochloric acid rapidly dissolves tin, but precipitation of the partially soluble lead chloride slows down the rate of attack; nitric acid dissolves lead, but forms insoluble metastannic acid with tin. Hwang and Sandonato [6] have used mixtures of fluoboric acid, nitric acid, and water for the dissolution of their samples. Bell [7] reported that the dissolution rate for lead alloys containing more than 50% lead is slow in these media, and has recommended mixtures of fluoboric acid, hydrogen peroxide, and disodium ethylenediaminetetraacetate (EDTA) at room temperature. The addition of EDTA is necessary to prevent the formation of insoluble salts during the dissolution step.

Yamazaki has used tin (II)-H_3PO_4 reagent to dissolve alumina containing small amounts of sulfur [8]. Bollman [8a] has found that this reagent is also excellent for the dissolution of many transition-metal nitrides without loss of nitrogen. The decomposition of the nitrides takes place upon heating for 2 to 3 min. Continued heating after the sample has reacted should be avoided, because phosphates are formed and become fused to the glass walls.

Whereas a large variety of solvents are available for the dissolution of organic mixtures, some solvent systems may be better than others. For instance, in a collaborative study of the partition chromatography of morphine in opium, Smith [9] has reported that more reproducible results are obtained with the use of dimethyl sulfoxide as solvent in sample preparation than with homogenization in water.

A number of solubilizing agents have been proposed to facilitate the dissolution of organic substances [9a]. Brock and co-workers [10] have found that aqueous solutions of purines dissolve aromatic hydrocarbons to a high degree. This is brought about by the formation of molecular complexes [11], and forms the basis of the dissolution by the aid of solubilizing agents. Stuchlik and co-workers [12] have performed the separation of polycyclic hydrocarbons and isoquinoline alkaloids [13] by paper chromatography after solubilization with purine bases. These workers have also separated caffeine, theobromine, theophylline, and their derivatives by thin-layer chromatography using arylsulfonic acid [14] or aryloxyacetic acids [15] as solubilizing agents. Durek-Kluczykowska and Krowczynski [16] have studied the effect of surfactants such as sodium lauryl sulfate on the extraction of alkaloids from drugs.

Wherever applicable, ultrasonic agitation is a useful technique to facilitate sample dissolution. The apparatus is commercially available [17]. Golden and Sawicki [18] have employed this method to dissolve polynuclear aromatic hydrocarbons from airborne particulates collected on filter paper into alcohol solutions at room temperature. Other solubilization media for organic materials that have been advocated include a quaternary ammonium hydroxide tissue solubilizer for use with toluene and xylene [19, 19a]. Gibbins [20] has constructed an apparatus for small-scale exploratory experiments with volatile organic solvents for solubility studies of air- and water-sensitive solids for prolonged periods. In dissolving single tablets for pharmaceutical analysis, Urbanyi and co-workers [20a] have reported that a greater rate of dissolution is achieved when the blender of the automated solid sampler is replaced by an ultrasonic device.

II. FRACTIONATION BY RECRYSTALLIZATION

A. PRINCIPLE

Fractionation by recrystallization is based on the difference in solubility of the individual compounds in a particular solvent system. Crystallization, in a broad sense, may be defined as the process of crystal formation from a saturated solution, melt, or syrup. This process usually is caused to take place by temperature changes. It should be recognized, however, that evaporation of a dilute solution may also lead to the formation of crystals. Recrystallization means the dissolution of solid mixtures in a suitable liquid medium, followed by the recovery of some or all compounds in form of crystalline solids.

Prior to the advent of chromatography, recrystallization was the experimental technique most frequently employed for the fractionation of solids. While the more effective methods of chromatographic separation (see Chapter 4) have displaced recrystallization to some extent, the fact that the crystallization technique is simple and requires no special equipment still maintains it as an essential operation in the laboratory. As a rule, the separation of the major components in a mixture is preferably achieved by recrystallization. Furthermore, when the amount of working material is more than 25 mg, it is usually easier to perform the separation by recrystallization than by chromatography.

Since the micromanipulative techniques, tools, and containers for recrystallization vary with the size of the working material, the methods discussed below are divided into two categories. Roughly, we treat the methods for separating more than 100 mg of solid mixtures in one category, and those for fractionating less than 100 mg in the other.

B. METHODS FOR HANDLING 100 MG OR MORE OF SOLID MATERIAL

Whereas macroscale crystallization [21, 22] may be inconvenient, time-consuming, or hazardous with flammable solvents [23], these objections do not present themselves when the working material is in the range between 100 mg and 1 g. Methods have been developed for manipulation in this scale that are simple, safe, and rapid. The necessary equipment is readily available, since it consists of common laboratory supplies. The time required to recrystallize a sample is a matter of minutes; this includes (1) dissolving the sample, (2) filtering the hot solution, (3) adjusting the conditions for the formation of crystals, and (4) separating the crystals from the mother liquor. Generally speaking, the laborious part is in the selection of the proper solvent system for crystallization, and the most difficult operation is the formation of crystals from a syrup.

1. Selection of Solvent and Preliminary Crystallization Tests

a. Choice of Solvent System

The success of crystallization as a fractionation process depends primarily on the availability of a proper solvent system. To achieve the maximum efficiency of the separation process, the solvent chosen should fulfill two conditions: (1) the separation of the crystals of the major compound should be nearly quantitative; (2) the separation should be selective.

The first condition means that practically the entire amount of the major compound in the working material crystallizes out of the solution, leaving only a very small portion in the mother liquor. The formation of crystals is controlled by the difference in solubility of a compound in a solvent at two temperatures, usually the boiling temperature and at room temperature. Four situations are possible with respect to the characteristics of the saturated solution, as listed in Table 3.1. The most effective separation is type 3; the least effective is type 2. Quantitatively speaking, small $\Delta[A]_{sat(T_2,T_1)}$ indicates low efficiency, while large $\Delta[A]_{sat(T_2,T_1)}$ indicates high efficiency. Small $[A]_{sat\ T_1}$ means less loss, while large $[A]_{sat\ T_1}$ indicates that considerable proportion of compound A remains in the mother

TABLE 3.1. Characteristics of Saturated Solutions

Type	$\Delta[A]_{sat(T_2,T_1)}{}^a$	$[A]_{sat\ T_1}{}^b$
1	Small	Small
2	Small	Large
3	Large	Small
4	Large	Large

[a] $\Delta[A]_{sat(T_2,T_1)} = [A]_{sat\ T_2} - [A]_{sat\ T_1}$; $T_2 > T_1$
[b] $[A]_{sat\ T_1}$ = concentration of a saturated solution of compound A at temperature T_1.

liquor. In the latter case, laborious treatment is necessary in order to recover all of compound A from the solution through repeated evaporation of the mother liquor.

Selective separation means that only a particular compound in the mixture crystallizes out while the other compounds remain dissolved in the mother liquor. This situation occurs when the compound in question crystallizes under type 3 and the other compounds under type 2. This can be achieved by proper selection of the solvent system. In order to optimize the conditions for selective crystallization, preliminary crystallization experiments are performed, and the composition of the solid phase and that of the mother liquor are determined (e.g., with the aid of chromatography). For best results, it is often necessary to carry out repeated crystallization using at least two different systems. Switching from nonpolar to polar solvent [24] can be the most successful fractionation procedure for compounds of medium polarity. The process permits the separation of less polar compounds in one solvent system, and more polar compounds in another. If this procedure fails to give satisfactory results, then it is advisable to employ other separation techniques such as sublimation or chromatography. In this connection, it may be mentioned that the conventional systematic fractional crystallization procedure [25] is not practiced in microscale manipulations. As a matter of fact, this systematic scheme has been displaced almost completely by the modern efficient separation techniques.

There are no definite rules for the selection of the best solvent for recrystallization. It is useful, however, to study the structures and properties of the solvents. Solvents are characterized by their polarity and specific solvation properties. The polarity of the solvent is generally given by the value of its dielectric constant [26]. The specific solvation property [27] is dependent on the presence of an active hydrogen atom that is able to form hydrogen bonding (e.g., alcohol), or a basic atom that can act as hydrogen acceptor (e.g., ether). The high dielectric value of molecules with polar structure, and the difference in the availability of negative and positive centers in the molecule for specific interactions, are factors to be considered in the application of dipolar aprotic solvents such as acetone, acetonitrile, nitromethane [28], dimethylformamide [29], and dimethylsulfoxide [30].

The solubility of an organic molecule depends on the overall polarity of the molecule. Basically, this molecular property involves two structural parameters: the size of the molecule and the number of polar groups present. It should be noted that these two parameters influence the polarity of the molecule in opposite directions. First, the solubility of organic compounds decreases with molecular weight; this principle holds for any type of solvent used. Second, the solubility of hydrocarbons (nonpolar molecules) in water and other polar solvents is low, and it increases when polar groups are introduced to form molecules of medium and high polarity. The polar groups are functional groups such as —OH, $\diagdown$CO, —COOH, —NHR, —NHCOR, —NR$_3^{\oplus}$, —SH, $\diagdown$SO, —SO$_2$H, —SO$_3$H, —SO$_2$NH$_2$. Since the molecular weight and the number of polar groups operate in opposite directions, a simple criterion of the overall polarity of the molecule is given by the ratio of carbon atoms in the molecule to the heteroatoms (O, N, S, etc.) present in the above-mentioned functional groups. A small value of C_m/Het_n ratio characterizes highly polar molecules that are readily soluble in water and other polar solvents (e.g., methanol, acetic acid). Contrastingly, a large value of C_m/Het_n ratio characterizes molecules that are easily soluble in nonpolar solvents (e.g., hydrocarbons, carbon tetrachloride, carbon disulfide).

In general, for compounds that are readily soluble, solvents of opposite character should be chosen (i.e., polar solvent for nonpolar compounds and vice versa). For compounds that are relatively insoluble, the specific interaction in the solvation process can be utilized, as well as affinity due to similarity in chemical structure (e.g., polar solvents for polar compounds, hydrocarbons for nonpolar compounds). There are correlations between solubility on the one hand and paper-chromatographic mobility [31], thin-layer-chromatographic mobility, and gas-chromatographic retention time [32] on the other hand, since all involve the solvation phenomena.

Sometimes significant enhancement of the solubility can be achieved by slight overheating of low-boiling solvents in a closed system. This has been confirmed by extensive laboratory tests [33]. For instance, diethyl ether becomes an extremely efficient solvent when it is heated slightly above its boiling point. For microscale manipulations, this technique can be used safely as follows. The sample and diethyl ether are placed in a test tube or round-bottom flask fitted with a ground-glass joint. A glass stopper is put in place and held firmly by means of the fingers or springs. The vessel is heated slowly in a water bath while being swirled, and repeatedly removed from the bath. After the heating is terminated, the solution is allowed to cool below its boiling point before releasing the stopper. Working behind a safety shield is recommended, and no open flame should be in the vicinity.

b. Preliminary Crystallization Experiments

The preliminary crystallization test is first carried out with single solvents. If the test is unsuccessful (i.e., no solvent of type 3 can be found), a mixture of two solvents is tried. One solvent should have the characteristics of type 1 as given in Table 3.1, and the other type 4. The two liquids should be miscible in all proportions. It is expected that mixing these two liquids will result in a solvent system with the characteristics of type 3. Usually one liquid is polar (e.g., methanol, acetone) and the other nonpolar (e.g., benzene, heptane). The common procedure is to add the less efficient solvent stepwise to the solution containing the sample in the other solvent while the solution is being continuously heated. If cloudiness appears, addition of the second solvent will clarify the solution.

In contrast to crystallization in the macroscale, the ratio of solute to solvent should be small in microscale manipulations. Difficulties in handling small volumes of dense suspensions make it imperative to work with systems having a large proportion of the solvent. The economic aspect is not important, since the quantity of solvent consumed is small.

As mentioned in Chapter 1, chemicals of high purity [34] are recommended in microscale experiments; this also applies to the use of solvents. Solvent selection has two facets, one being the problem of solubility as well as polarity of the sample, and the other being the efficiency of the solvent. Both can be simultaneously elucidated by a single experiment.

The solubility of the sample in a given solvent can be tested directly or indirectly. The direct method is carried out as follows. About 1 mg of the sample is transferred into a test tube of 4-mm i.d. The solvent is added dropwise by means of a pipet, graduated to 0.2 ml, while the test tube is being heated in a water bath or metal block. The sample is considered sufficiently soluble for recrystallization purposes if it dissolves in a boiling liquid to yield a 0.5% solution. If the sample does not dissolve in 0.2 ml of the solvent, the latter is removed by evaporation (see Section II.B.4) and another solvent is tried. For the systematic testing of solubility, it is advisable to start with low-boiling nonpolar solvents and change to liquids with increasing polarity.

The indirect method for testing solubility utilizes "schlieren" observation [35] in the following manner. A micro test tube filled with the solvent is illuminated under a dim light. In another test tube is prepared a saturated solution of the sample in this solvent at room temperature. The two test tubes are placed in the same water bath, so that their contents are at the same temperature. Using a micropipet, one drop of the solution is carefully added to the solvent under illumination, and the "schlieren" is observed, which is based on refractive-index differences. A positive test indicates that the sample is soluble in the solvent.

A semiquantitative crystallization test is carried out as follows. To a previously weighted test tube of 6-mm i.d. containing the sample (about 5 mg), the solvent is added dropwise until complete dissolution in the boiling liquid is achieved. The test tube is left to cool. Crystals are then separated by centrifugation, and the supernatant liquid is removed by means of the capillary pipet (see Chapter 2, Section III). The test tube containing the crystals is dried in a vacuum desiccator, and the weight of crystals recovered is determined. The ratio between this weight and the sample weight indicates the efficiency of the crystallization process. A value larger than 0.9 is characteristic of a good solvent system. It may be mentioned that an experienced chemist is able to estimate the crystallization ability of a solvent without weighing the sample and crystals.

If it is required to determine the crystallization efficiency precisely, manuals on quantitative analysis [36, 37] should be consulted in order to choose a method to establish the concentrations of the standard solutions at different temperatures. The following ebulliometric technique [38] can be employed: In the microebulliometer the difference between the boiling point of pure solvent and that of the saturated solution is measured at atmospheric and lower pressures. It is necessary to have an excess of sample in the apparatus so that saturation is assured. Ebulliometric constants of liquids can be found in the literature or obtained in a separate experiment using the pure solvent [38] and a known compound.

2. Dissolution of the Working Material for Recrystallization

On the basis of the information obtained from solubility tests and preliminary crystallization experiments, an appropriate technique is chosen to prepare a solution of the working material. Depending on the volume of solvent used, either a test tube (up to 10 ml) or a flask is employed as the container. Recording the weight of sample and amount of solvent is advisable, because these data can serve as reference for repeated experiments. Attention is called to the fact that fully saturated solutions are often difficult to handle. For instance, evaporation of solvent or slight cooling of the solution may cause separation of crystals and complicate the filtration process. Therefore, 5 to 10% excess of the solvent is recommended.

a. Preparation of Solution in a Test Tube

Attention should be given to the preparation of the solution in a test tube in order to prevent the heated liquid from running out. This can be done by covering the mouth of the tube with an inert plastic sheet and closing it with a finger or rubber band. No boiling chips are used; smooth boiling of the solution is aided by careful agitation with the hand. Heating is preferably effected by momentarily immersing the test tube in a water bath.

In another procedure the test tube is placed in the well of a heating block (see Chapter 2, Fig. 2.17), or inside an Erlenmeyer flask containing a boiling liquid (see Fig. 3.1). Since the test tube cannot be shaken, it is necessary to use a boiling capillary as shown. An electric hotplate should be used when organic liquids are involved.

b. Preparation of Solution in a Flask

The pear-shaped flask (see Chapter 2, Fig. 2.2) is recommended when the volume of solvent employed exceeds 10 ml. The flask is preferably fitted with a reflux condenser through a ground-glass joint. The solvent is added through the condenser while the flask is heated by a suitable device (e.g., a heating mantle).

For compounds that are slightly soluble, the working material is repeatedly treated with fresh portions of solvent after the saturated solution has been separated by filtration. This procedure is advantageous if most of the impurities (minor components) are removed in the first portion of the solvent. The technique of hot extraction (see Section II.B.6, Fig. 3.12) also can be used. In the latter case, however, the separation and purification effects are much less than in a true recrystallization procedure.

3. Filtration of the Hot Solution for Crystallization

An important step in the recrystallization process is the filtration of the hot solution in order to remove insoluble materials and/or excess sample if insufficient solvent has been used. We discuss the general techniques in this section. These filtration techniques are also employed in microscale manipulations other than crystallization. The volume of solution to be

Fig. 3.1. Heating a test tube in a liquid bath.

handled is in the range between 1 and 50 ml. It should be emphasized that the danger of fire and explosion exists in this scale of experimentation, albeit considerably less than in macroscale laboratory operations.

When the volume of hot solution to be filtered is below 50 ml, the use of scaled-down macroapparatus (e.g., a miniature Büchner funnel) is not recommended. On the other hand, a number of procedures have been developed for microscale hot filtration that require only very simple devices. The choice of the procedure depends on the volume of the solution, the amount and nature of the undissolved material, and the rate and conditions of crystal formation.

Preheating the whole apparatus is an essential part of hot filtration. For volumes larger than 10 ml, the following technique is recommended: The flask containing the hot solution is connected to an adapter of the kind designed by Nemec [39] (see Fig. 3.2, middle section). The porosity of the filtration bed is important, since a prolonged filtration process may cause separation of crystals and stoppage of the flow of the solution. The common packing material is cotton, which should be replaced by cellulose powder or asbestos for filtration with high efficiency. The chemical inertness of the material also should be ascertained. The material is packed into the apparatus by means of forceps and pressed down with a glass rod. The filtration assembly is warmed up by boiling the solution while the middle stopcock is open. Then the latter is closed and the receiver is

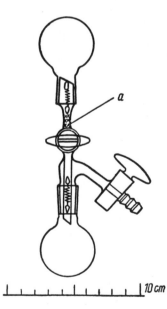

10 cm

Fig. 3.2. Adapter for hot filtration: *a*, cotton or other filtration bed.

evacuated. Subsequently the assembly is inverted, as shown in Fig. 3.2. When the side-arm stopcock is closed and the middle stopcock opened, the solution automatically passes through the filtration bed into the receiver.

The filter-stick technique is usually employed for hot filtration of volumes smaller than 20 ml. Of the various types of filter sticks (see Chapter 2, Fig. 2.38), those having fritted discs are inconvenient to use because of the difficulty in cleaning. Figure 3.3 depicts a technique which involves filtration under pressure. The filter stick is inserted through the rubber stopper into the test tube containing the solution to be filtered. At first the end of the filter stick is placed above the liquid surface while the solution is being heated. After the vapors have warmed up the capillary, the end of the filter stick is pushed down to the bottom of the test tube as shown in Fig. 3.3. By pressing the rubber bulb that is affixed to the side arm, the solution is filtered into the receiving tube. The advantage of applying pressure to filter hot solutions is the prevention of clogging. In contrast, vacuum filtration tends to cause precipitation due to vaporization of the solvent and lowering of the temperature of the solution. For this reason, the water aspirator is not recommended for the filtration of small volumes of hot solution. It is possible, however, to carry out suction filtration without the use of an aspirator. One technique [40] is illustrated in Fig. 3.4.

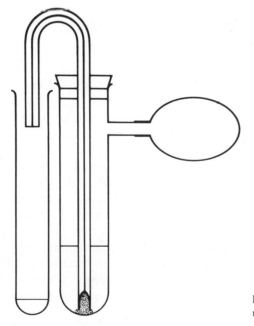

Fig. 3.3. Filtration using filter stick under pressure.

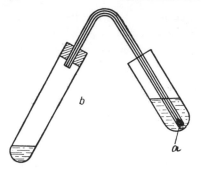

Fig. 3.4. Automatic filtration of hot solution [40]: *a*, hot solution; *b*, receiver.

Before starting the experiment, a few drops of the solvent are added to the receiver, and the rubber stopper carrying the filter stick is inserted. The receiver is heated so that the solvent vapor expels the air therein and at the same time warms up the capillary. The filter stick is immediately plunged into the hot solution, as shown in Fig. 3.4. The solution is spontaneously sucked into the receiver while it cools down. Another technique [41] utilizes an ampul with a capillary tip. The ampul is constructed as illustrated in Fig. 3.5; the tip is cut at the slightly enlarged part *A-B* where cotton is inserted to form the filter bed. The ampul is heated slightly and then dipped into the solvent, whereupon a little solvent is sucked into the ampul. On reheating the ampul to boil the solvent, air is driven out, and the tip of the ampul is now plunged into the solution as shown. After filtration, the capillary is sealed for crystallization and then cut off to facilitate the transfer of the material from the ampul.

4. Concentration of a Solution for Recrystallization

Concentration of the sample may be necessary when the working material received for recrystallization is a cold saturated solution or a very dilute solution. The mother liquor from a previous crystallization is an example of the former, and the eluate from a chromatographic column an example of the latter. In some cases the solution is evaporated to dryness to recover the solid mixture; this technique is applied when the solvent previously used is not suitable for the subsequent crystallization procedure. In other cases, the sample is concentrated until a hot concentrated solution results.

When the volume of solution is larger than 50 ml, concentration by means of rotatory evaporator is recommended. A simple apparatus is shown in Fig. 3.6. It consists of a round-bottom flask connected with an adapter through a ball joint. The flask is filled to about one-fourth of its

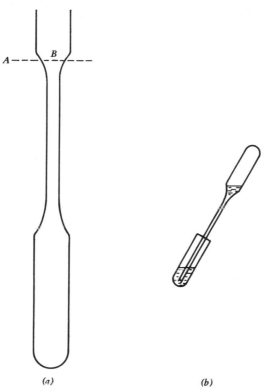

(a)   (b)

Fig. 3.5. Filtering hot solution into ampul. (*a*) Constructing ampul: *A*, where to cut off capillary; *B*, where to insert cotton wad. (*b*) Filtration in progress.

capacity. It is evacuated through the adapter and heated by placing in a water bath. The flask is rotated manually so that the evaporation of the solvent proceeds from the film on the walls of the flask. Commercial rotatory evaporators (see Chapter 8, Fig. 8.1) are motor driven.

When the volume of the solution is less than 50 ml, it is expedient to remove the solvent by applying a stream of gas over the liquid surface. One device is shown in Fig. 3.7. The vessel holding the solution is placed in a heating bath. The side-arm of the adapter is connected to the water aspirator so that air enters through the center tube. A cotton wad is inserted at the top of the center tube to prevent particulates in the air from contaminating the working material. A similar apparatus, described in Chapter 8, Fig. 8.6, is suitable for concentrating less than 5 ml of solution. Another technique for the removal of solvent consists of passing a current of gas (e.g., from a cylinder of nitrogen) over the liquid surface.

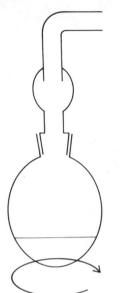

Fig. 3.6. Simple rotatory evaporator.

Fig. 3.7. Device for concentrating solution.

This method is applicable to less than 1 ml of solution contained in a capillary tube, as shown in Chapter 2, Fig. 2.47. When the gas inlet is slightly bent, it works like a spring holding up the capillary tube.

The apparatus shown in Fig. 3.8 was designed by Perold [42] for concentrating a solution in a test tube at constant temperature. The temperature is determined by the boiling point of the liquid in the flask. This apparatus also permits the concentration of a solution under controlled atmosphere.

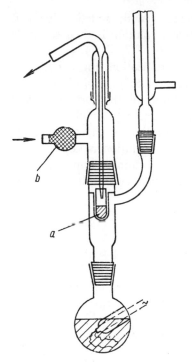

Fig. 3.8. Concentrating solution at constant temperature: *a*, test tube holding solution; *b*, bulb for purification of incoming gas with cotton.

5. Crystallization From the Solution

The reader is referred to Section II.B.1 for the discussion of crystallization. In microscale manipulations, the most common method of inducing crystal formation involves making a hot saturated solution and then lowering its temperature. As is well known, the size of the crystals formed is dependent on the rate of their formation. It should be recognized that crystal size is an important factor in the success of the subsequent filtration operation, and hence the fractionation process. Large aggregates, however, may contain impurities due to the enclosed mother liquor.

The rate of crystal formation can be slowed down by preventing the rapid cooling of the hot solution. This can be achieved in several ways: (1) The vessel containing the hot solution can be placed in a beaker packed with cotton, plastic foam, or cellulose material. (2) The vessel can be immersed in a hot water bath that has a film of paraffin oil on its surface; the water bath cools down much more slowly under this condition. (3) The most efficient way to grow large crystals is to put the vessel in a Dewar flask filled with hot water.

During the process of hot filtration (see Section II.B.3), if crystals begin to form, the filtrate should be reheated (with added solvent if necessary) until all crystals disappear. The clear solution is then set aside for crystallization.

When a mixture of solvents is used for crystallization, the experiment is carried out by adding the less effective solvent to the solution of the sample in the more effective solvent. The volume of the former solvent should not exceed the amount that causes the sample to precipitate. The procedure is as follows: To the hot filtered solution of the sample, the less effective solvent is gradually added while the solution is kept at boiling temperature. The addition is stopped when cloudiness appears. Then a little of the effective solvent is added just to clear up the turbidity.

6. Special Crystallization Apparatus and Techniques for Dissolution and Hot Filtration and Crystallization

A number of devices have been proposed for small-scale operations involving dissolution and hot filtration and/or crystallization. The apparatus designed by Cockburn [43] is illustrated in Fig. 3.9. The vessel has a short side arm sealed to a fritted disc and another side arm that serves as handle and condenser. The solid material or its solution is introduced through the open side arm. The vessel is heated and solvent added until the sample dissolves (Fig. 3.9a). Then the long side arm is closed with the thumb, whereupon vapors escape through the filtration arm to warm it up. Then the vessel is inverted (Fig. 3.9b) and the solution is filtered into the receiver. An improved version of this apparatus has been described by Martin-Smith [44] (see Fig. 3.10).

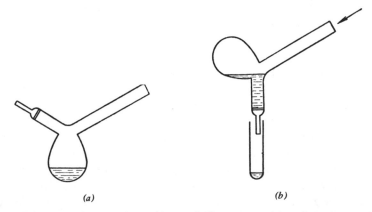

(a)   (b)

Fig. 3.9. Filtration device of Cockburn [43]: (a) position for preparation of solution and heating frit; (b) position for filtration under gas pressure.

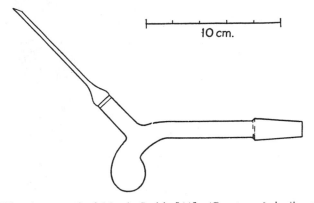
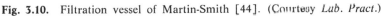

Fig. 3.10. Filtration vessel of Martin-Smith [44]. (Courtesy *Lab. Pract.*)

Shiba [45] has constructed an apparatus for hot filtration, extraction, and recrystallization using volatile and inflammable solvents without loss of solvent or deposition of crystals on the filter paper. Horak [46] has modified the Pregl hot-filtration apparatus in which the filter tube, connected by means of a ground-glass joint to an external jacket, is heated by the vapors of the solvent. Owen [47] has described a simple crystallization apparatus for application to small amounts of sample Cannon [48] has proposed a device for recrystallization that consists of a bulb (4.5-ml capacity) fitted with a three-bulb air condenser. The bottom of the container bulb has an orifice of 3-mm i.d. in which is inserted a ground-in plug with a stem projecting above the rim of the container. An enlargement on the side of the bulb serves as a heating area. To dissolve the material in a solvent, the apparatus is manipulated as a test tube; after cooling, the condenser is removed and the crystals are collected on a filter by manipulation of the stem of the stopper.

Kowala [49] has described an apparatus for the crystallization of air-sensitive compounds from both cold and hot solutions. Carel [50] has constructed an apparatus for filtration under nitrogen pressure at temperatures down to $-80°C$. A low-temperature crystallization apparatus designed by Friedrich [51] for semimicro quantitative work is shown in Fig. 3.11. The operation is as follows: The sample and solvent are introduced to the jacketed filtering flask O, and the stirring motor P is started. If the working material is insoluble in the solvent at room temperature, valve G is closed and valve F opened. The circulating pump L is turned on and steam introduced through the steam-jacketed section of the circulating system D. The circulating liquid (95% ethanol) is thereby warmed, in turn warming the mixture in the flask. To cool the solution, the steam

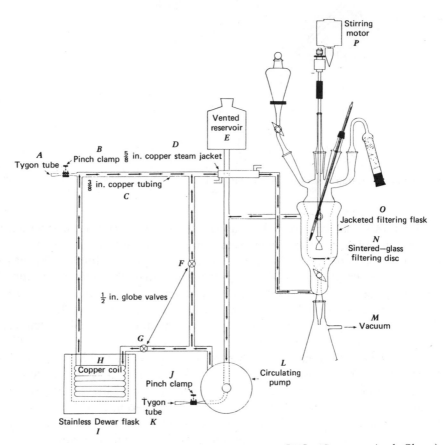

Fig. 3.11. Low-temperature crystallization apparatus [51]. (Courtesy *Anal. Chem.*)

is turned off, *G* is opened slightly, and *F* closed until the flow through the cooling bath (isopropyl alcolhol–Dry Ice) is large enough to give the desired cooling rate. Stirring is continued 15 to 30 min after the desired temperature has been reached. Stirring is then stopped, and the crystals are allowed to settle. The stopcock is then opened, and a vacuum, applied to the receiver *M*, removes the solvent and impurities and leaves the crystals on the fritted disc *N*.

Feldman and Ellensburg [52] have proposed a two-piece centrifuge crucible for handling precipitates so that they can be subsequently processed without changing containers. Bowers and Haschenmeyer [53] have described a small-volume ultrafiltration cell that has a sample chamber, of capacity up to 3 ml, fitted with a membrane supported by a porous polyethylene disc. Stajner [54] has reported that a dialysis membrane placed in a standard laboratory centrifuge is suitable for ultracentrifuging solutions under sterile conditions. For the filtration of difficult suspensions, Soudek [55] has described a dynamic filter that works under pressure by continuous stirring of the suspension.

The apparatus for hot extraction is used for the preparation of a solution of relatively insoluble materials. As illustrated in Fig. 3.12, the sample is placed in a paper basket or paper roll, which hangs on the coil condenser. As the solvent runs down from the condenser, it is heated by the vapor of the boiling liquid.

C. METHODS FOR HANDLING LESS THAN 100 MG OF WORKING MATERIAL

Generally speaking, the techniques for recrystallization of less than 100 mg of material are different from those described in Section II.B. For instance, the solubility test and preliminary experiments are usually not carried out. Normally the whole sample is used, and the experiment should be so arranged that the system can be changed from one solvent to another without affecting the compounds in the original mixture. Loss of material should be held at a minimum, and a high efficiency of separation is expected.

Two guidelines are in order: (1) The transfer of a concentrated solution from vessel to vessel should be avoided, except in special devices such as the hot-filtration apparatus. (2) Procedures should be simplified

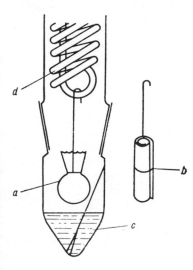

Fig. 3.12. Preparation of solution by hot extraction: *a*, paper basket containing the working material; *b*, paper roll holding the material; *c*, solvent; *d*, coil condenser.

wherever feasible. Centrifugation for separation of crystals from mother liquor, and concentration of solution by gas-flow evaporation, are recommended.

1. Manipulating Small Volumes of Solutions

It should be noted that transferring small volumes of concentrated solution containing less than 100 mg of working material is often beset with difficulties. If the container is not washed out carefully, a significant portion of the sample may be left behind. On the other hand, washing with a few drops of solvent may dilute the solution considerably. For this reason, it is advisable to use dilute solutions from the beginning of the experiment.

Pipetting is the common technique for transferring small volumes of solvent or solution. A commercial or home-made pipet with a long, fine capillary tip and a rubber bulb is suitable. Glass or polyethylene material can be employed to make the pipet. The advantage of a polyethylene pipet is that it is unbreakable, flexible, soft, and chemically inert. It cannot be used, however, to handle hot liquids. The polyethylene pipet is fabricated in the following manner: A transparent polyethylene tube (4- to 6-mm i.d.) is heated about 3 cm above the fine flame of a microburner. When the tubing becomes sufficiently soft, it is drawn out and held in a vertical position until the plastic hardens. (Note that polyethylene softens suddenly in a flame.)

Another technique for transferring solvent or solution is to use a bulb-pipet, which is constructed from glass tubing of 6- to 10-mm i.d. in the following manner (Fig. 3.13). A section (about 20 mm) of the glass

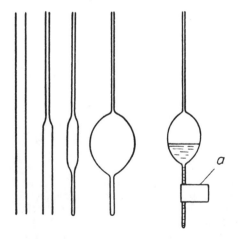

Fig. 3.13. Construction of the bulb-pipet: *a*, label.

tubing is heated and, after removing it from the flame, drawn out to form a capillary of 1- to 2-mm i.d. and 300-mm length. This process is repeated on the opposite side so that 10 to 30 mm of the tubing remains between the two capillary ends. Then one end is sealed to make a capillary about 100 mm long, and the opposite end is cut off to make a capillary about 300 mm long. Now the center part is heated and blown out to produce a bulb of suitable size, usually a few milliliters. While the glass is still hot, the round bulb is pulled out slightly. This tapered bulb permits the easy flow of liquid. The bulb-pipet is operated by warming the bulb over a small flame and then immersing the open tip of the capillary into the solvent or solution, whereupon the liquid is sucked in automatically. If the amount of liquid that has entered is not sufficient for the experiment, the bulb-pipet is inverted (Fig. 3.13a), and the bulb is warmed again to vaporize a small portion of the liquid. When the open capillary is now immersed in the solvent or solution, efficient suction will be achieved by the condensation of the vapor. For a very low-boiling solvent like diethyl ether, it is expedient to induce the liquid to enter the bulb-pipet by placing a piece of Dry Ice on the bulb instead of warming over a flame. Dispensing the whole volume from the bulb-pipet is effected by cutting the sealed end of the capillary; the liquid will come out by gravity or by applying slight pressure. Needless to say, a liquid can be stored in the bulb-pipet by sealing both ends, and the working material is thus protected from moisture and air contamination.

2. Recrystallization in a Test Tube

Depending on the volume of solvent, a test tube of 4- to 16-mm i.d. is used to prepare the solution. Centrifugation is a convenient method to separate the insoluble impurities in the working material. The hot supernatant liquid is transferred by means of a pipet into another test tube. The pipet should be previously warmed on a hot plate and operated with the aid of a rubber bulb. Centrifuging and pipetting, however, is not advisable if crystallization takes place rapidly from the hot solution. In this case, the hot-filtration technique should be employed as described in the next paragraph.

The technique mentioned in Section II.B.3 (see Figs. 3.3, 3.4, 3.5) may be used. The ampul method (Fig. 3.5) is particularly suited for working with small volumes of low-boiling solvents. The ampul is made from a test tube (6- to 16-mm i.d.) by drawing out the neck to form a capillary of 1-mm i.d. and about 100-mm length. The capacity of the ampul should be 2 to 3 times the volume of liquid to be filtered. After cutting at the line A-B (Fig. 3.5), a few drops of the solvent are added to the ampul, and a wad of cotton is placed inside the conical opening. The ampul and the test tube containing the solution are warmed up simultaneously. The capillary end of the ampul is then immersed in the solution, which is drawn into the ampul while the solvent vapors condense to produce a vacuum. After filtration, the cotton wad is removed, and the capillary can be sealed. The ampul is set aside for crystallization. Finally, the ampul is cut open near the constricted part, and the crystals can be transferred into the filter funnel shown in Chapter 2, Fig. 2.33a. Alternatively, the crystals can be left in the ampul while the mother liquor is removed by inverted filtration as illustrated in Chapter 2, Fig. 2.37.

Wright [56] has described an apparatus for the fractional crystallization of 1 to 100 mg of material. It comprises a glass tube with side arms that are plugged with cotton wool, one plug serving as a safety precaution and the other as a filtering medium. The sample and solvent are placed in the tube, which is then corked, and the solute obtained is ejected through the filter side arm by air pressure entering through the other arm.

3. Recrystallization in a Capillary

The recrystallization of less than 10 mg of solid material is carried out in capillary tubes. The capacity of the capillary ranges from about 100 to 500 μl. In handling such volumes, consideration should be given to capillary force, surface area, and the large ratio between length and diameter of the container. The narrow end (sealed) of a disposable pipet also can serve as the vessel, while the wide section serves as a funnel for introducing the sample.

The solid sample may be introduced into the capillary by dipping its open end into the working material, like filling a melting-point tube. For quantitative transfer of solid material, however, the plunger technique described in Chapter 9, Fig. 9.17 is recommended. Alternatively, the sample can be introduced into the capillary in the form of a solution. For this purpose the solid is dissolved in a low-boiling solvent with the characteristics of type 2 given in Section II.B.1, Table 3.1. The solution is transferred by means of a syringe. The needle is inserted into the capillary as deep as possible, and the liquid is expelled slowly. Owing to the capillary force, the solution may not flow down to the bottom of the capillary spontaneously. The capillary is tapped by hand or placed in a centrifuge to force the solution down.

A simple technique for filling the capillary with liquid is shown in Fig. 3.14. The open end of the capillary is submerged in the liquid contained in a test tube, which is placed in a chamber connected to the aspirator. Evacuating the chamber causes air to escape from the capillary. When

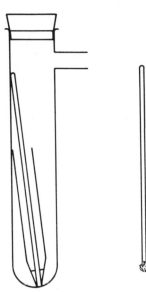

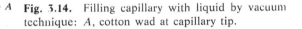

A **Fig. 3.14.** Filling capillary with liquid by vacuum technique: A, cotton wad at capillary tip.

pressure is reestablished in the chamber, the liquid will automatically enter the capillary. A cotton wad can be inserted at the capillary tip A to remove insoluble matter. For obvious reasons, this technique is not suitable for transferring solid material dissolved in low-boiling solvents.

If it is necessary to evaporate the solution to dryness, as in the case when a solvent of type 2 (Table 3.1) is replaced by one of type 3, the device shown in Fig. 3.15 can be used. The capillary is placed in a test tube, which serves as water bath. A stream of nitrogen gas is conducted above the surface of the solution by gradually lowering the inlet tube until all the solvent has evaporated. Then the needle and nitrogen source are removed. The next solvent is added to form a new solution. Since it is difficult to obtain a homogeneous solution in the narrow capillary while it is fixed in an upright position, the capillary is taken out of the water bath, sealed, and inverted (or centrifuged) several times in both directions.

The sealed capillary containing the hot solution is set aside for crystallization. This is preferably done in a warm water bath so that the temperature of the solution drops slowly. The growth of large crystals is advantageous, since they are easier to separate from the mother liquor.

The capillary may be modified to perform operations besides crystallization. For example, Kojola [57] has described a technique to separate the mother liquor as shown in Fig. 3.16. After crystallization is complete, one end of the capillary is constricted and then inserted into another capillary, which has an indentation. When the coupling is centrifuged, the crystals

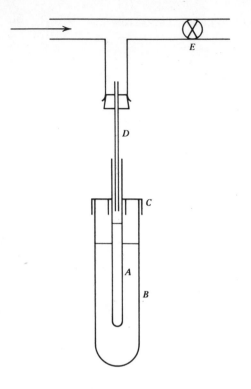

Fig. 3.15. Evaporating of solution in capillary: *A*, capillary containing solution; *B*, test tube as water bath; *C*, rubber cap; *D*, stainless-steel needle or glass capillary; *E*, nitrogen-gas valve.

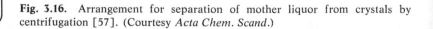

Fig. 3.16. Arrangement for separation of mother liquor from crystals by centrifugation [57]. (Courtesy *Acta Chem. Scand.*)

are retained in the original capillary. New solvent can be added by the technique shown in Fig. 3.14, and the recrystallization experiment repeated in the same capillary.

Fig. 3.17 shows the capillary at various stages of operation. A glass tube of 3-mm i.d. and 100-mm length is closed at one end (*a*). The solid sample is introduced into the bottom using the plunger technique (see Chapter 9, Fig. 9.17). By means of a sharp flame, a constriction is made at near the center of the capillary (*b*). Asbestos is added (*c*), and another constriction is made (*d*). The solvent is added and centrifuged to the bottom. The capillary is then sealed and, after warming, set aside for crystal formation (*e*). After crystallization is complete, the capillary is centrifuged in the reversed position. Crystals remain on one side while the mother liquor passes into the other end. The capillary is cut at the two constrictions to recover the crystals and solution, respectively. This method may be tested by recrystallizing 3 mg of crude *m*-dinitrobenzene using 95% ethanol as solvent.

4. Recrystallization on a Microscope Slide

When the working material is less than 1 mg, a microscope slide serves as the vehicle for crystallization. When water is used as solvent, the ordinary flat slide is suitable. A drop of water about 5 mm in diameter and 1 mm deep is deposited on the slide; the solid sample is added and stirred with a glass needle. The mixture is heated cautiously, in order to avoid evaporation, by moving the slide in a circular fashion over the micro-flame. Alternatively, the slide can be placed on a heating block kept at

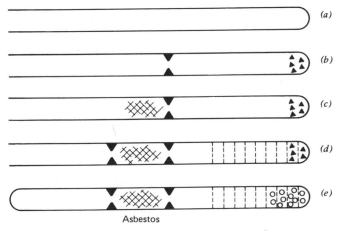

Asbestos

Fig. 3.17. Recrystallization in a capillary (see text).

95°C, and the solid added to the warm water. The saturated solution is filtered into a pipet (see Chapter 2, Fig. 2.40a) the tip of which has been warmed, and the droplet is deposited on another part of the slide, covered with a watch glass, and allowed to cool.

Microscope cavity slides [58] (Fig. 3.18) should be employed when organic solvents are used for recrystallization. Warming on metal block is preferred to heating the slide over a flame. A cover slip should be placed over the cavity to minimize loss of solvent.

Separation of the mother liquor from the crystals can be achieved by means of a capillary pipet, or a piece of filter paper (see Chapter 2, Fig. 2.40). In the latter case, the slide is slightly tilted to let the liquid move away from the crystals. Obviously, this technique is not recommended when the mother liquor is to be recovered.

III. FRACTIONATION BY SUBLIMATION

A. CHARACTERISTICS OF THE SUBLIMATION PROCESS

Sublimation utilizes the physical phenomenon of direct conversion of a solid into the gas phase; condensation of the vapor recovers the compound in crystalline form. The sublimation process is particularly suited for microscale manipulation. As a matter of fact, in contrast with crystallization or distillation, it is much easier and more effective to carry out sublimation on quantities of material below the decigram region than above this level. Fractionation by sublimation can be performed by means of relatively simple equipment (temperature-gradient apparatus), and the efficiency is comparable to that of chromatography.

Certain characteristics of the sublimation process should be recognized in order to utilize it fully for the separation of solid mixtures. (1) Sublimation takes places from the surface of the solid sample. This phenomenon is an important factor to be considered when working with materials that contain both volatile and nonvolatile components. As the sublimation progresses, the surface becomes richer in the nonvolatile component, and the process will slow down or even stop. Then it is necessary

Fig. 3.18. Microscope cavity slides. (Courtesy Metro Scientific, Inc.)

to regrind the residue or dissolve it to give a fresh surface before continuing the experiment. (2) Unlike boiling, which occurs at a constant temperature under a given pressure, sublimation of a substance occurs within a range of temperatures. Consequently, poor or no fractionation will result from a simple sublimation (e.g., in the apparatus described in Section III.B and C) if the working material is a mixture of compounds whose vapor pressures are close together. On the other hand, sublimation is the best technique for the separation of a volatile solid from the solvent used in crystallization, and from nonvolatile contaminants such as filter-paper fibers, as when one is preparing a sample for organic elemental analysis.

A significant advantage of the sublimation process is that no extraneous chemicals need be added, thus eliminating one source of contamination. The main limitation of sublimation lies in the fact that many compounds have very low vapor pressures. By applying a high vacuum (less than 10^{-2} torr), however, normally nonvolatile compounds such as amino acids and carbohydrates have been sublimed. Graves and Vincent [59] have discussed molecular sublimation, with reference to the possible sublimation of compounds that have hitherto resisted this method of fractionation. Loew and co-workers [60] have described an apparatus for vacuum sublimation of a thin film of material.

The collection of sublimates can be achieved below the microgram region. Janak [61] has described a technique to collect nanogram to picogram amounts of sublimates from a gas chromatograph in form of microcrystals. Sublimation is used extensively in the purification of isotopically labeled compounds [62].

Occasionally the condensate in a sublimation experiment becomes syrupy. When this happens, heating should be stopped, and a few crystals of the authentic compound inoculated on the surface of the condenser to induce crystal formation.

B. SUBLIMATION IN THE COLD-FINGER APPARATUS

1. General Considerations

The common device for the separation of 10 to 200 mg of solid material by sublimation is the cold-finger apparatus. Fig. 3.19 shows an example, which is designed for cooling with running water. Instead of using the ground-glass joint, the apparatus can be simply constructed by fitting a finger-condenser into a side-arm test tube through a rubber stopper. Fig. 3.20 shows an apparatus described by Marberg [63], which is fitted with a vacuum-jacket condenser that can accommodate Dry Ice, liquid nitrogen, or other fluid at the desired temperature for condensing the vapors. The

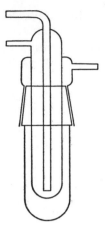

Fig. 3.19. Sublimation apparatus with water cooling.

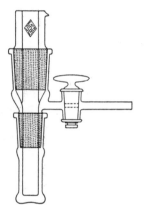

Fig. 3.20. Sublimation apparatus with Dry Ice cooling [63]. (Courtesy SGA Scientific Inc.)

sublimation vessel usually has an inside diameter of 30 to 60 mm, and the cold finger is 100 to 150 mm long, with its end about 10 mm above the bottom of the vessel. Since sublimation occurs only on the surface of the working material, it is not advisable to reduce the dimensions of the apparatus even for handling milligram amounts of solid mixture. For the same reason, the metal heating block (see Chapter 2, Fig. 2.17) is preferred to the liquid bath, so that the sublimation assembly can be set up horizontally, instead of vertically, and the sample can be spread out along the length of the vessel. The side arm of the vessel is connected to the vacuum line. For the sublimation of a compound that exhibits sufficient vapor pressure at a temperature about 20°C below its melting point, a water aspirator usually suffices [64]. When the apparatus shown in Fig. 3.20 is used, cooling the sublimate with the aid of Dry Ice permits sublimation to be performed on heat-sensitive compounds. On the other hand, placing a warm fluid in the vacuum-jacket condenser well assures the removal of high-boiling solvent in order to obtain a pure sample for subsequent micromanipulations such as infrared spectroscopy or quantitative organic analysis.

2. The Procedure

Sublimation using the cold-finger technique is carried out in the following manner: The cold finger (Fig. 3.19 or Fig. 3.20) is disengaged from the assembly. The solid material to be sublimed is spread evenly in the form of a thin layer in the vessel. When the working material is in the milligram range, it is expedient to introduce the sample in the form of a dilute solution, which is then evaporated to dryness in the sublimation vessel, employing the technique discussed in Section II.B.4 (Figs. 3.7, 3.8). The apparatus is reassembled and placed in the metal heating block or liquid bath.

While the sublimation vessel is being heated gradually to within 10°C of the melting point of the desired compound, the cold finger is watched for the appearance of crystals. If no sublimate deposits on the condenser, the apparatus is evacuated by connecting the side arm to the water aspirator (about 20 torr), and then to the oil pump (0.02 torr) or higher vacuum to vaporize the solid. In the case of thermally unstable substances, however, it is recommended to perform the sublimation at the lowest possible temperature; then the apparatus should be evacuated by means of the oil pump before the vessel is (carefully) heated. If the pressure inside the vessel is sufficiently low, and Dry Ice–acetone mixture is used as coolant, many compounds can be obtained as sublimates at ambient temperatures.

When the sublimation slows down or stops, the solid material remaining in the vessel can be taken out, pulverized, and returned for repeated heating. If the amount of the remaining solid is very small, it is better to introduce a few drops of a suitable solvent and evaporate the solution.

The sublimate can be recovered from the condenser by scraping with a spatula. Prior to this operation, the coolant should be removed and the apparatus brought to ambient temperature in order to prevent moisture from condensing on the sublimate. Another technique for removing the sublimate is to use a solvent, as illustrated in Fig. 3.21. The condenser is held upside down, the solvent is added from a syringe, and the solution drains through the side arm into a test tube. This method should be used for quantitative recovery of the material, such as in the preparation of radioactive compounds [62].

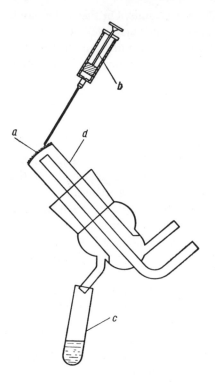

Fig. 3.21. Recovering sublimate with a solvent: *a*, sublimate; *b*, syringe for delivering solvent; *c*, receiving tube; *d*, finger condenser.

3. Miscellaneous Sublimator Designs

Appel and Romero [65] have improved the cold-finger apparatus by placing a metal wire mesh between the sample and the condenser. When the right conditions (bath temperature, vacuum), which vary with the compound, are chosen, most of the product will deposit on the mesh. The speed of sublimation is increased and the crystals are much better formed than when the mesh is absent.

A device for microsublimation proposed by Bhati [66] comprises an annular oil bath, into which the sublimation tube is inserted, heated by a Thermocord element. This obviates charring, which can occur with direct electric heating. The working range of this apparatus is 45 to 200°C.

Nemec [67] has constructed a sublimation apparatus that consists of a flask connected by a ground-glass joint to a still head, through which a cold-finger condenser extends vertically into the flask. The condenser is sealed to the upper end of the still head by a rubber sleeve, and has at its lower end a spherical recess. A simple sublimation apparatus [68] can be fabricated from a Pyrex tube that is connected by pressure tubing to a vacuum pump, and heated by a metal block encircling the tube. Schrecker [69] has used a bulb-tube assembly for vacuum sublimation. Pino and

Zehrung [70] have described an apparatus that comprises a sublimation chamber fitted inside a heating tube, which is wound with chromel wire as the heating coil, to which the power input is governed by a variable transformer. One end of the sublimation tube just fits into a condenser tube, and at the other end is fitted an inlet tube containing a drying agent between two asbestos plugs. An aspirator connected to the exit end of the condenser chamber provides an air current to transport the vapors of the sublimate.

C. SUBLIMATION ON THE METAL HEATING STAGE

The sublimation of material below mg-quantity should be performed on the metal heating stage and not in a liquid bath. A simple technique using capillary tubes is illustrated in Fig. 3.22. The solid to be sublimed is spread out inside a capillary of about 2-mm i.d., which is placed in the well of the heating block (Chapter 2, Fig. 2.17) in a horizontal position. Another capillary with taper end is fitted snugly at the opening. While the metal block is being heated, condensation of the sublimate under the Dry Ice is observed. The Dry Ice can be supported on a piece of plastic tape affixed to the capillary receiver. The collection of different fractions can be conveniently achieved by changing the taper capillary receivers, which can be connected to the water aspirator.

The sublimation of material placed on a glass slide can be carried out on a metal block (Chapter 2, Fig. 2.17) or microscope heating stage. As shown in Fig. 3.23, a sublimation chamber encloses the solid to be sublimed. This chamber may be an inverted glass dish; it also may consist of a glass or metal ring covered with a glass plate. The distance between the top and bottom glass plates varies from less than 1 mm to several millimeters, depending on the experimental conditions. The formation of crystals on the condensing surface is observed through the microscope objective, or with the aid of a magnifier. Evacuation is carried out by

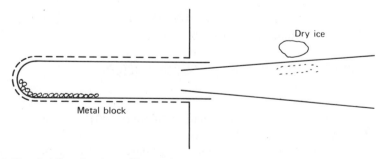

Fig. 3.22.   Sublimation in capillary tube.

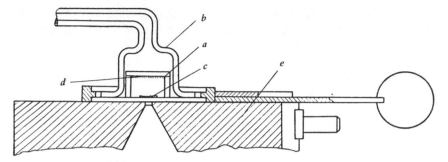

Fig. 3.23. Sublimation from glass slide: *a*, sublimation chamber; *b*, evacuation attachment; *c*, sample placed on glass slide; *d*, sublimate; *e*, heating stage.

means of the attachment (*b*), which is connected to the aspirator. A commercial vacuum sublimation chamber that fits the Kofler heating stage [71] is shown in Fig. 3.24.

The heating-stage technique was developed by Kofler [72] and has been extended by Kuhnert-Brandstätter [73] and other workers. For instance, Kunze [74] has described a special microscope with a stage that can be heated and cooled from −20° to 360°C, which is used to sublime dye stuffs. Kassau has separated by sublimation 1- to 5-µg amounts of pharmaceuticals [75] and carcinogens [76]; the compounds are obtained as characteristic crystals. Weisz and Schapky [77] have described a device for the simultaneous vacuum sublimation of organic substances with a Kofler heating bench.

Baehler [78] has reported successful sublimation of 1 to 5 µg of caffeine, theobromine, and barbiturates from silica gel plates; but 20 to 50 µg of material is required in order to get good results for atropine and morphine. Fawkes and co-workers [79] have described a method for working with thin-layer spots as follows: All adsorbent except that in the vicinity of the spots is scraped from the plate to reduce the moisture content; then the compounds are sublimed on a glass plate (cooled by a superimposed Plexiglas tray of ice water) by heating the thin-layer plate at 265°C on an aluminum hot plate maintained at 280 to 290°C. The compounds are subsequently resublimed onto a microscope cover glass.

D. TEMPERATURE-GRADIENT SUBLIMATION

1. Principle

The temperature-gradient sublimation method [80] is a refinement of the sublimation technique that permits the separation of several components in one operation. The principle is based on the condensation of

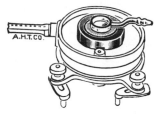

Fig. 3.24. Kofler heating stage and vacuum sublimation chamber. (Courtesy A. H. Thomas Co.)

the individual compounds at different positions in a sublimation tube in which there is decreasing temperature gradient. The rate of migration of each compound along the temperature gradient is dependent on the ratio between its rate of volatilization and rate of condensation at each temperature. After certain time necessary for equilibration, the zones of the respective compounds become practically stationary. In contrast with simple sublimation (Section III.B), temperature-gradient sublimation provides a powerful fractionation technique.

2. Apparatus

The apparatus usually consists of a glass tube where sublimation takes place, and a heating mantle, with an adjustable temperature gradient, in which the glass tube is accommodated. The heating mantle is basically another tube wound with resistance wire. The temperature gradient is achieved by putting a decreasing number of windings along the length of the tubing. The solid mixture to be fractionated is placed in a glass capsule which is positioned in the warmest part of the sublimation tube. After the experiment, the sublimation tube is cut into sections to recover the separated compounds.

3. Operation

Two procedures can be employed to facilitate the movement of the separated zones along the temperature gradient. In one procedure the sublimation tube is evacuated to 10^{-3} torr or lower pressure and sealed [81] (Fig. 3.25), or evacuated continuously [82] (Fig. 3.26). In another procedure, a constant flow of carrier gas is conducted through the sublimation tube at a predetermined rate, for example, 10 ml per minute [83] (Fig. 3.27).

The assembly described by Bates [81] is shown in Fig. 3.25. The horizontal glass tube B (50 cm×2 cm) has its inner and outer surfaces insulated with asbestos paper, while a spiral of nichrome wire is wound around the outer surface to produce a uniform temperature gradient along the tube. The sample (10 to 100 mg placed in a small test tube) is

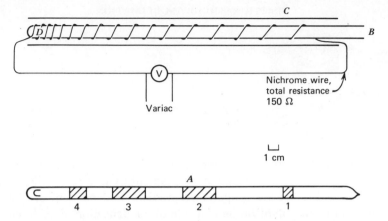

Fig. 3.25. Temperature-gradient sublimation in sealed tube [81]: *A*, Evacuated tube containing sample (in a small test tube); *B*, heating tube with temperature gradient; *C*, glass draft shield; *D*, highest temperature. (Courtesy *Chem. Ind.*)

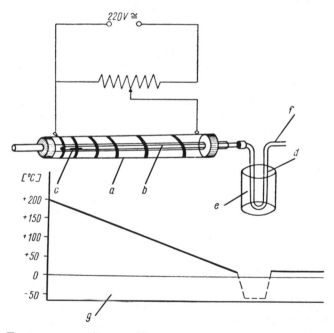

Fig. 3.26. Temperature-gradient sublimation with continuous evacuation [82]: *a*, outor tubing of the heating mantle with the temperature gradient; *b*, sample tubing connected to vacuum; *c*, sample; *d*, U tube for freezing out volatile sample; *e*, cooling bath; *f*, connection to vacuum; *g*, temperature-gradient graph.

110

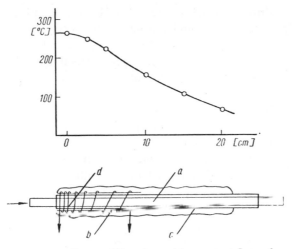

Fig. 3.27. Temperature-gradient sublimation with constant flow of carrier gas [83]. Graph represents temperature gradient inside tubing a. a, tubing containing sample with the inert-gas inlet; b, heating mantle with temperature gradient; c, insulation; d, location of sample.

inserted into another glass tube A (55 cm × 1 cm), which is then evacuated to a pressure of less than 10^{-3} torr, sealed off, and placed in the wider tube B. The temperature is raised gradually by increasing the voltage at 10-min intervals in 10-V steps. For example, for the fractionation of a mixture containing biphenyl and the three terphenyl isomers, the required temperature is 170°C and is kept constant for 1 hr. The inner tube A is then removed and cooled in a horizontal position, and the four compounds recovered by cutting the tube.

4. Applications and Modifications

Shigetomi and co-workers have fractionated quinones [84] and polycyclic aromatic hydrocarbons [85] by the following arrangement: A porcelain outer tube (5 cm × 60 cm) is surrounded by a spiral coil of nichrome wire, and a spiral of copper tubing (0.5-cm diameter) fits snugly within it. The sample is placed in a glass tube (8-mm i.d.), which is evacuated to 1 torr and sealed. The glass tube is inserted in the copper spiral, and air is passed through the copper tubing at an appropriate rate to produce a temperature gradient (1 to 3°C/cm) in the glass tube. The temperature gradient reaches equilibrium in about 3 hr. Flaschenträger and co-workers [86] have used a Pyrex-glass tube, divided into four parts that are joined by ground joints, to perform high-vacuum fractionation sublimation in a heating block, through which the sections of the tube are pushed in turn. When the fractionation is complete, the fractions may be weighed in their individual tube sections.

Gosling and Bowen [87] have designed a fractional-sublimation apparatus that permits the introduction of air-sensitive compounds without exposure to air and provides a variable temperature gradient, facilitating the separation of compounds that differ only slightly in their vapor pressures. As illustrated in Fig. 3.28, the apparatus consists of a 1000-mm-long Pyrex sublimation tube (11-mm o.d.), which is sealed at the lower end; a small indentation is created to facilitate the breaking of the sample ampul. The upper end is connected to a vacuum line via a right-angled

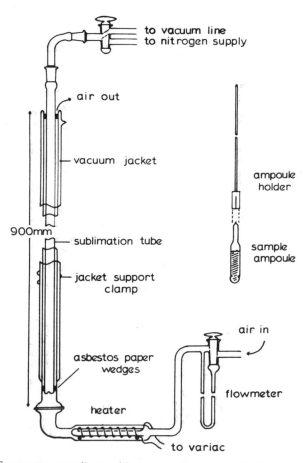

Fig. 3.28. Temperature-gradient sublimation apparatus of Gosling and Bowen [87]. (Courtesy *Anal. Chem.*)

elbow. The vacuum line carries a three-way stopcock to permit the evacuation or admission of dry nitrogen. A 900-mm-long unsilvered vacuum jacket (15-mm i.d.) surrounds the sublimation tube and is centered by three asbestos wedges (each 3 mm wide) at top and bottom. The assembly is completed by a nichrome heater coil enveloped in a Pyrex tube and attached to the lower end of the vacuum jacket by a spherical ground joint. Compressed air is led via a flowmeter to the heater tube, then up the annular space between the sublimation tube and the vacuum jacket, and exhausted at the top. The sample ampul is made from 8-mm-o.d. Pyrex tubing with a 6-mm-o.d. neck. The lower end is sealed and blown out to form a thin-walled bulb. The insertion of the sample ampul into the sublimation tube is accomplished with an ampul holder, which is fabricated from a stainless steel tubing brazed to a brass rod. The tubing is slit longitudinally to provide a clamping action on the glass ampul. In operation, the sublimation tube is flamed out under vacuum, filled with nitrogen, and removed from the vacuum line. The sample ampul is introduced and broken on the protrusion at the bottom of the tube, and the sample shaken out. With the ampul holder removed, the sublimation tube is attached to the vacuum line and pumped out immediately. The vacuum jacket and heater tube are then fitted, and connections made to the air supply and the Variac. The air supply is turned on fully to give a constant reading on the flowmeter, and the temperature of the air stream is increased gradually by means of the Variac control until sublimation begins. A further slight increase in temperature will cause steady movement of the sublimate up the tube. When sublimation is complete, the heater is turned off and the air supply interrupted. Nitrogen is introduced into the sublimation tube. The tube is now sealed off with a torch at convenient places to include separate compounds for transfer to a dry-box. This sublimation apparatus has been employed for the separation of a mixture of geometric isomers that cannot be fractionated by crystallization because of their isomerization in solution.

Berg and co-workers [88, 89] have described a method, termed "fractional-entrainment sublimation," which utilizes a horizontal sublimation tube, with provision for a thermal gradient to be maintained along the tube and for a carrier gas to be drawn across the sample (placed at the hot end of the apparatus) at low pressures (about 1 torr). The apparatus [89] is shown in Fig. 3.29. It has been employed for the separation of 10 to 60 mg of solid materials. The components are obtained in sharply defined crystal zones and are pure enough for identification by their melting points and ultraviolet spectra. The temperature limit of any zone is reproducible and is probably comparable with the R_f value in chromatography although the degree of separation is not as great. Crosmer and co-workers [89a] have

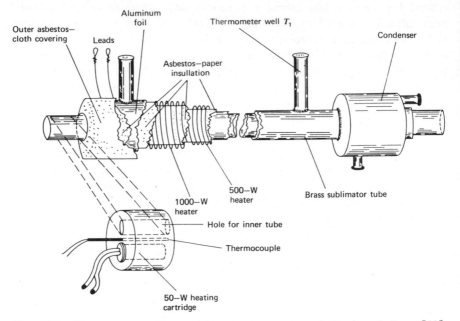

Fig. 3.29. Temperature-gradient sublimation apparatus of Reed and Berg [89]. (Courtesy *Anal. Chem. Acta.*)

proposed a cold-trap fractionation technique for a gas-chromatography–mass-spectrometry combination. The trap containing condensed organic mixture is gradually warmed, whereby the components separately vaporize and pass into the mass spectrometer.

E. LYOPHILIZATION

1. Principle

Lyophilization [90] is the process in which one portion of a mixture is separated by sublimation from the frozen material at reduced pressure, and the residue is obtained as porous solid. It is generally known as "freeze-drying" or "drying by sublimation." This technique is particularly useful for the removal of water from thermolabile substances and biological materials. Thus, it has been utilized to study free fatty acids in cheese [91] and amino acids in tissues [92]. The sublimate is usually water vapor; the residue, if sealed *in vacuo*, retains its biological attributes for long periods. When reconstitution is desired, the porous residue readily readmits the added solvent; hence the appropriateness of the term lyophile, or solvent loving.

5. Apparatus and Operation

The lyophilization of small quantities of material can be achieved with very simple equipment. For example, Young and Christian [93] have freeze-dried 1 ml of serum as follows: A Dry Ice–acetone bath is prepared using a flat-bottomed rectangular dish (10×12 inches) with a glass cover. Then 1 ml of serum is pipetted into a 10-ml beaker, which is immersed in the cooling bath for 5 to 10 minutes. The beaker is swirled gently while immersed, so that the serum is frozen on the glass in as thin a film as possible. Following the freezing step, the beaker is placed in a vacuum desiccator (160-mm o.d.), and a vacuum is applied for about 30 min. The serum is dried efficiently without losses, and can subsequently be used for neutron-activation analysis or other analytical procedures.

A lyophilization apparatus that can handle several samples simultaneously is shown in Fig. 3.30. It consists of a cold-finger condenser with side arms to which the flasks are attached through ground-glass joints. Before connecting the flasks to the body, the samples in the flasks should be solidified. This is effected by swirling the flask in a Dry Ice–acetone bath to produce a thin film over a large area of its inner surface. The well of the apparatus is filled with Dry Ice–acetone mixture, and a vacuum of 1 to 10^{-2} torr is applied through the exit tube. During the sublimation of ice, the flasks do not require external cooling, since this is accomplished by the evaporation process. About 6 g of Dry Ice is consumed for the sublimation of 1 g of ice in the flask. Liquid nitrogen may be placed in the finger-condenser well, and its level can be automatically controlled [94].

Habeeb and De Luca [95] have described a flask that serves for lyophilization without transfer of the heat-labile material or interruption of the experimental process. Iijima [96] has designed freeze-drying flasks that

Fig. 3.30. Lyophilization apparatus. (Courtesy Kontes Glass Co.)

incorporate filters. Gabella and Costa [97] have constructed an apparatus for freeze-drying tissues without damaging their structures. The assembly consists of a cryostatic cell with compressor and ancillary equipment from which a refrigerated liquid is circulated through two refrigerated plates placed in chambers connected to a vacuum pump. Lyophilization apparatus that utilizes thermoelectric cooling and permits simultaneous treatment of several specimens is commercially available [98].

IV. RING-OVEN TECHNIQUE

A. PRINCIPLE OF THE METHOD

The ring-oven technique [99] was developed by Weisz [100] for the analysis of metals and salt dissolved in aqueous solutions. In principle it combines the radial elution of the soluble portion of a mixture from the center of a filter paper with the concentration of the migrating solution by evaporating the solvent at elevated temperature. The ring-oven controls the evaporation, and the product appears in the form of a narrow circular tracing on the filter paper. This method thus utilizes the techniques of electrography [101] and the paper chromatography (see Chapter 4, Section II).

B. APPARATUS AND PROCEDURES

The ring-oven apparatus and ancillary tools are very simple. A diagrammatic sketch is shown in Fig. 3.31. It comprises an aluminum block of cylindrical shape (55-mm o.d., 22-mm i.d., 35-mm height) resting on a tripod provided with an electric bulb, and a clamp for fixing a capillary tube containing the washout liquid exactly above the center of the bore of the aluminum block. The centering technique is shown in Fig. 3.32. In operation, a filter-paper disc (55-mm diameter) of the best analytical quality is placed on top of the aluminum block and covered with a porcelain or metal ring of 25-mm i.d. The sample solution (1 to 3 μl) is spotted on the paper disc by means of the capillary sample pipet of 0.1-mm i.d. (Fig. 3.33), precisely at the center of the aluminum cylinder. In the case of aqueous solutions, the temperature of the aluminum block is maintained at 105 to 130°C so that it is above the boiling point of the washout liquid. The washing pipet is now positioned above the ring oven. As its tip touches the filter paper, the liquid drains onto the sample and migrates outward radially. When the solution reaches the area heated by the aluminum block, the solvent evaporates, and the nonvolatile material is left on the paper disc in form of a narrow ring tracing. The washing

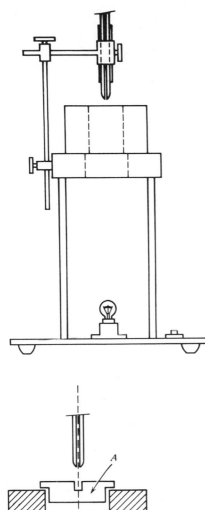

Fig. 3.31. Ring oven [99]. (Courtesy Pergamon Press.)

Fig. 3.32. Centering of capillary pipet [99]: *A*, centering device. (Courtesy Pergamon Press.)

procedure is repeated until all soluble substances in the original mixture have moved to the ring tracing. In order to prevent the ring tracing from running off the edge of the aluminum cylinder, or the solution from drying out before reaching the edge, an appropriate volume of washout liquid should be used each time, and the interval between two washings should be controlled. In addition, a suitable temperature of the metal block must be selected for the particular solvent.

In a modified version of the above procedure, the sample is transferred

Fig. 3.33. The capillary sample pipet [99]. (Courtesy Pergamon Press.)

from a small filter-paper disc to the filter paper resting on the aluminum block, as illustrated in Fig. 3.34. The small disc may be mounted with tiny drops of plastic cement [102]. The washout liquid dissolves the soluble substances in the small disc and carries them to the large disc.

A reverse operation has been developed [103] by means of which the sample is caused to converge from the outer ring into a small area in the center of the filter-paper disc. The device for this "washing-in" proceure is shown in Fig. 3.35.

C. APPLICATIONS

Owing to the simplicity of the apparatus as well as its operation, the ring-oven technique has found many applications in the microanalysis of inorganic and also organic materials [99]. It has been utilized for air-pollution studies [104], a commercial unit for which is shown in Fig. 3.36. It also complements the spot-test methods of Feigl [105, 106]. For the analysis of thermolabile substances, evaporation of the solvent by heat can be circumvented by the use of an adsorption barrier [107].

No separation is involved if all working material on the filter paper is transferred from the center to the ring tracing. Separation into two fractions is achieved, however, when a portion of the original sample is insoluble in the washout liquid. For example, in the analysis of dust mixture collected on a small disc of an air filter (see Chapter 7, Section II.B), the small disc is placed on the ring oven as illustrated in Fig. 3.34 and is

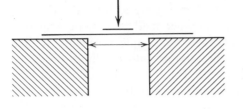

Fig. 3.34. Transfer of substances from little filter paper to ring-oven paper [99]. (Courtesy Pergamon Press.)

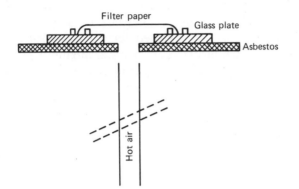

Filter paper

Glass plate

Asbestos

Hot air

Fig. 3.35. Device for "washing-in" technique [99]. (Courtesy Pergamon Press.)

Fig. 3.36. Air-pollution trace oven. (Courtesy A. H. Thomas Co.)

washed with 3 N HCl. The components that migrate to form the ring are separated from the insoluble material. Another method for effecting separation is carried out as follows: The spotted sample is treated with a reagent that converts one portion of the mixture into insoluble products, and the soluble component is washed out. The insoluble products can be further separated by cutting out the center area of the filter paper. The cut-out portion is placed on top of another filter paper as shown in Fig. 3.34, and a second reagent is added to dissolve the material for another washing-out operation. The use of gaseous reagent is advantageous in this method. A gas generator designed for the treatment of a sample on the filter paper is shown in Fig. 3.37. Thus, a mixture containing Fe(III) and Cu(II) ions is treated with H_2S gas to precipitate CuS. After Fe(III) is washed out in the ring oven, the center disc is cut out and treated with H_2O_2 to convert CuS into water-soluble $CuSO_4$. The disc is then placed on the ring-oven, and Cu(II) is washed out.

For the analysis of the substances in the ring tracing, the reagent can be applied by spraying or dipping the filter paper into the reagent solution. The paper can be cut into several sections to be tested with different reagents. Semiquantitative estimation can be made by comparison against standards.

A glass ring oven has been constructed by Heath [108], and glass-fiber paper [109] has been used on the ring oven [110]. A combination of the

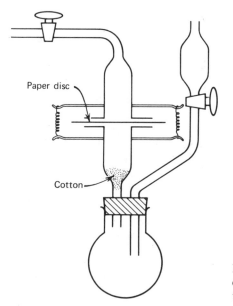

Fig. 3.37. Gas generator and device for clamping filter-paper disc [99]. (Courtesy Pergamon Press.)

ring-oven technique with thin-layer chromatography for separation has been suggested by Scherz and co-workers [111].

V. ZONE REFINING AND RELATED METHODS

A. PRINCIPLE

Zone refining, also called zone melting [112, 113], is a technique for separating minute constituents (usually impurities) from solid materials by a melting-solidification process. The material is packed in a glass tube, and a narrow molten zone is produced by a heating coil, which is slowly moved from one end of the tube to the opposite end. Each complete run is known as a "pass," and the passes are repeated until the desired separation is achieved.

The primary factor in determining how the solute redistributes during zone melting is the distribution coefficient K, which is the ratio of the solute (minute constituent) concentration in the solid phase to that in the liquid phase at a constant temperature. The components with distribution coefficients less than one are more soluble in the liquid than in the solid; consequently they are deposited toward the finishing end. Conversely, the components with distribution coefficient greater than one are more soluble in the solid; therefore they tend to be rejected by the liquid and deposited at the starting end.

The zone melting method is used to produce semiconductor materials and metals with extremely low impurities, to the level of parts per billion. Many inorganic and thermally stable organic compounds of high purity (better than 99.99%) have been prepared by this technique [113–117]. The accumulation of trace impurities by zone melting makes possible the analytical evaluation of these impurities.

B. APPARATUS FOR ZONE MELTING

A schematic diagram of a zone-refining apparatus proposed by Schildknecht and Vetter [118] for small samples is shown in Fig. 3.38. The glass tubing (2-mm i.d.), in which the working material is tightly packed, moves at the rate of 10 mm/hr through two systems of wire loops, one for heating and the other for cooling. As a result, several narrow molten zones migrate simultaneously in the same direction. After one passage of zones from one end to the opposite end, the operation is repeated as many times as required to complete the separation of impurities.

The commercially available apparatus for zone refining [119] is designed for handling gram-quantity and large-scale operations. Knypl and

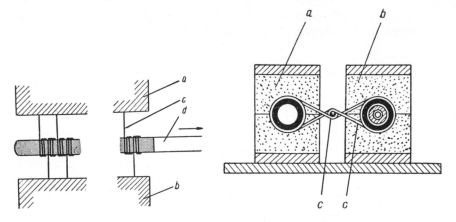

Fig. 3.38. Schematic diagram of zone-refining apparatus [118]: *a*, heating block; *b*, cooling block; *c*, wire loops for heat transfer; *d*, capillary containing sample.

Zielenski [120] have described a simple automatic apparatus for the zone refining of organic substances as shown as Fig. 3.39. The essential element for moving the tube is the floating disc, which floats upward with precision while water rises in the vessel. When the water level reaches the highest bend of the siphon tube, the water pours off to the initial level. The zones then begin repeated migration. The heaters are made from resistance wire wound 50 times around a 3-mm-thick glass ring, and they are supplied with 200-V ac through a transformer. A temperature of 350°C is easily obtained in the center. To demonstrate this apparatus, naphthalene mixed with 0.1% bromoscresol purple was packed in the tube. After two passages of the molten zone (10-mm width) at a velocity of 23 mm/hr on the 80-mm length from the initial position of the heater, there was no visual trace of the dye. Only transparent, crystallized ultrapure naphthalene was left, and the dye was collected.

Kennedy and Moates [121] have constructed a horizontal continuous zone-refining apparatus, shown in Fig. 3.40. The enriching section consists of a 22-mm-o.d. tube 43 cm long joined to a 17-mm-o.d. tube 8.9 cm long. When the apparatus was tested with benzoic acid that contained 1% of iodine, after 24 passes, iodine could be detected visually in only the last 2.5 cm of the 22-mm portion of the enriching section. The amount of purified product collected per pass was 0.6 to 0.7 g.

Generally speaking, a vertical tube with zones moving downwards is the most efficient arrangement. However, its efficiency deteriorates if the compactness of the solid column is disturbed by the formation of gas pockets. Then a horizontal arrangement is preferred, because the vapor

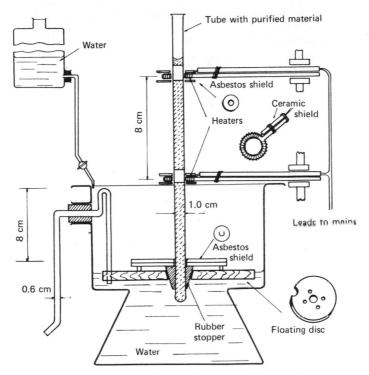

Fig. 3.39. Automatic zone refiner [120]. (Courtesy *J. Chem. Educ.*)

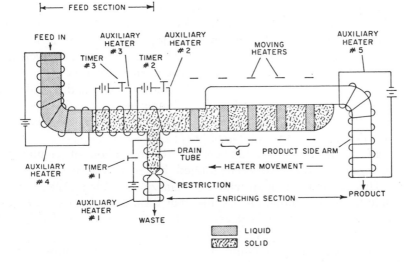

Fig. 3.40. Continuous horizontal zone-refining apparatus [121]. (Courtesy *Rev. Sci. Instrum.*)

123

can escape upwards without demaging the column, although some vapor may condense on the purified material.

Schildknecht [122] has described a device that is a modification of zone refining on the microscale (see Section C below). The material is melted, and the crystals that are gradually formed are transported by means of a spiral. A long path through the apparatus is obtained by inserting a metal spiral into the annular space between two glass tubes. If a temperature gradient is maintained along the length of the tube, it is possible to feed a mixture in at the midpoint of the tube and to withdraw material of low melting point at one end, and of high melting point at the other end. This technique has been employed for the separation of o-, m-, and p-terphenyls. Martinek [123] has purified 50-μg amounts of organic compounds using the Kofler heating stage (see Fig. 3.24). The material is placed between two microscope slides, which are heated on the stage. Gentle pressure is applied at one point on the upper slide, so that the fraction with the lower melting point is extruded. Alternate heating and cooling is continued until a constant melting point is reached.

C. APPARATUS FOR CONTINUOUS COLUMN CRYSTALLIZATION

Column crystallization from a melt can be utilized as a multistage fractionation method by countercurrent contacting of melt and crystals in a temperature gradient [124]. A continuous-column crystallization apparatus [125] consists of two concentric tubes between which is a rotating metal spiral that transports the crystals along the length of the tube; the lower part of the tube is heated while the top is maintained at a lower temperature [126] (Fig. 3.41). A pulsating column is used to overcome the problem of crystal transport. For microscale manipulation in the milligram region, the apparatus is fitted with quickly rotating threaded pins instead of spirals [126] (Fig. 3.42). By employing a rotating speed of 500 to 2000 rev/min, 10 mg of stilbene mixed with azobenzene was separated in crystallization time of 20 sec [127].

The effects of various parameters on the operation of continuous-column crystallization have been studied by Schildknecht and Breiter [124]. This technique is more efficient than zone melting [128], and is applicable to the separation of mixtures that form eutectics (e.g., azobenzene with acetanilide) or mixed crystals (e.g., azobenzene and stilbene) [129].

A solid in solution or a liquid mixture at room temperature can be fractionated by a similar method in which crystallization is effected by gradually freezing the material [130]. The apparatus proposed by Schild-

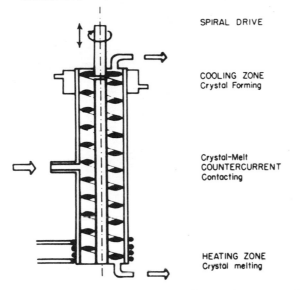

SPIRAL DRIVE

COOLING ZONE
Crystal Forming

Crystal-Melt
COUNTERCURRENT
Contacting

HEATING ZONE
Crystal melting

Fig. 3.41. Schematic diagram of a continuous-flow column crystallizer [126]. (Courtesy *Sep. Sci.*)

knecht and co-workers [131] is shown in Fig. 3.43. The solution is placed in a glass tube, which is gradually submerged (6 cm/hr) in an efficient cooling bath while the mixture is stirred at 1400 to 3000 rev/min. For aqueous solutions, the cooling temperature should be about $-40°C$. Lukhovitskii and co-workers [132] have described an apparatus that consists of a vertical fractionating column having a metal heat exchanger above a jacket containing Dry Ice, producing a temperature gradient. The substance to be separated is passed down the column and is continuously melted and then solidified, thus being separated into zones according to the melting points of the components. The apparatus was used for the isolation of thiophene from a 5% solution in benzene. This method is particularly useful for thermally unstable substances such as enzymes and for compounds that decompose above their melting points.

VI. SEPARATION BY PARTICLE SIZE OR WEIGHT

A. MECHANICAL SEPARATION, SIEVING, AND SCREENING

In microscale manipulations it is often expedient to separate solid mixtures by particle size under a magnifier or microscope when the working material is below the milligram range. Screening is thus performed man-

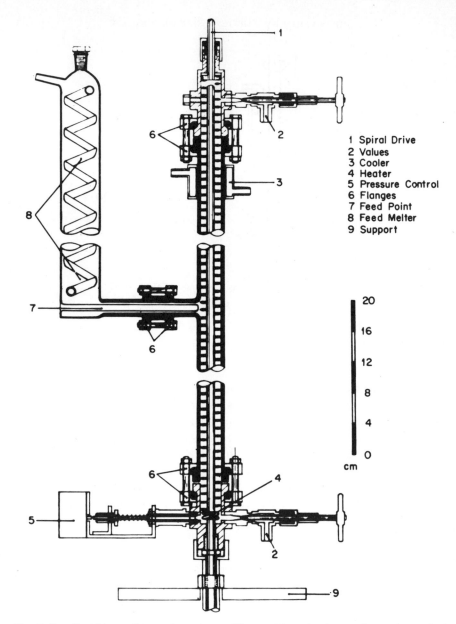

1 Spiral Drive
2 Values
3 Cooler
4 Heater
5 Pressure Control
6 Flanges
7 Feed Point
8 Feed Melter
9 Support

Fig. 3.42. Continuous-flow column crystallizer with pulsating and rotating spiral [126]. (Courtesy *Sep. Sci.*)

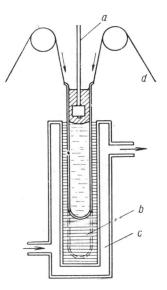

Fig. 3.43. Column crystallization by freezing [131]: *a*, stirrer; *b*, cooling bath; *c*, cooling mantle; *d*, device for lowering the test tube.

ually with the aid of a microspatula or needle. This method has the advantage of being free from the danger of contaminating the working material.

The common laboratory technique for separating particles by size is sieving [133, 134]. The sieves are graded by mesh size; the higher the mesh size number, the smaller is the sieve opening (e.g., No. 60 has a sieve opening of 0.250 mm, and No. 5 an opening of 2.000 mm). Copper screens are generally used, and the sieves can be conveniently made as shown in Chapter 2, Fig. 2.49. Since copper screens may contain other metallic elements, polyethylene sieves with polyester screens, now commercially available [135], are recommended for sieving samples intended for trace-metal analysis. It is obvious that a screen separates the working material into two parts: One portion of the particles passes through, and the rest remains on top of the screen. Therefore, fractionation of the mixture is achieved by using screens of various mesh sizes. When it is desired to obtain particles within a specific size range (e.g., the solid supports in chromatography), two screens are required.

Sieves are not effective in separating particles when their dimensions are smaller than 0.1 mm. While filter paper and discs made of fritted glass or unglazed ceramics can be use for this purpose, these screening devices suffer from the drawback that their openings are not uniform. In order to discriminate the particulates and separate them into known size ranges, membrane filters such as Millipore Filters [136] with specific pore size should be employed. For example, Hannah and Dwyer [137] have used Millipore Filters to separate suspended particles (e.g., paint pigments, aerosol mists, bacteria, colloidal trace contaminants in tea) and accomplish both the separation and sample preparation for ATR infrared analysis. Spurny and co-workers [138, 139] have studied Nuclear Pore Filters for the separation of ariborne particulates. Because these membrane filters are transparent, particles collected thereon may be measured and counted in an electron microscope. A two-layer filter has been fabricated that can collect material in a range of particle sizes [140]; it comprises a layer of membrane to collect particles 1 μm in diameter and a second layer to intercept particles down to 0.1 μm in diameter. When membrane filters and cellulose fibers (filter paper) are employed, attention should be given to the adsorption of certain substances from the system. Chiou and Smith [141] have reported that the extracted materials may vitiate the analytical results. Bowen [142] has separated traces of metals by absorption on polyurethane foams. Absorption equilibrium is complete after shaking the foam with the solution for 1 to 1.5 hr, but there is appreciable absorption after squeezing the foam in the solution for 30 sec.

B. SEDIMENTATION AND FLOTATION

Solid particles of different densities and sizes can be separated by sedimentation [143]. The mixture is suspended in a liquid (e.g., water) placed in a tall vessel. The supernatant, which contains very fine particles, is carefully separated from the sediment. Fractional sedimentation can be achieved by letting the suspension stay in the vessel for a fixed period of time and separating the supernatant. Fresh liquid is added to the vessel, the sediment is resuspended, and the separation process is repeated. Giddings and co-workers [143a] have developed the theory for sedimentation field-flow fractionation.

Flotation is based on the same principle as sedimentation, except that a gas is imparted to the solid particles to make them lighter than the liquid medium. An apparatus designed by Keil [144] is shown in Fig. 3.44. The solid mixture is placed on the fritted glass disc. A stream of water or air is flushed from the bottom. As the water reaches the overflow side arm, the solid particles that float on the liquid surface are drained off with the current. With increasing flow rate, larger particles are suspended.

C. CENTRIFUGATION

1. Separation of Crystals and Colloidal Particles

The centrifugation technique is used extensively in microscale manipulations. Based on separation by specific gravity, it can be employed to sep-

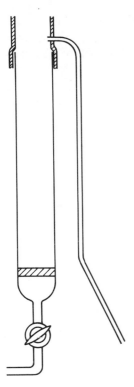

Fig. 3.44. Apparatus for flotation [144].

arate a precipitate from the mother liquor, two or more solids in the same liquid medium, or two liquid phases (see Chapter 6, Section III.C).

The apparatus for centrifugation is the centrifuge [145]. The efficiency of a centrifuge depends on the diameter of the centrifugal motion and on the number of revolutions per minute (tangential velocity of the centrifugal motion), and is expressed in g (gravity) units. In the centrifuge tube, a centrifugal-force gradient exists between the points closest and furthest from the axis of rotation.

The common laboratory centrifuges are constructed to carry 6 to 12 centrifuge tubes and operate at a speed of 3000 to 9000 rev/min. The capacity of the centrifuge tube is 5 to 20 ml. In the angle-type rotor, the tubes are accommodated in the metal holders of the rotor, and they remain in a fixed oblique position (about 45-degree) during centrifugation. In the swinging-tube rotor, the heads of the holders are suspended by two points; therefore the angle between the axis of the centrifuge tube and that of the centrifugal motion will vary with the speed of rotation. These centrifuges are used to separate crystals from the liquid phase. After centrifugation, the crystals are collected at the bottom of the centrifuge tube, except in the case where the liquid (e.g., carbon tetrachloride) is heavier than the solid. Centrifuge tubes have been designed to permit the mother liquor to pass into anothr vessel through an opening [146] or channels [147], as shown in Fig. 3.45.

A centrifuge fitted with a heating mantle, so that the material in the tube can be warmed up to 100°C during the operation, is commercially available [148]. On the other hand, thermally sensitive substances should be centrifuged at low temperatures.

The separation of colloidal particles requires highly efficient centrifuges,

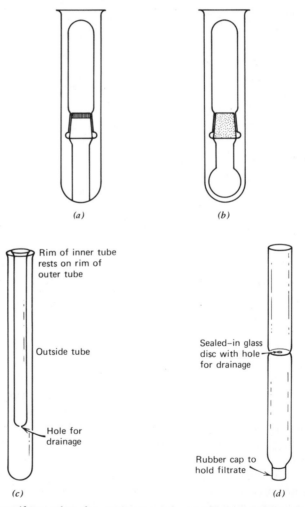

(a)   *(b)*

Rim of inner tube
rests on rim of
outer tube

Outside tube

Sealed–in glass
disc with hole
for drainage

Hole for
drainage

Rubber cap to
hold filtrate

(c)   *(d)*

Fig. 3.45. Centrifuge tubes for separation of crystals from mother liquor: *a*, *b*, liquid drains through channels; *c*, *d*, liquid drains through hole.

known as ultracentrifuges, which operate at 20,000 rev/min or more. Because of the large centrifugal forces involved, these instruments should be handled with extreme caution. The weight of the oppositely positioned centrifuge tubes must be exactly balanced. Polyethylene and cellulose nitrate centrifuge tubes are preferred, since glass tubes may not stand the enormous pressures. Neuhoff [149, 150] has described a technique to centrifuge small capillaries containing biological fluids without breaking at 60,000 to 70,000 rev/min by placing them in a medium of approximately the same density as the sample. A simple apparatus for centrifuging 10 μl to 1 ml of liquid up to 20,000 rev/min has been constructed; it consists of a suitable-size borosilicate-glass capillary tube, containing the sample, sealed with Parafilm and supported in a centrifuge tube filled with fine sand, with a cotton-wool wad (3 mm thick) placed 3 to 4 mm from the bottom of the tube. Schumaker and co-workers [151] have described a pressure chamber in which conventional ultracentrifuge cells may be filled with nitrogen at pressures of up to 135 atm.

2. Separation by Density-Gradient Centrifugation

In density-gradient centrifugation [152] the mixture of particles is suspended in a medium of low specific gravity and layered on top of a preformed density gradient. When centrifuged, the components form discrete zones. Hence this technique is also called zonal centrifugation. Separation is primarily achieved according to the sedimentation rate, which depends predominantly on the particle size.

If a density gradient is used that covers the complete range of all fractions, particles reach levels of equal density after a sufficient period of time. In ultracentrifuges at sufficiently high speeds the density gradient does not need to be preformed, but is formed by the centrifugal field itself.

Eichenberger [152] has reviewed the applications of this method to the analytical and preparative separation of macromolecules and biological materials. Unlike differential centrifugation, zonal methods allow the quantitative recovery of the separated constituents. Zonal separation in centrifuge tubes with swinging-bucket and fixed-angle rotors and in sector cells has been used for the determination of physical characteristics, such as molecular weight, sedimentation coefficients, density, and heterogeneity.

Bonner and co-workers [153] have separated minerals by centrifuging them in a medium of continuous density gradient (1.8 to 2.8 g/ml) prepared by mixing 10% ethanolic poly(vinyl-pyrolidinone) with 1,1,3,3-tetrabromacetone in the required varying proportions. The sample (about 0.15 g), pretreated to remove air and water and to prevent cementation, is ultrasonically dispersed in 5 ml of low-density solution ($d = 1.8$) and then layered onto 35 ml of the density-gradient solution. After centrifuging for 16 hr at 1500 rev/min and 20°C, the separated zones are collected and mineral components are identified by X-ray analysis.

Anderson and co-workers [154] have constructed a zonal-centrifuge rotor for center or edge unloading of the density gradient. Siakotos and Wirth [155] have devised a method for the mass production of density gradients. A gradient pump is modified to allow the mixed outflow to run into a special loading head fastened to the 12-place head of a centrifuge and rotating at 1200 rev/min. Centrifuge tubes are thus loaded under increased gravity to stabilize the gradient being formed. Morton and Hirsch [156] have described a high-resolution apparatus for the analysis and fractionation of density-gradient preparations with a volume of less than 5 ml (Fig. 3.46). The ultracentrifuge is fitted directly under the cell

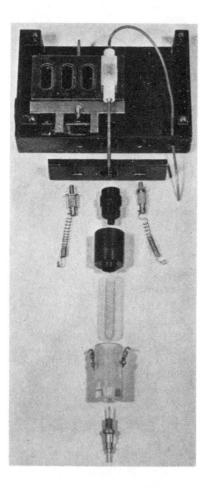

Fig. 3.46. Components of apparatus of Morton and Hirsch [156] for analysis and fractionation of density-gradient preparations. (Courtesy *Anal. Biochem.*)

compartment of a spectrophotometer. A dense solution from a syringe pump displaces the sucrose gradient solution upwards through a cap, then through a vertical straight flow-path cell, and, after measurement, downwards to a fraction collector. Manley and Murray [157] have proposed a method that combines sedimentation (see Section B) and density-gradient centrifugation.

REFERENCES

1. J. C. Giddings (Ed.), *Separation Science*. Marcel Dekker, New York, 1966—.
2. E. S. Perry and C. J. van Oss (Eds.), *Separation and Purification Methods*. Marcel Dekker, New York, 1973 -.
3. B. L. Karger, L. R. Snyder, and C. Horvath, *An Introduction to Separation Science*. Wiley, New York, 1973.
4. E. W. Berg, *Physical and Chemical Methods of Separation*. McGraw-Hill, New York, 1963.
5. A. Paris, *Indus. Chim.*, **52**, 353 (1965)
6. J. Y. Hwang and L. M. Sendonato, *Anal. Chem.*, **42**, 744 (1970)
7. H. F. Bell, *Anal. Chem.*, **45**, 2296 (1973).
8. Y. Yamazaki, *Jap. Anal.*, **19**, 187 (1970)
8a. D. H. Bollman, *Anal. Chem.*, **44**, 887 (1972).
9. E. Smith, *J. Assoc. Off. Anal. Chem.*, **53**, 603 (1970)
9a. H. B. Klevens, *Chem. Rev*, **47**, 1 (1950).
10, N. Brock, H. Druckery, and H. Hamperl, *Arch. Exp. Pathol. Pharmakol.*, **189**, 709 (1938)
11. L. F. Fieser and M. S. Newman, *J. Amer. Chem. Soc.*, **57**, 1602 (1935).
12. M. Stuchlik, L. Krasnec, and I. Csiba, *J. Chromatogr.*, **30**, 543 (1967).
13. M. Stuchlik and L. Krasnec, *J. Chromatogr.*, **36**, 522 (1968).
14. M. Stuchlik, I. Csiba, and L. Krasnec, *Czech. Farm.*, **18**, 91 (1969); **16**, 123 (1967).
15. I. Csiba, L. Krasnec, and M. Stuchlik, *Czech. Farm.*, **17**, 28 (1968).
16. B. Durek-Kluczykowska and L. Krowezynski, *Dissnes Pharm. Warsz.*, **20**, 221, 229 (1968).
17. Tekmar Co., Cincinnati, Ohio.
18. C. Golden and E. Sawicki, reported at Am. Chem. Soc. National Meeting, Dallas, Tex., Apr., 1973.
19. A. J. Jackson, L. M. Michael, and H. J. Schumacher, *Anal. Chem.*, **44**, 1064 (1972)
19a. L. Wheeler and A. Strother, *Anal. Biochem.*, **53**, 42 (1973)
20. S. G. Gibbins, *Anal. Chem.*, **43**, 1348 (1971).
20a. T. Urbanyi, W. T. Brunskill, and M. Lin, *J. Assoc. Off. Anal. Chem.*, **56**, 1069 (1973).
21. R. S. Tipson, "Crystallization and Recrystallization," in A. Weissberger (Ed.), *Technique of Organic Chemistry*. 2nd ed., Vol. III. Wiley-Interscience, New York, 1966.

22. A. Lüttringhaus, "Krystillisieren," in E. Müller (Ed.), *Houben-Weyl Methoden der organischen Chemie*, Vol. I, Part 1, p. 343. Thieme, Stüttgart, 1958.

23. J. A. Riddick and E. E. Toops, Jr., "Organic Solvents," in A Weissberger (Ed.), *Technique of Organic Chemistry*, Vol. VII, p. 40, Wiley-Interscience, New York, 1970.

24. H. G. Cassidy, *Adsorption and Chromatography*, Wiley-Interscience, New York, 1951, p. 101.

25. Ref. 21, p. 490.

26. Ref. 23, p. 39.

27. O. Fuchs, "Zwischenmolekulare Kräfte," in E. Müller (Ed.), *Houben-Weyl Methoden der organischen Chemie*, Vol. III, *Part 2, p.* **399.** Thieme, Stüttgart, 1955.

28. A. J. Parker, *Quart. Rev. (Lon.)*, **16**, 163 (1960).

29. R. S. Kittila, *Dimethylformamide, Chemical Uses*. DuPont, Wilmington, Del., 1967.

30. N. Kharasch (Ed.), *Quarterly Reports on Sulfur Chemistry*, Vol. 3, No. 2. Intra-science Research Foundation, Santa Monica, 1968.

31. I. M. Hais and K. Macek (Eds.), *Handbuch der Papierchromatographie*, p. 125. Fischer, Jena, 1963.

32. V. Horak and J. Peeka, in *Mechanisms of Reactions of Sulfur Compounds*, Vol. 4, p. 43, Intra-science Research Foundation, Santa Monica, 1969.

33. J. Fajkos, private communication.

34. Ref. 23, p. 279.

35. A. A. Benedetti-Pichler, *Identification of Materials*, p. 44. Springer-Verlag, New York, 1964.

36. N. D. Cheronis and T. S. Ma, *Organic Functional Group Analysis*. Wiley, New York, 1964.

37. F. T. Weiss, *Determination of Organic Compounds*. Wiley, New York, 1970.

38. W. Swietoslawski, *Ebulliometric Measurement*. Reinhold, New York, 1945.

39. J. Nemec, *Chem. Listy*, **56**, 1196 (1962).

40. V. Horak, *Chem. Listy*, **54**, 723 (1960).

41. M. Svoboda, *Chem. Listy*, **53**, 31 (1959).

42. G. Perold, *Mikrochim. Acta*, **1959**, 251.

43. W. F. Cockburn, *Can., J. Chem.*, **29**, 715 (1951).

44. M. Martin-Smith, *Lab. Prac.*, **7**, 572 (1958).

45. H. Shiba, *Anal. Chem.*, **26**, 943 (1954).

46. V. Horak, *Chem. Listy*, **48**, 616 (1954).

47. T. C. Owen, *J. Chem. Soc.*, **1964**, 3486.

48. J. H. Cannon, *J. Assoc. Off. Agric. Chem.*, **38**, 844 (1955).

49. C. Kowala, *Chem. Ind.*, **1966**, 1029.

50. A. B. Carel, *Lab Prac.*, **19**, 1239 (1970).

51. J. P. Friedrich, *Anal. Chem.*, **33**, 974 (1961).

52. C. Feldman and J. Y. Ellensburg, *Anal. Chem.*, **29**, 1557 (1957).

53. W. F. Bowers and R. H. Haschemeyer, *Anal. Biochem.*, **25**, 549 (1968).

54. A. Stajner, *Chem. Listy*, **65**, 653 (1971).

55. J. Soudek, *Chem. Prum.*, **21**, 241 (1971).

56. G. F. Wright, *Can. J. Technol.*, **32**, 250 (1954).

57. N. Kajola, *Acta Chem. Scand.*, **8**, 698 (1954).

58. Metro Scientific, Inc., Farmingdale, N. Y.

59. B. B. Graves and B. F. Vincent, *Chem. Ind.*, **1962**, 2137.

60. B. Loev, K. M. Snader, and M. F. Kormendy, *J. Chem. Educ.*, **40**, 426 (1963).

61. J. Janak, *J. Chromatogr.*, **16**, 491 (1964).

62. N. D. Cheronis, A. R. Ronzio, and T. S. Ma, *Micro and Semimicro Methods*, p. 380. Wiley-Interscience, New York, 1954.

63. C. M. Marberg, *J. Am. Chem. Soc.*, **60**, 1509 (1938).

64. T. S. Ma and D. Spiegel, *Microchem. J.*, **10**, 62 (1966).

65. H. H. Appel and P. A. Romero, *Chem. Ind.*, **1970**, 92.

66. A. Bhati, *Lab. Prac.*, **15**, 1141 (1966).

67. J. Nemec, British Patent, 1,156,115 (1968).

68. *M. & B. Lab. Bull.*, **3**, 26 (1958).

69. A. W. Schrecker, *Anal. Chem.*, **29**, 1113 (1957).

70. L. N. Pino and W. S. Zehrung, *J. Chem. Educa.*, **31**, 476 (1954).

71. A. H. Thomas Co., Philadelphia, Penn.

72. L. Kofler and A. Kofler, *Mikromethoden zur Kennzeichnung organischer Stoffe und Stoffgemische*. Universitätsverlag, Innsbruck, 1948.

73. M. Kuhnert-Branstätter, in N. D. Cheronis (Ed.), *Microchem. J., Symp. Ser.*, Vol. 2, p. 221. Interscience, New York, 1962.

74. K. Kunze, *Tex.-Prax.* **11**, 160 (1956).

75. E. Kassau, Dtsch. *Apoth.-Ztg.*, **111**, 1197 (1971); **110**, 1375 (1970).

76. E. Kassau, Detch. *Apoth-Ztg.*, **109**, 1290 (1969).

77. H. Weisz and G. Schapky, *Mikrochim. Acta*, **1967**, 310.

78. B. Baehler, *Helv. Chim. Acta*, **45**, 309 (1962).

79. J. Fawkes, R. O. Thomas, and L. Fishbein, *J. Chromatogr.*, **31**, 576 (1967).

80. M. Behrens and A. Fischer, *Naturwiss.*, **41**, 13 (1954).

81. T. H. Bates, *Chem. Ind.*, **1958**, 1319.

82. G. Schmidt, *Mikrochim. Acta*, **1959**, 406.

83. W. H. Melhuish, *Nature*, **184**, 1933 (1959).

84. Y. Shigetomi and K. Yoshizumi, *J. Chem. Soc. Jap., Pure Chem. Sect.*, **89**, 530 (1968).

85. Y. Shigetomi and S. Yamamoto, *Jap. Anal.*, **17**, 1477 (1968).

86. B. Flaschenträger, S. M. Abdel-Waheb, and G. Habib-Labib, *Mikrochim. Acta.*, **1957**, 390.

87. K. Gosling and R. E. Bowen, *Anal. Chem.*, **45**, 1574 (1973).

88. E. W. Berg and F. R. Hartlage, Jr., *Anal. Chim. Acta*, **33**, 173 (1965); **34**, 46 (1966).

89. K. P. Reed and E. W. Berg, *Anal. Chim. Acta*, **37**, 472 (1967).

89a. W. E. Crosmer, N. C. Thomas, P. H. Tsang, and R. Duckett, *Rev. Sci. Instrum.*, **44**, 837 (1973).

90. G. Broughton, "Solvent Removal, Evaporation, and Drying," in A. Weissberger (Ed.), *Technique of Organic Chemistry*, Vol. VIII, Part 1, p. 831. Wiley-Interscience, New York, 1966.

91. E. Hote-Baudert, *Bull. Rechs. Agronom. Gembloux*, **3**, 689 (1968).

92. K. Adriaenssens, R. Vanheule, D. Karcher, and Y. Mardens, *Clin. Chim. Acta*, **18**, 351 (1967).

93. J. W. Young and G. D. Christien, *Anal. Chem.*, **45**, 1296 (1973).

94. Y. Alon, *Rev. Sci. Instrum.* **40**, 20 (1969).

95. A. F. S. A. Habeeb and C. De Luca, *Chem.-Anal.* **55**, 92 (1966).

96. H. K. Iijima, *Anal. Biochem.*, **23**, 350 (1968).

97. G. Gabella and M. Costa, *Boll. Soc. Ital. Biol. Sper.*, **43**, 1159 (1967).

98. Freeze-Dryer/3, Compagnia Europea Apparecchi Scientifici, Torino, Italy.

99. H. Weiss, *Microanalysis by the Ring-Oven Technique*, 2nd ed. Pergamon, Oxford, 1970.

100. H. Weisz, *Mikrochim. Acta*, **1954**, 140.

101. H. W. Hermance and H. V. Wadlow, "Electrography and Electro-Spot Testing," in W. G. Berl (Ed.), *Physical Methods of Chemical Analysis*, Vol. 2, p. 156. Academic, New York, 1951.

102. P. W. West and S. L. Sachdow, *J. Chem. Educ.*, **46**, 96 (1969).

103. C. J. van Niewenburg, and J. W. L. van Ligten, *Qualitative Chemische Analyze*, p. 271. Springer-Verlag, Wien, 1959.

104. P. W. West, in A. C. Stern (Ed.), *Air Pollution*, Vol. II, 2nd ed., p. 147. Academic, New York, 1968.

105. F. Fiegl and V. Anger, *Spot Tests in Inorganic Analysis*, 6th ed., Elsevier, Amsterdam, 1972.

106. F. Feigl, *Spot Tests in Organic Analysis*, 6th ed., Elsevier, Amsterdam, 1960.

107. S. Abe and H. Kikuchi, *Mikrochim. Acta*, **1973**, 615.

108. P. Heath, Anlyst, **90**, 175 (1965).

109. T. S. Ma and A. A. Benedetti-Pichler, *Anal. Chem.*, **25**, 999 (1953).

110. D. T. Burns, *Mikrochim. Acta*, **1964**, 687.

111. H. Scherz, E. Bancher, and K. Kaindl, *Mikrochim, Acta*, **1965**, 255.

112. H. Schlidnecht, *Zone Melting.* Academic, New York, 1966.

113. N. Y. Paar, *Zone Refining and Allied Techniques.* Newness, London, 1960.

114. R. Polland, *Chim Anal.* **53**, 759 (1971).

115. W. G. Pfann and H. C. Theurer, *Anal Chem.*, **32**, 1574 (1960); *Science*, **135**, 1101 (1962).

116. W. R. Wilcox, *Chem. Rev.*, **64**, 187 (1964).

117. A. Gauman, *Chimia*, **18**, 300 (1964).

118. H. Schildknecht and H. Vetter, *Agnew Chem.* **71**, 723 (1959).

119. Zone Refiner, Fischer Scientific Co., Pittsburgh, Penn.

120. E. T. Knpyl and K. Zielenski, *J. Chem. Educ.*, **40**, 352 (1963).

121. J. K. Kennedy and G. H. Moates, *Rev. Scien. Instrum.*, 37, 1530 (1966).

122. H. Schildknecht, *Chemia*, **17**, 145 (1963); *Angew. Chem.*, **73**, 612 (1961).

123. A. Martinek, *Mikrochim. Acta.*, **1971**, 877.

124. H. Schildknecht and J. Breiter, *Chemztg.-Chemappar.*, **94**, 3 (1970).

125. H. Schildknecht and J. E. Powers, *Chemztg.-Chemappar.*, **90**, 135 (1966).

126. H. Schildknecht, J. Breiter, and K. Maas, *Sep. Sci.*, **5**, 99 (1970).

127. K. Maas and H. Schildknecht, *Z. Anal. Chem.*, **236**, 451 (1968).

128. H. Schildknecht and J. Breiter, *Chemztg.-Chemappar.*, **94**, 81 (1970).

129. H. Schildknecht, V. Reimann-Dubbers, and K. Maas, *Chemtg.-Chemappar.*, **94**, 437 (1970).

130. H. Schildknecht and A. Mannl, *Angew. Chem.*, **69**, 634 (1957).

131. H. Schildknecht, G. Rauch, and F. Schlegelmilch, *Chemztg.*, **83**, 549 (1959).

132. V. I. Lukhovitskii, Y. A. Chikin, and V. L. Karpov, *Dokl. Akad. Nauk S.S.S.R.*, **176**, 1075 (1967).

133. J. W. Axelson and W. C. Streib, in a Wciszberger (Eds.), *Technique of Organic Chemistry*, Vol. III, Part II, p. 199. Wiley-Interscience, New York, 1957.

134. H. Rumpf and T. Lange, in E. Müller (Ed.), *Houben-Weyl Methoden der organischen Chemie*, Vol I, Part 2, p. 31. Thieme, Stüttgart, 1959.

135. Polyethylene sieves, Metro Scientific, Inc., Farmingdale, N Y.

136. Millpore Corp., Bedford, Mass.

137. R. W. Hannah and J. L. Dwyer, *Anal. Chem.*, **36**, 2341 (1964).

138. K. R. Spurny and J. P. Lodge, Jr., *Coll. Czech. Chem. Comm.*, **33**, 3679, 4385 (1968).

139. K. R. Spunry, J. P. Lodge, Jr., E. R. Frank, and D. C. Sheesley, *Environ. Sci. Technol.*, **3**, 453, 464 (1959).

140. K. R. Spunry and E. Wiesner, British Patent 1,222,638 (1965).

141. W. L. Chiou and L. D. Smith, *J. Pharm. Sci.*, **59**, 843 (1970).

142. H. J. M. Bowen, *J. Chem. Soc., A*, **1970**, 1082.

143. Ref. 134, p. 41.

143a. J. C. Giddings, F. J. F. Yang, and M. N. Myers, *Anal. Chem.* **46**, 1917 (1974).

144. B. Keil, *Laboratory Technique in Organic Chemistry* (in Czech), p. 67. CSAV Publisher, Praha, 1963.

145. A. M. Ambler and F. W. Keith, Jr., in A. Weissberger (Ed.), *Technique of Organic Chemistry*, Vol. III, Part I, p. 563. Wiley-Interscience, New York, 1966.

146. Ref. 62, p. 31.

147. T. S. Ma and J. M. Tien, *Microchem. J.*, **2**, 254 (1958).

148. National Scientific Co., Cleveland, Ohio.

149. V. Neuhoff, *G.I.T.*, **13**, 86 (1969).

150.   V. Neuhoff, *Arzneim.-Forsch.*, **18**, 629 (1968).
151.   V. N. Schumaker, A. Wlodawer, J. T. Courtney, and K. M. Decker, *Anal. Biochem.*, **34**, 359 (1970).
152.   W. Eichenberger, *Chima*, **23**, 85 (1969).
153.   W. P. Bonner, T. Tamura, C. W. Francis, and J. W. Amburgey, Jr., *Environ. Sci. Technol.*, **4**, 821 (1970).
154.   N. G. Anderson, C. T. Rankin, Jr., D. H. Brown, C. E. Nunley, and H. W. Hsu, *Anal. Biochem.*, **26**, 415 (1968).
155.   A. N. Siakotos and M. E. Wirth, *Anal. Biochem.*, **19**, 201 (1967).
156.   B. E. Morton and C. A. Hirsch, *Anal Biochem.*, **34**, 544 (1970).
157.   T. R. Manley and B. Murray, *Br. Polym. J.*, **4**, 291 (1972).

CHROMATOGRAPHIC AND ELECTROPHORETIC TECHNIQUES

I. GENERAL

A. CHARACTERISTICS AND BASIC PRINCIPLES OF CHROMATOGRAPHIC SEPARATION

Chromatographic separation [1–5] is based on the differential migration of the various compounds of a mixture in a specific environment. It is interesting to note that all chromatographic methods are microtechniques in their conception. Two reasons may be presented: (1) Efficient chromatographic separation requires a very high ratio between the number of molecules in the substrate and that in the stationary phase and the carrier; this results in low capacity of the chromatographic methods. (2) Very sensitive detection techniques are available; thus only minute amounts of the separated components are need in the experiment. For example, it is simple to separate microgram quantities of mixtures by paper or thin-layer chromatography, but separation on the macro scale becomes extremely laborious. Whereas the analytical gas chromatograph is at present a common piece of laboratory equipment for the separation of volatile compounds, techniques for injecting large samples into the preparative apparatus in order to obtain milligram or larger amounts of products are still under investigation [6].

Generally speaking, chromatographic separation is a process in which molecules (or ions) are distributed between a stationary phase and a mobile phase, and migrate in the direction of the flow with a certain characteristic velocity. The stationary phase may be solid or liquid. If it is the latter, the liquid is fixed on the surface of an inert solid adsorbent. The mobile (dynamic) phase is either liquid or gaseous. Partition (distribution between two liquids, or between gas and liquid), or adsorption (on a solid surface), or sometimes both phenomena simultaneously are involved in the separation process. The partition process is characterized by (1) symmetrical distribution curves, (2) the mobility of each compound being independent of the presence of other substances, and (3) low capacity of the system. Contrastingly, the adsorption process is characterized by (1) more or less asymmetrical distribution curves, (2) the

139

mobility of each component being affected by the presence of other compounds because of the competition for the active sites of the adsorbent, and (3) higher capacity than the partition system.

Besides adsorption and partition, other principles are utilized in chromatographic separation methods. For example, molecular dimensions and shapes are the dominant characteristics for the separation process that takes place in gel permeation chromatography (on molecular sieves, zeolites, Sephadox gels, etc.). Coulombic interactions are involved in ion-exchange chromatography. Interactions due to hydrogen bonding are the special feature of chromatography on polyamides. Complexation is the basis of certain chromatographic separation techniques employing selectively reactive adsorbents.

B. HISTORICAL

The technique of differential migration has been used for the separation of chemical species for over a century. General adoption of chromatographic separation in the laboratory, however, began less than three decades ago. Perusal of the old chemical literature will reveal that on special occasions the flow of solutions through porous sorptive media such as paper and soil was employed for the resolution of certain mixtures, and the methods were called capillary analysis, breakthrough analysis, retardation analysis, selective filtration, etc. The petroleum chemists were using this technique prior to 1900, and Day of U.S. Geological Survey published a paper in 1897 describing the fractionation of a crude oil by forcing it through a column of powdered limestone [7].

The discovery of chromatography is generally credited to Tswett, a botanist who succeeded in separating leaf pigments by passing the green solution through a column of adsorbents [8]. Surprisingly, whereas Tswett's findings were published in 1903, his method remained unrecognized for nearly thirty years until Kuhn and other workers utilized it extensively for the investigation of carotenoids and various natural products [9]. Then Martin and Synge developed partition chromatography in 1941, leading to the popularity of separation on filter paper [10]. Gas chromatography had a modest start in the 1950s, followed by rapid expansion. Thin-layer chromatography became popular in the 1960s, whereas column chromatography has been revived in recent years in form of high-speed, high-resolution liquid chromatography.

It is somewhat unfortunate that the term "chromatography" has been accepted in the field of separation science to represent all separation processes that involve differential migration. This word is derived from the Greek word chroma ($\chi\rho\omega\mu\alpha$), meaning color. Historically speaking, colored compounds were separated and color tests were used in the detection of the separated components, but these techniques are equally applicable to colorless materials. Furthermore, separation in the gas chromatograph is not concerned with colored species and does not use any color detection methods.

C. MYRIAD VARIETIES OF CHROMATOGRAPHY

Chromatography being defined so very broadly as described above, it is understandable that there appear in the current literature myriad variations of chromatographic separation processes. Thus, based on the principles of adsorption and partition, combinations of different stationary and mobile phases produce the following types:

1. Liquid-solid adsorption chromatography
2. Gas-solid adsorption chromatography
3. Liquid-liquid partition chromatography
4. Gas-liquid partition chromatography

When the separation is dependent on partition, the adsorption phenomenon should be excluded as much as possible, since it operates as a disturbing factor in the system. This is usually achieved by employing highly indifferent stationary carriers (e.g., Celite, cellulose) for the liquid phase.

From the viewpoint of apparatus and experimental procedure, adsorption and partition chromatography can be classified into a number of categories, such as

1. Thin-layer chromatography
2. Paper chromatography
3. Classical adsorption liquid-column chromatography
4. Modern high-pressure liquid chromatography
5. Dry column chromatography
6. Partition liquid-liquid column chromatography
7. Gas chromatography

These and several other methods that are based on the electronic structures and chemical properties of the separated species are discussed in detail in subsequent sections. It should be mentioned that the list of chromatographic techniques is still expanding. For example, there are countercurrent chromatography [11], centrifugal chromatography [12], mass chromatography [13], plasma chromatography [14], affinity chromatography [15], thin-film chromatography [16], molecular-sieve thin-layer chromatography [17], hotplate chromatography [18], phosphate-induced protein chromatography [19], and so on. Haber [20] has developed a differential migration method which is based on the use of high voltages to control certain electrochemical properties of molecules and their environment, and called it the electromolecular propulsion technique.

There are fundamental differences between chromatographic operations using a liquid mobile phase and those using a gaseous mobile phase. Even among the methods employing a liquid phase, however, there are variations in technical concepts and applications. Different degrees of efficiency, sophistication, and automation can be achieved, depending on the chromatograph and chromatographic aids available.

Ever since the general acceptance of chromatography in the chemical laboratory, there have been continuous developments in chromatographic techniques. Symposia are frequently organized. For instance, in 1974 the 13th Annual Meeting on the Practice of Chromatography was held in October in Philadelphia, Pennsylvania by the American Society for Testing and Materials, and the 9th International Symposium on Advances in Chromatography took place in November in Houston, Texas.

D. SELECTION OF CHROMATOGRAPHIC TECHNIQUES

The selection of the correct chromatographic technique for a particular case is a complex problem and depends on many factors, such as the nature of the individual components in the mixture, the technical level of the laboratory, the efficiency and speed required, whether routine operation or research separation, and so on. Besides, a decision should be made between chromatographic and nonchromatographic methods. For instance, it would be unwise to use chromatography if the same effect could be achieved by means of simple recrystallization or partition between two liquids in a separatory funnel. On the other hand, experience has shown that a simple elution through a short adsorption column causes the working substance to crystallize after the solution is evaporated even if it failed to crystallize before. Combination of chromatographic and nonchromatographic techniques often can facilitate the separation of complicated mixtures. This is particularly true of mixtures containing compounds with large differences in polarity. Thus, partial separation can be first performed with a nonchromatographic technique into two or more groups of compounds with similar polarity, followed by final resolution into individual compounds of each group with the aid of chromatography.

Within one type of chromatographic technique, the experimental conditions vary with the working material to be fractionated. In general, the polarity of the molecules is of chief concern, both in adsorption and in

partition chromatography. The polarity of a compound can be characterized by its dipole moment, or for simple molecules, by the C_n-versus-
(heteroatom)$_m$ ratio. The polarity variation with respect to functional
groups can be expressed roughly in the following series:

R—H [alkane < alkene < aromatic] < R—Hal < R—COOR' < R$_2$CO
< RCHO < R—OH [alcohol < phenol] < R—NH$_2$ < R—COOH

It should be noted, however, that this sequence of increasing polarity
represents only the first approximation in the differentiation of the various
types of molecules in their interactions with the active centers of adsorbent
or in the solvation process. In the case of polyfunctional molecules and
cyclic structures, other principles also participate. Factors such as flexibil-
ity and rigidity of the molecular structure, spatial arrangement, proximity
of functional groups for intramolecular chelation, and so on, add to the
multiparameter character of the interactions in the system.

Efforts have been made to elucidate the theoretical aspects of chromatography in order to facilitate the selection of the proper method and
experimental conditions, although these ends are not yet in sight. Keller
[21] has reviewed the selectivity and polarity parameters on the basis of
the equation of general migration proposed by Martin [22] in 1947. The
migration parameter R, defined as the fraction of any time period an
average molecule spends in the mobile phase where it moves with the
fluid carrier velocity, depends on the three-way interactions between the
solute, stationary, and mobile phases. The various chromatographic systems exploit different emphases on one or more of these interactions.
Hence it is difficult to predict the three-way interaction on the basis of
the properties of the isolated entities or combinations of two of them
without some regard for the interactions important in each system. Rouser
[23] has studied the elution selectivity principles as applied to quantitative liquid-column and thin-layer chromatography. Majors [24] has
reported on the effect of particle size on column efficiency in liquid-solid
chromatography.

Ettre [25] has summarized the factors affecting the speed of gas-
chromatographic separation and the ways to predict the retention time in
a given system. Haken [26] has studied retention prediction with respect
to molecular structure. Maynard and Greeshka [27] have investigated the
effect of dead volume on the efficiency of a gas chromatograph. Rogers
and co-workers have published methods for predicting the resolution [28]
of mixtures of n-alkanes, as well as the variation in peak shape resulting
from changes in concentration and in the overlap of two peaks [29].

Farrell and Pescok [30] have presented a computer-generated table for
the selection of chromatographic column systems to separate organic

bases. Underhill and co-workers [31] have developed a procedure for calculating spatial movements of a velocity-programmed chromatographic column. Bowen and co-workers [32] have evaluated high-precision sampling techniques for chromatographic separations. Chromatographic input peak profiles from computer-controlled sampling values are characterized in terms of their precision and statistical moments. It should be mentioned that high-precision sampling in gas and liquid chromatography is a prerequisite for the development of new techniques for the measurement of fundamental parameters in chromatographic systems and for quantitative analysis.

E. USES OF CHROMATOGRAPHY

Notwithstanding the empirical nature of the experimental procedures, chromatography has found very wide applications in chemistry, biochemistry, and biology; in applied sciences such as medicine, toxicology, and forensic science; in the food, cosmetics, pharmaceutical, and petroleum industries; and elsewhere. The American Society for Testing and Materials has been holding annual meetings on the "Practice of Chromatography" for over a decade [33].

Chromatographic separation of the working material is performed for the purpose of either microanalysis or micropreparation. The recovery of the separated compounds is usually not required in analytical applications, where sharp separation and quantitative measurement are the main objects. In contrast, quantitation is seldom needed for preparative purposes, since the chief interest is in separating one or more components, as cleanly as possible, from the mixture and obtaining maximum yields of the desired compounds. Obviously chromatographic separation must be accompanied by a suitable detection device to locate the positions of the respective separated compounds. Some detection techniques are applicable to several types of chromatography [34]; others are restricted to one category of chromatographic system (e.g., gas chromatography [35, 36]). The detection device may be nondestructive (e.g., fluorescence, refractive index) or destructive (e.g., H_2SO_4 spray, flame ionization). Understandably, nondestructive methods are chosen if possible for preparative chromatography. If such a method is not available for the particular system, only a small fraction of the separated compound should be exposed to the detection device.

Sometimes it is advantageous to combine two chromatographic procedures in order to achieve the desired separation of the complex working material. For example, Hesselberg and Johnson [37] have analyzed pesticides in fish by (1) blending the dry sample with 2 g of $NaSO_4$, placing the mixture in a column, and eluting with a suitable solvent, (2) using partition column chromatography on Florisil to clean up the eluate, and (3) determining the pesticides by gas chromatography. Brewington and co-workers [38] have isolated steroids from milk using ion-exchange chromatography, passed the extract through 1 g of alumina in a pipet, and analyzed the eluate in a gas chromatograph. Fritz and Latwesen [39] have developed a sequential scheme for the quantitative separation of 27 different metal ions. Five different chromatographic and ion-exchange columns are employed to separate the metals into groups; the metal ions are then eluted separately from each column by means of effective eluents. Cram and Chester [40] have described the coupling of high-speed plasma chromatography with gas chromatography. In the experiments for preparative chromatographic separations, it is a common practice to monitor the separated compounds with the aid of an analytical chromatographic procedure.

The scope of a particular chromatographic technique depends both on the special features of the method and on the characteristics of the compounds to be separated. Thus, insolubility of a compound in the liquid system can be a limitation on its chromatographic separation with a liquid mobile phase. However, owing to the dynamic character of the method, it may be possible to separate compounds that, according to the static solubility test, are only slightly soluble. Volatility of compounds is a requirement for gas chromatography; hence it is a general practice to increase the volatility of large molecules by employing high temperatures, and to enhance the volatility of compounds containing polar groups through derivatization to mask the polar functions. In contrast, volatility of the working material can be a limiting factor in applying thin-layer or paper chromatography. As a rule, compounds with boiling points below 160°C are not suitable for detection on the thin-layer plate, since they will evaporate from the plate during the developing and drying processes. Therefore, derivatization is utilized to convert these compounds into non-volatile substances (e.g., 2,4-dinitrophenylhydrazones of aliphatic aldehydes). It should be noted that the derived compounds may influence the chromatographic separation in a negative way (i.e., making the components more difficult to separate), and experimental conditions must be optimized in order to circumvent the difficulty.

Derivatization of difficultly detectable compounds by means of "chromatotags" is very useful for facilitating detection and quantification of such compounds in thin-layer chromatography and particularly in high-resolution liquid chromatography employing specific detectors. The chromatotag may be a reagent that carries an intensive chromophore in the visual or ultra-violet region, one that is fluorescent or radioactively labeled, or one that can be electrochemically reduced easily.

Solid, liquid, and gaseous working materials can all be handled by chromatography. Whereas gases can be separated by means of gas chromatography only, liquid and solid samples are amenable to liquid- and gas-phase chromatography without technical difficulty. For this reason, chromatography will be discussed in one chapter; but additional comments will be given in the appropriate sections in other chapters.

F. PREPARATIVE CHROMATOGRAPHY

Preparative chromatographic separation is usually based on the results obtained from analytical chromatographic experiments. Such experiments are necessary in order to optimize the experimental conditions for the preparative runs. It should be noted that results obtained from analytical experiments can be applied to the preparative scale only when identical chromatographic methods are used. Therefore, results from thin-layer chromatography on silica-gel plates cannot be used for preparative elution chromatography on a silica-gel–packed column without certain modifications (e.g., change of the composition of the elution liquid mixture, or alteration of the activity of the silica gel). One of the major problems in transferring analytical results to the preparative scale is the low capacity of almost all chromatographic systems. This factor must be considered carefully if complete separation is to be achieved economically. Overloading a chromatographic system results in tailing, broadening of zones, and lengthening of spots with increasing danger of overlapping. Also, to increase the capacity of the chromatographic system by increasing the width of a column (or the thickness of the layer in thin-layer chromatography) frequently worsens the separation. Generally speaking, for preparative separations, it is preferable to slow down the migration of zones or spots by using less polar solvent systems, or less active adsorbents, than those used in analytical experiments.

II. THIN-LAYER CHROMATOGRAPHY AND PAPER CHROMATOGRAPHY

A. PRINCIPLES AND EQUIPMENT

Thin-layer chromatography (TLC) and paper chromatography are most frequently utilized for the separation of solid mixtures because these methods are simple and inexpensive. These two techniques can be discussed together owing to their close similarity. As a matter of fact, when Martin and co-workers [10] investigated the methods for separating amino acids which led to the development of paper chromatography [41], silica-gel plates were first employed [42]. It should be noted, however, that separation by means of paper chromatography depends primarily on the principle of partition, while the adsorption phenomenon predominates in thin-layer chromatography [42a].

The basic techniques can be described as follows: In TLC, a plate or a sheet of supporting material is coated with a layer of chromatographic material (e.g., silica gel), and the layer is activated. The sample to be separated is spotted close to one end of the plate, and a liquid is made to migrate a certain distance from the end carrying the working material toward the opposite end of the plate. In paper chromatography the support consists of cellulose fibers, which need not be activated.

The supporting media for TLC may be glass, plastic, or aluminum sheets. Glass plates of uniform thickness are usually employed for laboratory-made TLC plates. The most common dimensions are 5×20 cm and 20×20 cm; some commercial plates of the latter size can be used as such or split into four 5×20 cm segments. Recently plastic [43, 44] and aluminum supports have been advocated. The advantages of such plates are their light weight, lesser fragility, easy storage, and convenience for cutting into any size or shape.

The layers of adsorption material can be either compact or loose. In the compact type the pulversized adsorbent is spread over the plate in the form of a thick paste, which, after drying, adheres to the support, and being mechanically stable, can even be held in a vertical position. In the case of a loose layer, the dry powder is simply spread over the plate; hence the layer is mechanically unstable, and the plate can be positioned only slightly off the horizontal level. Precoated TLC plates [46] and sheets [47], which are made by many firms, belong to the compact type.

The dimensions of a TLC plate depends on the kind of work for which the TLC is specifically intended. For some analytical tests and preliminary experiments, 3×10 cm strips may be used. The thickness of an analytical layer is 100 to 250 μm. In preparative TLC for which the capacity of the system has to be increased, large plates with layers several millimeters thick are employed [48]. Normal coating is applicable to layers of 100 μm to 3 mm; for thicker layers, cracking should be prevented by special manipulation and by using special compositions of the adsorption material. Fig. 4.1 shows a commercial spreader that can be used to prepare a layer up to 3 mm thick. Fig. 4.2 illustrates the preparation of a loose layer and the position of the finished plate in the developing tank; a glass rod with two rubber sleeves serves to adjust the thickness

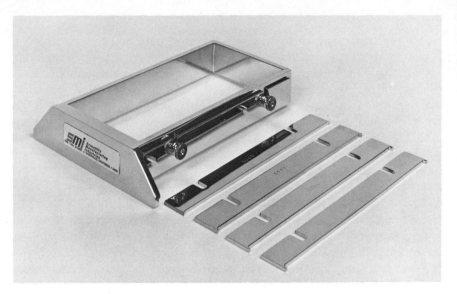

Fig. 4.1. Spreader for preparative TLC, adjustable up to 3-mm thickness [44]. (Courtesy Scientific Manufacturing Industries.)

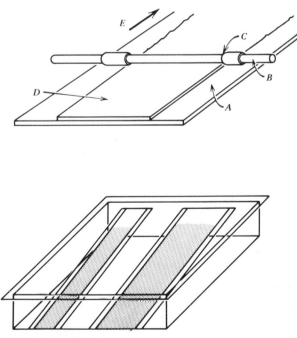

Fig. 4.2. Preparation of loose layer for TLC. Top: Rolling glass rod over pulverized adsorbent. Bottom: Position of TLC plate in developing tank. *A*, glass plate; *B*, glass rod; *C*, rubber tubing which serves a spacer; *D*, alumina or other adsorbent; *E*, direction of the motion of the glass rod.

and width of the layer. Glass plates with elevated edges are commercially available [48a].

A general procedure for preparing TLC plates with a compact layer is as follows: A number of thoroughly degreased (by washing) and dried glass plates are coated with a slurry of the selected composition using a spreading device. The ratio of adsorbent and liquid (water, ethanol, etc.) is given in Table 4.1. For thick layers of over 3 mm, a binder (e.g., 2% CaSO₄) is recommended, even for those adsorbents that normally do not require one [48]. The freshly prepared plates are air dried for 24 hours, and can be used without activation (dehydration) by heating. A drying rack is shown in Fig. 4.3 [49]. A plate scriber [50] can be used to divide the layer into separate segments for the simultaneous separation of many samples, as depicted in Fig. 4.4. An improved scriber has been designed by Herbert and co-workers [50a]. An inexpensive spreader resistant to corrosive slurries has been described by Lowry [51].

Unlike TLC, paper chromatography requires no preliminary work to prepare the paper. The equipment for paper chromatography [52] is shown in Fig. 4.5.

B. SELECTION OF EXPERIMENTAL CONDITIONS

1. Adsorbents

Currently most TLC separations are carried out on silica gel or alumina; the former is frequently used in a compact layer; the latter, in a loose layer. The activity of these adsorbents is related to the degree of dehydration. The water content in the layer is adjusted during the preparation of the TLC plate and depends primarily on the duration and intensity of the exposure to heat. Lowering the water content, within certain limits, results in increased adsorption power and a slowing down of the migration rate. Silica gel of medium activity is obtained after heating at 105°C for 30 min, whereas high activity requires activation for several hours. The use of TLC plates of very high activity is recommended only when the compounds to be separated have low polarity (e.g., hydrocarbons). In general, better separation can be achieved with layers of medium or low activity. This is particularly true for preparative TLC.

TABLE 4.1.   Composition of the Slurry for Spreading on Five Plates 20×20 cm Covered with a Layer of about 300 μm

30 g Silica gel, Woelm TLC	+45 ml water
25 g Silica gel G, Woelm TLC	+50 ml water
35 g Aluminum oxide (basic, acidic or neutral), Woelm TLC	+40 ml water
35 g Aluminum oxide G. Woelm TLC	+40 ml water
5 g Polyamide, Woelm TLC	+45 ml ethanol

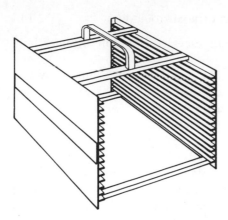

Fig. 4.3. Drying rack for TLC plates [49]. (Courtesy Scientific Manufacturing Industries.)

Fig. 4.4. Plate scriber for dividing TLC layer into 20 areas [50]. (Courtesy The Lab Apparatus Co.)

Fig. 4.5. Equipment for paper chromatography [51]. (Courtesy *Mikrochim. Acta.*)

The common brand of silica gel used in TLC, identified as Silica gel G, contains about 10% anhydrous $CaSO_4$ as binder, and the particles are 5 to 25 μm in diameter. Because of the presence of $CaSO_4$, the slurry of Silica gel G must be applied to the plate within a time limit, after which the mass hardens. Another brand, Silica gel H, has no binder, but it forms a compact layer with the proper solvent.

Alumina for chromatography is classified as basic, neutral, or acidic; the designation refers to the pH of the water extract and is pertinent when certain classes of compounds are separated. While basic and acidic alumina may cause hydrolysis of esters, dehydration, dehydrohalogenation, adol condensation, and molecular rearrangement, neutral alumina of low activity is usually indifferent. Compact layers prepared with alumina usually have lower activity and smaller capacity than those prepared with silica gel. If, instead of TLC alumina, regular alumina for chromatography is used to prepare the loose layers, the efficiency of separation decreases drastically because of the much larger size of the particles (100 μm versus 10 μm).

TLC layers may be modified for specific purposes. For instance, a fluorescent indicator can be mixed with the adsorbent. The presence of this indicator facilitates the detection of many colorless and nonfluorescent compounds. After separation, examination of the TLC plate under a short-wavelength ultraviolet lamp (254 nm) reveals dark spots indicating that the particular compounds have a quenching effect on fluorescence. Layers loaded with scintillators are useful for scanning radiochromatograms [53].

Dohmann [54] has reviewed the selection of adsorbents for TLC. Besides silica gel and alumina, various materials have been used to prepare TLC plates, among which may be mentioned cellulose and its derivatives [55–58], sintered glass powder [59, 60], polyamide [61, 62], polyacrylonitrile [63], ion exchangers [64, 65], and Sephadex [66]. Since the techniques of chromatographic separation on these layers are based on different principles, they are discussed in other sections.

Enhanced selectivity can be achieved by modifying the common adsorbents. Different classes of compounds show different migratory velocity on neutral, acidic, and basic alumina. Pretreatment of silica gel with silver nitrate solution in acetonitrile produces an adsorbent with new properties, such as increased affinity toward ethylenic compounds and certain aromatic structures like polyphenols [66a].

2. Solvent System

The solvent system for TLC can be selected only after careful consideration of the mixture to be separated, the TLC procedure, and the aim of the separation experiment. The mixture for separation is characterized by the number of components and their proportions, the polarity of the individual compounds, and the range of polarities. The TLC procedure is characterized by the type of adsorbent layer and by the technical conditions of the experiment (e.g., vapor-programmed TLC). For a particular TLC procedure, the relationship between the polarity of the substrate molecules, the activity of the adsorbent used, and the polarity of the solvent system is demonstrated by Stahl's triangular indicator [67], shown in Fig. 4.6. The aim of the separation may be the complete quantitative resolution of all compounds, or partial fractionation into groups, or the selective separation of only one component. Massart and DeClercq [67a] have proposed the application of numerical taxonomy techniques to the selection of suitable solvents in TLC.

The migration rate and the efficiency in separating different compounds in the same layer of adsorbent depends primarily on the differences in their polarities. The types of interaction involved are hydrogen bonding, charge transfer, chelation, the dipole-dipole interaction, and so on. In the simplest approach to selecting the solvents, their dielectric constants [68] are used as a measure of their polarity. In such an eluotropic series (Table 4.2), solvents are arranged by increasing dielectric-constant values, representing their increasing polarity and desorption power. It should be emphasized that the numerical value of the dielectric constant characterizes the

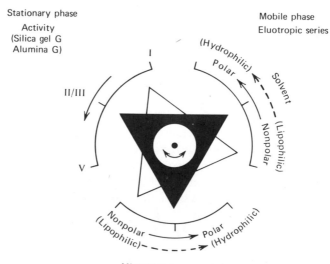

Fig. 4.6. Stahl's indicator of relationship between polarity of substrate and solvent and activity of adsorbent [67]. (Courtesy *Pharm. Rundsch.*)

TABLE 4.2. Solvents and Their Dielectric Constants

Solvent	Dielectric Constant
Hexane	1.88
Benzene	2.29
Ether	4.47
Chloroform	5.2
Ethyl acetate	6.11
Dichlorethane	10.4
Butanol-2	15.5
Acetone	21.5
Ethanol	26
Methanol	31.2

desorption power of the solvent only qualitatively, and may even lead to erroneous predictions. For instance, chloroform and ethyl acetate differ only slightly in dielectric constant, but remarkably in desorption power. Comparing acetic acid ($\varepsilon = 6.15$) with methylene chloride ($\varepsilon = 9.08$), one sees that the former has strong desorbing power, while the latter is very weak. The effect of a specific solvent-substrate interaction [69] on the migration of some compounds can be seen in Table 4.3.

Generally speaking, mixed solvents are more effective than single solvents, owing not only to the possibility of fine adjustment of the desorption power but also to the gradient of the composition of the solvent mixture along the axis of migration of the solution. A summary of the well-tested solvent systems for various classes of organic compounds on silica-gel layers is given in Table 4.4. Probably the most common solvent systems for separating compounds of medium polarity are mixtures of cyclohexane (or chloroform) with ethyl acetate (or methanol) in ratios of 20:1 for the least polar up to 1:1 for the most polar substances. In order to optimize the conditions for preparative TLC, the solvent system is selected after preliminary tests in the analytical scale, varying both the

Table 4.3. Effect of Solvent-Substrate Interaction on Migration on Alumina Layer [69]

Compound	R_F in solvent	
	Chloroform	Diethyl ether
o-Nitrophenol (pKa 7.23)	0.28	0.05
m-Nitrophenol (pKa 8.35)	0.03	0.26
o-Hydroxybenzaldehyde	0.26	0.17
m-Hydroxybenzaldehyde	0.11	0.42
Brucine	0.85	0.08
Cinchonine	0.32	0.45

TABLE 4.4.   Solvent Systems for the Separation of Various Classes of Components

Class of Compounds	Solvent system (volume ratio)[a]	
Hydrocarbons	1.	Pentane
	2.	Cyclohexane
	3.	Cyclohexane (3) : ethyl acetate (1)
	4.	Benzene (2) : ether (1)
Ketones	1.	Hexane (9 : ether (1)
	2.	Petroleum ether (2) : ether (1)
	3.	Cyclohexane (1) : ethyl acetate (1 to 2)
Alcohols	1.	Pentane (10) : ether (1)
	2.	Cyclohexane (1) : ethyl acetate (1 to 2)
Esters, lactones	1.	Petroleum ether (2) : ether (1)
	2.	Cyclohexane (1) : ethyl acetate (1 to 2)
	3.	Ethyl acetate
Enolic diketones	1.	Ethyl acetate
	2.	Ethyl acetate (4) : methanol (1)
Carboxylic acids	1.	Chloroform saturated with 90% formic acid
Amides	1.	Ethyl acetate (5) : methanol (1)
Amines and aminoalcohols	1.	Chloroform (3) : methanol (2) containing 1% of conc. aqueous ammonia
Quaternary ammonium salts	1.	Chloroform (6) : methanol (6) : conc. aqueous HCl

[a] In order of increasing polarity within each class of compounds.

adsorbents and solvent systems. If the activity of the adsorbents decreases, the polarity of the solvent system should be decreased accordingly. Sometimes a solubilizing agent [70] (see Chapter 3, Section I.B) is incorporated in the solvent system to facilitate the migration of the separated species. The use of homogeneous azeotropic solvent mixtures has been reported to aid in the attainment of constant R_F values [71].

Most of the above discussion on solvent selection is applicable to paper chromatography, which is based on the partition principle. It should be noted, however, that two or more liquids of limited miscibility are involved in paper partition chromatography. Normally the polar liquid (usually water) is adsorbed on cellulose fibers and serves as the stationary phase, while the "organic phase" (e.g., butanol–acetic acid–water, 4:1:5) is used for developing the chromatogram. Contrastingly, if a water-repellent support (e.g., silicone-impregnated paper) is employed, the nonpolar organic solvent is the stationary liquid phase. The latter technique, known as "reversed-phase paper chromatography," is suitable for the separation of lipophilic (i.e., water-insoluble) substances.

3. Standardization

The efficiency of TLC separation on adsorbents such as silica gel and alumina depends on many factors. Since the activation procedure is not fully reproducible, and since changes in activity may result from storage conditions, standardization methods have been advocated to monitor the activity of the TLC plates. The method using azo dyes as reference compounds is frequently employed. The data [72] are tabulated in Table 4.5; a higher activity (i.e., stronger adsorption) is designated by a lower number in the Brockmann-Schodder scale. Waksmundzki and Rozylo [73] have proposed the standardization of adsorbent layers in TLC by measurement of the flow rate of the liquid mobile phase. It is shown that under standardized conditions, the relationship between the square of the distance of solvent migration and time is linear, with the slope inversely proportional to the specific surface area of the adsorbent.

C. OPERATION TECHNIQUES

1. Spotting

A spotting line is marked with a fine pencil point on the layer (or chromatographic paper), parallel to the edge of the plate at a distance from it (about 15 mm), so that the spotted material is never in direct contact with the developing liquid. Analytical samples are spotted by means of a microliter pipet [74] (Fig. 4.7) or microsyringe. The spotting should be done in such a manner that the spot does not spread beyond 5 mm in diameter. With dilute solutions, the spotting is repeated after waiting for the solvent to evaporate, preferably with the aid of a hot-air dryer. The sample is usually spotted as a solution, even if the working material is a liquid. Understandably, easily volatile solvents of low polar-

TABLE 4.5.   Standardization of TLC Plates Using Azo Dyes [72]

| | Activity on the Brockmann-Schodder Scale (R_F) | | | |
Dye	II	III	IV	V
Azobenzene	0.59	0.74	0.85	0.95
p-Methoxyazobenzene	0.16	0.49	0.69	0.89
Sudan Yellow	0.01	0.25	0.57	0.78
Sudan Red	0.00	0.10	0.33	0.56
p-Aminoazobenzene	0.00	0.03	0.08	0.19

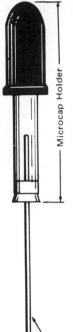

Fig. 4.7. Microliter pipet for spotting [74]. (Courtesy Boleb Inc.)

ity are preferred. Simple [75] and automatic [76–79] applicators have been proposed. Stahl [80] has described a thermal microprocedure to transfer solid material from a capillary tube onto the TLC layer by sublimation (see Chapter 9, Fig. 9.13).

For preparative separation using large samples, the starting line can be loaded with overlapping spots. Alternatively, the solution can be applied in the form of a continuous line, using the device shown in Fig. 4.8, which can deliver 100 to 200 μl over 20 cm by controlling the flow through the mouthpiece. Applicators [81, 82] for preparative TLC and an automatic streaking device [83] have been reported. The manual applicator [85] illustrated in Fig. 4.9 is made of plastic material. It has double rows of slots, each being filled with a precise reproducible volume of solution, and delivers the sample in the form of a line 100×3 mm or 20×3 mm with volume from 5 to 50 μl after touching the starting line. Since the sample is transferred from a plastic trough, this device is recommended for aqueous solutions only, in view of the creeping effect of organic liquids. Samuels [86] has designed an enclosure for rapid TLC spotting. An apparatus for spotting samples in a nitrogen atmosphere is commercially available [87].

One spot on the analytical TLC plate holds less than 0.1 mg of working

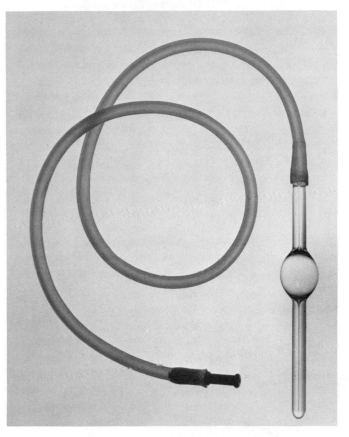

Fig. 4.8. Streaking pipet for preparative TLC. (Courtesy Scientific Manufacturing Industries.)

material, while the starting line of a 20×20 cm plate can be loaded with 1 to 10 mg, and a thick layer up to 100 mg. It should be recognized, however, that the efficacy of TLC separation is dependent on a large absorbent-to-sample ratio (1000 to 10,000). If this ratio is decreased by heavy loading of the plate, the zones are broadened and tailing occurs.

2. Developing

a. Conventional

The TLC plate is placed in a glass vessel containing the solvent system for the development of the chromatogram. As shown in Fig. 4.10, the developing tank is usually a rectangular chamber of heavy glass with a flat ground top and matching glass lid. For developing small plates or

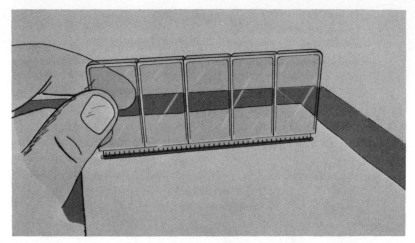

Fig. 4.9. TLC applicator for 5 to 50 μl of solution [85]. (Courtesy Cordis Laboratories.)

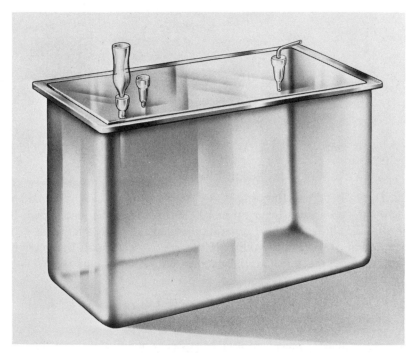

Fig. 4.10. Developing tank for preparative TLC. (Courtesy Scientific Manufacturing Industries.)

strips, glass containers like reagent bottles with a plastic cover are suitable. Chambers for holding up to five 20×20 cm plates are commercially available. Stainless-steel tanks have been constructed [88]. Developing vessels for paper chromatography are shown in Fig. 4.5, and can be used either for ascending or for descending development.

For efficient separation and to provide the highest possible reproducibility of R_F values, thorough saturation of the chamber with the vapors of the liquid system is essential. For this purpose a sheet of filter paper matching the dimensions of the chamber is adhered to its wall. Such provision is particularly important for low-boiling solvents such as petroleum ether or diethyl ether. Pittoni and Sussi [89] have constructed a chamber with provision for admitting a desired gas. Separation in unsaturated chambers by the technique of vapor programmed TLC has been reported [89a].

b. Sandwich Technique

In the sandwich technique for TLC development [90], the layer is placed between two glass plates, which are kept apart by means of spacers. This technique is recommended for handling two or more plates simultaneously and for plates larger than 20×20 cm. For the very large plate (up to 100×20 cm), only the cover plate is needed. The sandwich is held together with clips; for developing it is placed in a trough containing the solvent system and is supported in a nearly vertical position with a proper device (Fig. 4.11). Owing to the narrow spacing between the

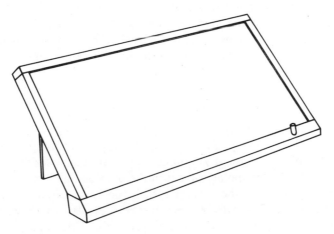

Fig. 4.11. Preparative sandwich chamber for 40×20 cm TLC plates [49]. (Courtesy Scientific Manufacturing Industries.)

adsorbent layer and the cover plate, no special provision for saturation is required.

c. Repetitive and Successive Developing Techniques

In most cases the separation process is completed when the solvent front approaches the upper end of the TLC plate. The movement of the spots system is stopped instantaneously when the plate is removed from the chamber or trough. However, for certain objectives, a variation of the developing process is utilized. Thus, repetitive development is carried out, which is equivalent to separation on a path several times longer than the plate. The spots are caused to migrate slowly (R_F lower than 0.3) by selecting a less polar solvent system and by repeated developments, and better separation is achieved. In preparative TLC, this technique also can be utilized to make the separation more economical by slightly overloading the layer. Further improvement of this version of TLC is represented by the programmed multiple development technique of Perry and co-workers [90a], for which a commercial apparatus is available.

Successive development in two different solvents is carried out as follows: The front of the first solvent is allowed to reach the midpoint of the pathway; after drying the plate, the development is completed in the second solvent. If the first solvent is polar, a reasonable separation of polar compounds in the lower half of the plate is achieved, while the nonpolar compounds are at the front of the solvent. The second development in the nonpolar solvent does not move the polar species but separates the nonpolar species in the upper half of the plate. The same principle can be utilized to form a new narrow starting line for preparative TLC, if the sample is streaked on the starting line by using a nonpolar solvent for dissolving the sample. This is accomplished by dipping the plate into a polar solvent and allowing the front to migrate approximately 10 mm above the upper edge of the original spot.

d. Gradient Developing Technique

The separation of a mixture consisting of a wide range of polarities can be facilitated by applying the gradient developing technique. A solvent system with polarity gradient is obtained by continuously adding the more polar solvent to the less polar solvent in the developing chamber, provided that the homogeneity of the solvent system can be maintained by stirring. For preparative TLC, however, preseparation into two or more fractions with a narrower range of polarities is preferred. The preseparation can be realized by forming mixed zones of low, medium, and high polarity in one chromatographic process. Samples recovered from the respective zones are then treated under specifically controlled conditions to achieve complete separation. In another version of preseparation, the TLC plate is developed successively with solvents of increasing polarity; in each step the least polar species is isolated from the zone that migrates with the solvent front. The limitation of the latter method is that the operative length of the plate is diminished with the removal of each fraction.

e. Two-dimensional Developing Technique

For two-dimensional developing, the sample is spotted on one corner (about 15 mm from the edges) of the TLC plate (or chromatographic paper). After developing with one solvent system, the plate is removed from the tank and the solvent is expelled. The plate is then turned 90 degrees and placed in another solvent system to be developed. Obviously, only one original spot can be treated by this technique on the same plate.

3. Detection

a. Nondestructive Methods

After developing on the adsorbent layer or chromatographic paper and removing the solvent, the locations of the separated species must be revealed by an appropriate method. Procedures that are nondestructive to the compounds are ideal for preparative TLC and convenient for analytical purposes. Thus, colored substances can be seen directly, and fluorescent compounds under ultarviolet light [91] (Figs. 4.12, 4.13), the long-wave ultraviolet being preferred. Conversely, the quenching effect on fluorescent plates is frequently utilized. Radioactive species can be recognized after autoradiography upon contact with a photographic film. An instrument has been described that permits chromatograms to be scanned in an X-ray fluorescence spectrometer for elements such as phosphorus, sulfur, chlorine, or iodine [92].

Exposure of the dry TLC plate to iodine vapors produces dark brown spots on a yellow background, because of the dissolution of iodine mole-

Fig. 4.12. Ultraviolet lamp: long wave (366 nm) or short wave (254 nm) [91]. (Courtesy Laboratory Supplies Co.)

Fig. 4.13. Ultraviolet exposure cabinet [91]. (Courtesy Laboratory Supplies Co.)

cules in compounds with lipophilic character. The intensity of the color decreases with time as iodine evaporates from the plate. It should be noted that, while this test is based on a physical phenomenon, chemical changes may occur if the separated compounds are susceptible to oxidation.

b. Destructive Methods

Destructive methods are those that utilize chemical reagents to react with the separated species to produce visible colors on the layer (or chromatographic paper). Since they are usually used as sprays, these chemicals are called spray reagents. Some have broad applications (e.g., concentrated H_2SO_4 for organic substances); others (e.g., $KMnO_4$-H_3PO_4, ninhydrin, Ag_2O-NH_4OH) are more or less specific for particular functional groups. The latter criteria are more important for analytical purposes than for preparative separation.

When a destructive spray reagent is employed for the detection of zones in preparative TLC, only a small portion of the layer should be exposed. The main part of the layer is covered with a glass plate, leaving the edges to be sprayed. Since the zones do not always migrate exactly horizontally, the spray reagent should reach far enough to locate the position of the separated zones. For recognizing all irregularities of zones, it is useful to observe the TLC plate under ultraviolet light even though the separated compounds are not visible by themselves.

4. Recording

The results of separation of a mixture by TLC (or paper chromatography) are expressed in terms of "rate of flow" (R_F) for the respective components. The quantity R_F is a physical constant that is characteristic of each compound with respect to its rate of migration under the given experimental conditions. Mathematically R_F is expressed as the ratio of the distance of the center of the particular spot to the distance of the front of the solvent, both measured from the starting line (i.e., the position where the mixture was originally placed). Numerically the R_F values range between 0 and 1, and are dimensionless. A transparent overlay (Fig. 4.14) facilitates the measurement of distances on the plate.

It should be emphasized that the experimental conditions comprise all factors of the chromatographic system, such as type and activity of the adsorbent, solvent, temperature, saturation of chamber, and width of the strip. Because the system has many parameters and it is difficult to record and characterize all the factors involved, the reproducibility of the R_F values is limited. Therefore the use of an authentic compound for comparison is recommended. It is also advisable to make a graphic reproduction or photographic record of the chromatogram.

For the best analytical characterization of the compounds, the experimental conditions should be fixed in such a manner that the R_F values range between 0.25 and 0.75, the region in which the chromatographic separation is most efficient. R_F values higher than 0.8 are of little use, while low R_F values can be utilized with the repetitive developing technique. The same criteria hold for preparative TLC separations.

5. Recovery and Quantitation

Quantitative evaluation of a TLC (or paper chromatography) experiment is done by direct reading on the plate or by determining the separated

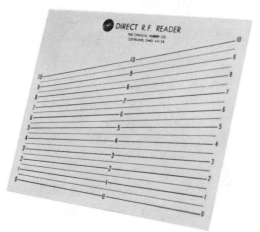

Fig. 4.14. Transparent overlay for measuring distance on TLC plate [50]. (Courtesy The Lab Apparatus Co.)

species after extraction from the respective spots. In the former case, visual comparison of the dimensions and intensity of the spot with a standard plate carrying a series of spots of comparable and known quantities processed under exactly the same conditions is the best and simplest method. A sophisticated scanning densitometer has been proposed to measure to spots at one or several wavelengths *in situ* [93].

For preparative TLC the last operation is the recovery of the desired separated species from the spot or zone. In the case of the compact layer, the adsorbent containing the particular compound is scraped off the plate and collected in an apparatus for extraction. The technique and equipment are discussed in Chapter 7, Figs. 7.6 and 7.7. A simple device to remove adsorbent from a loose layer by suction is illustrated in Fig. 4.15. A commercial instrument [94] for scraping off TLC zones into glass bottles is shown in Fig. 4.16. Desorption of the sample from silica gel or alumina is accomplished by using polar solvents such as methanol, pyridine, and acetic acid. These solvents may be mixed with less polar liquids (e.g., chloroform, benzene).

D. APPLICATIONS OF THIN-LAYER CHROMATOGRAPHY

Whereas the beginnings of thin-layer chromatography are in debate [95, 96], publications on this subject have numbered well over 10,000, not including papers that mentioned the use of TLC for separating certain compounds [96]. The reader is referred to the manuals [97–100] and reviews [101–104] for comprehensive surveys. Selected papers that appeared recently are presented below.

Scott [105] has investigated the theoretical and practical aspects of the stationary phase in TLC. Waksmundzki and Rozylo [106] have com-

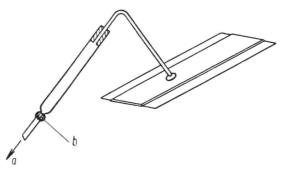

Fig. 4.15.   Removal of adsorbent and sample for loose layer: *A*, to suction; *B*, cotton wad.

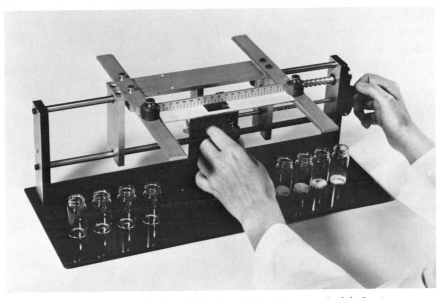

Fig. 4.16. Autozonal scraper for TLC plate [94]. (Courtesy Isolab Inc.)

pared the R_F values of organic compounds of various structures on Silica gel G, Alumina G, and Kieselguhr G using active solvents (electron donors and acceptors) and inactive solvents as mobile phases. Vanhaelen [107] has reported on the factors that affect the rate of saturation of a chromatographic tank, containing or not containing layers of adsorbents of various porosities, and the effect of repetitive developing on R_F values. It has been shown that separation of the components of the solvent system, such as chloroform-methanol (1:1), on silica-gel layers is reduced by equilibration and increased by increasing the thickness of the layer. Hurtubise, Lott, and Dias [108] have compiled the current information on instrumentation for TLC.

Understandably, most of the applications of TLC are in organic chemistry [99] and biochemistry [109]. Steinke and Schmidt [110] have advocated TLC as routine technique in the pharmacy laboratory. Egli [111] has recommended two solvent systems—namely, $CHCl_3–CH_3OH–$ conc. aq. NH_3 (85:14:1) and $CHCl_3–CH_3OH–85\%$ HCOOH (17:2:1)— for the TLC of drugs on Kieselgel F_{254}. Successful separations were achieved for all but seven of the 74 drugs tested. Goenechca [112] has described the preparative TLC of barbituate mixtures for infrared spectroscopy; the method is applicable to biological materials. Bujna and Machovicova [113] have utilized the microsublimation procedure [80] (see Chapter 9, Fig. 9.13) for spotting aromatic drugs, infusions and other galenical formulations on silica-gel layers. From 20 to 50 mg of sample is required. Haywood and Moss [114] have put forward a rapid system for the screening of alkaloids in toxicology.

TLC is a convenient technique for the microchemical investigation of medicinal plants [115] and other natural products [116]. Kolattukudy [117] has employed TLC to analyze the lipids on the surfaces of animals, plants, and insects. The lipids can be isolated from the surfaces by dipping the sample in chloroform for 30 sec. The result with a cabbage leaf is shown in Fig. 4.17. Schwartz and Virtanen [118] have used thin-layer partition chromatography to study the carbonyl compounds in butter oils. Separated as the corresponding 2,4-dinitrophenylhydrazones, 37 carbonyls were found in the butter oil from normal milk, whereas 28 were detected in the milk fat of synthetically fed cows.

A silica-gel layer can be used to carry out organic synthesis on the microgram scale [119]. The reactants are brought together in a spot on the layer. After the completion of the reaction, the products and unreacted starting materials are separated by TLC on the same layer. An example is shown in Fig. 4.18; various amounts of 2,5-dimethoxyphenyldiazonium salt are placed at different spots on the layer, and α-napthol solution is added. The coupling reaction is completed in 2 min, and TLC separation in benzene–ethyl acetate (2:1) takes 45 min [120]. This experiment demonstrates the advantage of using small samples for analytical TLC, and the nonhorizontal migration of the separated zone. Chlorination, nitration, hydrolysis, esterification, acylation, and other reactions can be performed in a similar fashion [121]. A simple and rapid technique for multiple synthesis on a 20×20 cm layer has been published [122].

Quantitative analysis of thin-layer chromatograms has been surveyed by

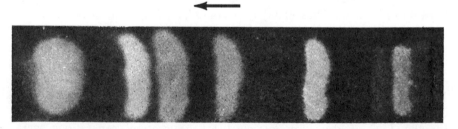

Fig. 4.17. Thin-layer chromatogram of cabbage-leaf surface lipids [117]. Chromatography on Silica gel G with benzene as the developing solvent. The visible components in the order of increasing polarity are hydrocarbons, wax esters, ketones, aldehydes, secondary alcohols, ketols, and primary alcohols. (Courtesy *Analabs, Inc. Res. Notes.*)

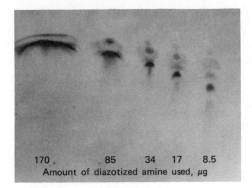

170   85   34   17   8.5

Amount of diazotized amine used, μg

Fig. 4.18. Synthesis of azo dyes on silica-gel layer and separation by TLC.

Novacek [123]. Chahi and Kratzing [124] have constructed a multi-sample applicator for quantitative TLC. Samuels and Fisher [125] have proposed direct quantification by measuring the diameters of the spot with a magnifying instrument, and determining the extinction with a fiber-optic light guide leading to a spectrophotometer. The amount of substance is directly proportional to the product of the largest and the smallest diameter of the spots and the extinction. Kyriakides and Balint [126] have described the quantitative recovery of lipids after argentation chromatography on thin layers. DeMedeiros and Simmons [127] have put forward a procedure for the determination of amino acid ratio by means of TLC. About 80 μl of blood is required, from which 40 μl of plasma is used in the actual analysis. Applications of preparative TLC have been reported by Stutz and co-workers [127a] and by Lerch and Moffatt [127b].

E. APPLICATIONS OF PAPER CHROMATOGRAPHY

The numerous applications of paper chromatography have been documented by Zweig and co-workers [128, 129] and by Hais and Macek [130]. Paper chromatography has been displaced by TLC in popularity, primarily because the latter technique is more rapid; but paper has some advantages over adsorbent layers. For instance, paper is more convenient to handle and can be cut and folded, so that the separated compound is easily recovered. Thus this separation technique has been recently employed in the analysis of natural products [131, 132], plant materials [133], amines in fish [134], and drug metabolites [135]. Ligney and Veen [136] have studied the peak broadening in paper chromatography. Mankinen and Fischer [137] have described an inexpensive device for semiautomated multiple spotting. Milborrow [138] has designed a simple apparatus for holding the paper during equilibration and development, and Wachler [139] a trough for descending paper chromatography with a solvent gradient. According to Bush and Crowshaw [140], the use of $(C_2H_5)_2O\text{-}CH_3OH\text{-}H_2O$ (20:4:1) for the pretreatment of chromatographic paper causes an even and reproducible change in the solvation of the cellulose and renders unnecessary the usual period of equilibration; for chromatograms of 35 to 50 cm, this technique halves the time needed for a single development.

The quantitative analysis of colored spots on a paper chromatogram can be performed by cutting the paper and affixing it to a glass slide, which is placed in a specially constructed aluminum block fitted into a Beckman DU spectrophotometer [141]. Jeroschewski and co-workers [142] have proposed the quantitative evaluation of metal ions on paper chromatograms by amperometric measurements *in situ*. Bush [143] has constructed an apparatus for the rapid photometry of paper chromatograms. Rockland [144] has described a technique for preparative paper chromatography by using a roll of paper; the chromatogram is developed by radial, horizontal migration. Recently spin thimbles specially designed for the elution of separated compounds from the spots of a paper chromatogram have become commercially available [145]. The area of the spot is cut out, rolled into a coil, and inserted into the thimble, which is placed in a centrifuge tube. The spot is moistened with the solvent and then centrifuged.

F. MODIFICATIONS OF TLC PROCEDURES

A number of procedures have been published for miniaturizing the TLC technique described in Section II.C. Instead of the 5×20 cm plates, thin layers can be prepared by dipping microscope slides into a slurry of the adsorbent, thus coating both sides. (Only one side is coated if two slides are held together.) Shevchenko and Favorskaya [146] have described a method to purify silica gel and prepare the layers on slides, which are ready for use after 2 min. Tyman and Higdon [147] have constructed a spreader that permits the coating (up to 1 mm) of TLC supports varying in size from microscope slides to 8×20 cm. DeThomas and co-workers have described a pre-equilibration apparatus for 1×12 cm plates [148], with the technique for descending [149] or horizontal sandwich [150] TLC. Martin [151] has designed a developing rack and tank to extend the benefits of miniature TLC. In order to confine the microspot, Edgar [152] has proposed placing the adsorbent in a scratch (0.2 mm wide and 0.1 mm deep) on the slide and applying the reagent as a smear. The scratch must be uniform throughout its length. For the separation of steroids, the detection limit is 0.01 μg. Lenk and Gruber [153] have per-

formed TLC on degreased yarn fibers wound around a glass frame (10×
14 cm) to form a strip containing 25 threads on each side spaced at 3 mm.

TLC operations have been modified in various ways. For instance,
Lepoivre [154] has performed centrifugal radial TLC using an apparatus
that consists of a chamber containing a TLC plate (up to 40×40 cm)
mounted on a rotating disc; solvent is continuously applied to the layer
through a capillary tube placed eccentrically. Hashmi and co-workers
[155] have studied the factors that affect circular TLC. Hutzul and
Wright [156] have advocated developing the TLC plate in a horizontal
position (see Fig. 4.19), as suggested by Mistryukov [157]; the layer
does not need to contain a binder or to be baked, and this allows a wider
choice of adsorbent and speeds up the separation.

Berger and co-workers [158] have reported on the applications of TLC
to successive juxtaposed layers containing different kinds of adsorbents.
Niederwieser [159] has surveyed gradient TLC on nonuniform layers.
Misra and co-workers [160] have used glass-fiber sheets to separate drugs
and metabolites. Turina and co-workers [161] have described an adapter
for solvent evaporation on the top of the TLC plate to prolong the developing time, which increases the efficiency of separation. Sandroni and
Schlitt [162] have modified a commercial TLC chamber [163] in order
to use it for the concurrent developing of a 20×20 cm layer with up to
five different single solvents (see Fig. 4.20). Miroslawa [164] has reviewed the vapor-programmed TLC technique for separating compounds

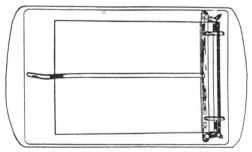

Top view: plate and its support in chamber

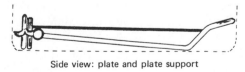

Side view: plate and plate support

Fig. 4.19. Developing TLC plate in horizontal position [156]. (Courtesy *Can. J. Pharm. Sci.*)

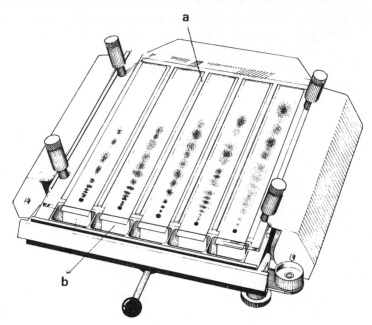

Fig. 4.20. TLC chamber for simultaneous developing with a solvent gradient [162]: *a*, barriers; *b*, solvent tanks. (Courtesy *J. Chromatogr.*)

of similar molecular structure. Perry and co-workers [165] have designed an instrument for programmed multiple developing, which makes TLC faster, more sensitive, and more versatile. Automated TLC machines are commercially available [166]. Another Commercial apparatus [163] is constructed for the simultaneous evaluation of ten TLC developing conditions.

III. DRY-COLUMN CHROMATOGRAPHY

A. PRINCIPLE AND CHARACTERISTICS

Dry-column chromatography (DCC) is a technique recently proposed by Loev and co-workers [167, 168]. Basically, the procedure is carried out by filling an empty column with adsorbent, depositing the working material on top of the adsorbent, and developing the chromatogram by allowing solvent to move down the dry adsorbent by capillary action and gravity. When the solvent reaches the bottom, the process is completed. The resemblance between DCC and TLC is apparent. Actually DCC evolved from TLC for preparative separation in the range for which TLC would not be economical. Whereas the capacity of a 20×20 cm TLC plate is 1 to 10 mg of working material, the DCC technique can separate 1 g or more in a single operation.

The relationship between DCC and TLC is also shown by the comparable speed of the separation, and by the almost identical R_F values that are obtained if the type and activity of the adsorbent and the solvent system are the same. Consequently, it is possible to optimize the conditions of preparative DCC by running preliminary tests with TLC in order to conserve the working material. A significant advantage of DCC is its simplicity in equipment and operation.

B. SELECTION OF ADSORBENTS

Alumina and silica gel are the common adsorbents for DCC, with slight preference for the former. The previous discussion on activity and solvent systems for TLC (Section II.B.1 and 2) is also applicable to DCC. Nevertheless, some features of DCC should be noted. First, better separations on dry columns have been experienced with a single solvent. Special pretreatment of the adsorbent is necessary for mixed solvents [169]. This fact imposes strict requirements on the activity of the adsorbent used. Secondly, the particle size of the adsorbent for DCC should be 80 to 200 mesh. If the fine particles for TLC are employed to pack the dry column, it results in substantial increase column resistance and of separation time, accompanied by a broadening of the bands and a decrease of the sharpness of the separation.

It is important to standardize each batch of adsorbent for DCC. The activity of the adsorbent is determined as follows: A 1×75 mm capillary is filled with the adsorbent. (This is conveniently achieved by tamping the particles by means of paper clips, as illustrated [170] in Fig. 4.21). One end of the capillary is placed in a vial filled to a depth of several milli-

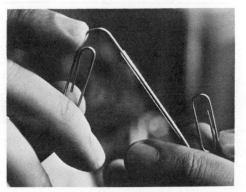

Fig. 4.21. Packing a capillary column [170]. (Courtesy Dr. D. P. Schwartz.)

meters with a 0.5% solution of the appropriate dye (*p*-aminoazobenzene for alumina, *p*-dimethylaminoazobenzene for silica gel) in benzene, until a drop of dye solution is adsorbed. The capillary is transferred to another vial containing a few milliliters of benzene, and the miniature column is allowed to develop. The R_F is calculated, and the corresponding activity is determined from Table 4.6. The first column in the table indicates the amount of water (in %) that is added to the fully activated adsorbent (activity I) in order to obtain an adsorbent of lower activity.

The ratio of adsorbent *vs* substrate is about one order of magnitude lower in DCC than in TLC. In general it ranges between 70 and 300 g of adsorbent per gram of substrate. Fig. 4.22 shows the relationship between sample weight and column length, and the diameter of the column for average separation (difference in R_F >0.4) and for difficult separation (difference in R_F ~0.1).

C. OPERATION

1. Packing the Column

Glass, silica, or plastic tubing may be used to prepare the dry column. Glass and silica columns are packed in an upright position by pouring the dry adsorbent in small portions and tapping each layer. Even packing can be achieved by holding a vibrator against the column. If the column has a stopcock, the latter should be opened to prevent the formation of air pockets. Packing is easier with alumina than with silica gel.

TABLE 4.6.   Standardization of Adsorbent [168]

% Water added	R_F of appropriate dye	Activity grade of adsorbent according to Brockmann scale
For alumina		
0	0	I
3	0.12	II
6	0.24	III
8	0.46	IV
10	0.54	V
For silica gel		
0	0.15	I
3	0.22	
6	0.33	
9	0.44	
12	0.55	II
15	0.65	III

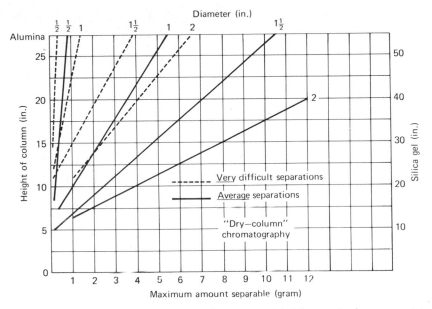

Fig. 4.22. Graph used for determining diameter and height required to separate a given quantity of "average" or "difficulty separable" mixture by "dry-column" chromatography. (Courtesy *Chem. Ind.*)

Flexible plastic (nylon) tubing is packed as follows: One end of the tube is sealed; the bottom is perforated with a few holes, and a wad of glass wool is inserted. Using a funnel, the adsorbent is poured into the tube to about one-third of its length. The adsorbent is compacted by allowing the tube to drop several times in an upright position from a height of about 15 cm. The packing is repeated with the second and third portions of the adsorbent. When finished, the column is firm and can be clamped. Commercially available [171] dry columns are in the form of compact rods or are provided with a glass housing that can hold pressure up to 50 psi.

2. Loading and Developing

The sample to be separated is introduced into the dry column mixed with dry adsorbent. A general procedure is to mix the solution of the sample with a small amount of adsorbent and then evaporate the solvent. The adsorbent loaded with the sample is transferred to the top of the column, and a layer of sand or glass beads is added to prevent the disturbance of the surface when the liquid is poured in. The column is developed by filtering the solvent through by gravity while keeping the solvent level at 3 to 5 cm above the top of the adsorbent. The developing is finished when the solvent front reaches the bottom of the column. The approximate volume of liquid required is given in Table 4.7.

3. Detection and Recovery

The use of plastic columns is advantageous for the detection of colorless compounds and the recovery of the separated substances. Observation with an ultraviolet lamp is a common practice; it should be noted that glass adsorbs ultraviolet light. The recovery of compounds from silica or glass tubes usually involves extruding the column of adsorbent, whereas a plastic tube can be cut into sections. Hodd and Caspi [172] have proposed the following procedure for preparative DCC: After developing, the column is laid horizontally. At intervals of 1 or 2 cm the tubing is punctured with the tips of disposable glass pipets; a 5-mm length of adsorbent is removed in each tip. The pipets are placed in test tubes, and the contents are expelled and then extracted with 5 drops of methanol. Aliquots of these extracts are subjected to TLC or gas chromatography. These analyses permit marking the plastic tubing appropriately so that separated zones can be sliced from the column. In order to prevent mixing by diffusion, the separated zones should be removed from the column as soon as possible. The usefulness of DCC has been demonstrated for the separation of five azobenzenes and of alkaloids that could not be separated in a liquid-flow column [167].

IV. ELUTION LIQUID CHROMATOGRAPHY

A. CHARACTERISTICS

Elution liquid chromatography (LC) consists of separating compounds in a packed column with a liquid that runs through the column by gravity or pressure, and collecting the effluents in portions from which the separated components are recovered [8, 173, 174]. Several types of LC are

**TABLE 4.7. Volumes of Solvent for Developing
in Dry-Column Chromatography [168]**

Column size (in.)	Volume of Solvent (ml)
20 × ½	20
× 1	90
× 1½	300
× 2	500

distinguishable according to the principle controlling the separation process. Adsorption and partition LC are treated separately below. Modern high-pressure LC, which differs from classical elution LC in instrumentation, is also discussed here. Elution chromatography, which depends on affinity or ion exchange, is treated in other sections. Some simple apparatus for elution chromatography are shown in Fig. 4.23.

B. ADSORPTION LIQUID CHROMATOGRAPHY

1. Principle

The separation of compounds between an adsorbent functioning as the stationary phase and a liquid as a mobile phase operating in a vertical column has been discussed in Section III (DCC). The difference between DCC and elution LC is twofold: (1) the column for elution chromatography is already equilibrated with the liquid phase before the sample is added, and (2) the individual components are gradually eluted from the

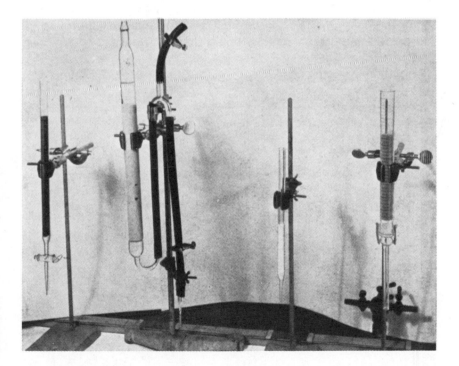

Fig. 4.23. Some simple apparatus for elution chromatography [51]. (Courtesy *Mikrochim. Acta.*)

column in separate fractions, from which they are recovered by evaporation. Consequently these two techniques differ in certain technical aspects.

2. Preparation of the Column

The columns utilized for adsorption elution chromatography are made of glass or silica, the latter having the advantage of permitting detection by means of ultraviolet light. The common type of columns (Fig. 4.24) carry a stopcock (a Teflon plug is recommended, since stopcock grease may contaminate the effluent). The bottom of the column is a fritted disc. Columns with negligible dead volume at both ends are preferred (Fig. 4.25); in this type the packing is squeezed between adjustable fittings that carry capillary connections. Some columns are intended for aqueous solutions; other columns are for organic solvents.

The column can be packed with dry adsorbent in the same manner as described for DCC. However, for a typical LC, before the sample is introduced into the column, the adsorbent should be saturated with the solvent by passing the liquid through the column by gravity. Alternatively, the column is packed with adsorbent suspended in the proper solvent by pouring the suspension rapidly through a wide funnel, as shown in Fig. 4.26. In the latter case, the excess liquid draining from the bottom facili-

Fig. 4.24.   Column for elution chromatography. (Courtesy Kontes Co.)

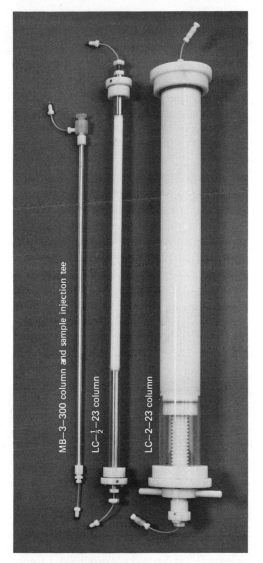

Fig. 4.25. Elution column with negligible dead space. (Courtesy Laboratory Data Control.)

tates homogeneous packing of the column. The horizontal homogeneity of the packing is extremely important for the compounds to migrate in parallel horizontal zones. The adsorbent packed in nonhorizontal and distorted layers causes deformation of moving zones and mixing of frac-

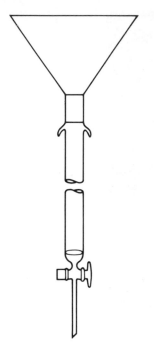

Fig. 4.26. Packing the elution column.

tions. Good separation also depends on the uniformity of the adsorbent with respect to particle size. Disturbances may arise from the presence of fine particles, which may be found even in an originally uniform batch of adsorbent because they are formed by mechanical abrasion during transportation. In such a case it is advisable to remove the fine powder by sieving. After equilibration with liquid, no part of the column must ever become dry.

3. Loading of the Column

The sample is transferred into the column in the form of a concentrated solution. Even if the working material is a liquid, it should be added to the column as a solution (at least $1:1$ mixture) to minimize the effect of the heat of mixing. Evidently, a sample that is slightly soluble in the selected solvent will have to be introduced in rather large volume. This may produce broad diffuse zones, and the efficiency of separation will suffer. Such a situation can be improved in several ways. (1) The sample is adsorbed on a small quantity of adsorbent as described for DCC (Section III.C.2). The adsorbent carrying the sample is spread evenly on the wet column. Then the solvent is added carefully to allow air bubbles to escape upwards without disturbing the column. (2) Another solvent, which improves the solubility of the sample drastically, is mixed with the liquid for elution. However, if that solvent is strongly polar, it will occupy part of the column, which will not be available for separation. (3) The solvent system is changed to one which is good for separation as well as for dissolving the sample. A sample injection valve (Fig. 4.27) which facilitates and improves the loading of the sample into the column, is commercially available [175].

4. Elution and Recovery

The elution process is different from the DCC developing procedure in that it is not stopped at the moment when the solvent reaches the opposite end of the column. The migration of zones continues, and finally the individual components leaves the column. Such an elution technique also has been used with "dry" columns. The problem in classical (wide-tube) elution LC is in recognizing the migration and collection of colorless compounds. In general, no detector is used. Fractions of equal volume are collected, evaporated, and weighed. The result of the separation is obtained by plotting the dry weight versus fraction number. From the pattern of such a graph the individual components of the mixture are recognized and collected from the corresponding fractions. Evaporation of each fraction, however, is time consuming, and complete evaluation of the separaiton is tedious, although this process can be facilitated by using a fraction evaporator [176] (Fig. 4.28). The elution process is usually automated by continuously feeding the column with the solvent and by utilizing a fraction collector [177] for separating equal-volume fractions. Instead of being evaporated, the fractions can be analyzed by TLC; the use of a multispotter (Fig. 4.29) facilitates the analysis [178]. On the basis of TLC or other analytical results (e.g., conductivity, spectroscopy), the fractions containing the same component are combined for recovery. Continuous monitoring of the effluent is performed by coupling the column with a detecting device. For instance, Schutte [179] has described the detection of radioactive effluents by heterogeneous or homogeneous scintillation counting. Since such detection techniques are generally employed in modern high-pressure LC, they are discussed in Section IV.D.

5. Selection of the Experimental Conditions

The selection of experimental conditions for adsorption LC is similar to that discussed in TLC and DCC (Sections II and III). Alumina and silica gel usually give better results than other adsorbents. The conditions with respect to the selection of a proper solvent system can be studied on

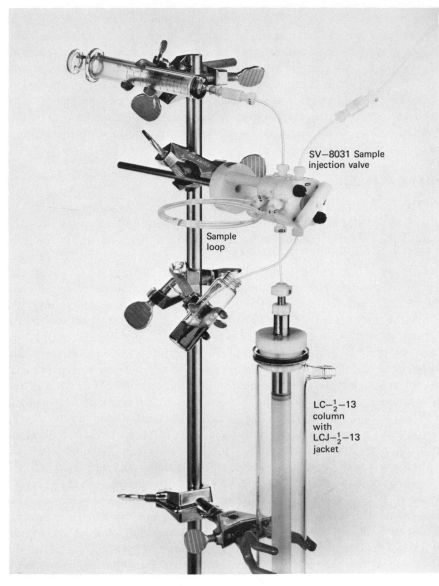

Fig. 4.27. Sample injection valve connected to elution liquid column [175]. (Courtesy Laboratory Data Control.)

Fig. 4.28. Fraction evaporator for simultaneous evaporation of 24 samples [176]. (Courtesy RHO Scientific, Inc.)

Fig. 4.29. Multispotter for TLC of 10 fractions [178]. (Courtesy Pierce Chemical Co.)

181

TLC plates, although the R_F values of TLC are not transferable to separations on LC columns without additional consideration. The extended developing procedure of elution is more efficient and versatile than the stop-flow technique in DCC; the capacity of the column also can be utilized more economically. Improper elution procedure, however, results in substantial extension of the elution time, which causes broadening of zones by diffusion and turbulent flow. Such disturbing factors reduce the efficiency of the column and increase expenses for the solvent and operation time. Fig. 4.30 shows the relationship between the position of a compound with a certain R_F value on the column and the volume of solvent required for elution. The volume V indicates the volume capacity of the column for liquids. At the unit volume ($V=1$), the R_F is evaluated for the compound. The numbers at 100% column length represent the volume of solvent necessary to elute out of the column a compound characterized by a certain R_F. From the graph it is apparent that compounds with small R_F values, which cannot be separated successfully by DCC, can be separated with the same solvent by LC. This fact becomes particularly useful for the separation of mixtures with large differences in migration rates. However, if such differences are very large ($V>20$), the simple elution technique is not suitable. In such a case, there are two ways to solve the problem: (1) The mixture is first separated into two or more fractions with narrower differences in migration rates. These fractions are subse-

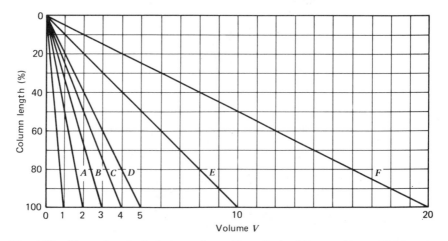

Fig. 4.30. Dependence of elution volume V on R_F in LC experiment. Holdup (capacity for liquid) of the column is used as volume unit V. At $V=1$, compounds A to F show position from which R_F is calculated. Readings at 100% length show that compound A with $R_F=0.5$ is eluted from column with $V=2$; compound F with $R_F=0.05$ is eluted with $V=20$.

quently separated individually under conditions that can be optimized for each fraction. (2) Elution is performed with a solvent gradient [179a], which is prepared batchwise or continuously by mixing two solvents of different elution power. For continuous mixing a simple two-compartment reservoir is employed (see Fig. 4.31). The gradients can be steep or flat, depending on the difference in polarity of the two components. In a flat gradient, the solvent mixture differs from the basic solvent only slightly; for example, the more polar component may be prepared by mixing 5 to 25% of the polar solvent with the basic solvent. A steep gradient may cause migrating zones to be too close, resulting in overlapping and consequent loss of discriminating power. Thorough TLC evaluation (using the loose-layer technique) of the compositions of solvents is very useful for the optimization of elution procedures.

C. PARTITION LIQUID CHROMATOGRAPHY

1. Principle

In partition liquid chromatography [3] the separation results from the distribution of molecules between two liquid phases, one stationary and the other mobile. The stationary phase is anchored to the surface of solid particles (solid "support"). The solid material should be inert, so that it does not interfere with the separation process (e.g., by adsorption). The interference of the solid is minimized by proper selection of the material, by chemical modification of the surface (e.g., treatment with acids), and by using sufficient quantities of the stationary liquid to coat the entire surface. Additional requirements for such solid support are uniform poros-

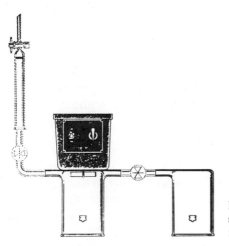

Fig. 4.31. Two-chamber reservoir for solvent gradient in elution chromatography. (Courtesy Kontes Co.)

ity of the surface and mechanical stability of the material. The most popular solids for the purpose of partition chromatography are Celite, (Kieselguhr) silica gel, starch, and cellulose.

The two liquid phases involved in a typical partition process are immiscible. Before use, the liquids are equilibrated (i.e., mutually saturated), or at least the mobile phase is saturated with the stationary phase. The common immiscible solvent pairs include water–n-butyl alcohol, water-phenol, water-collidine, and dimethylformanide–petroleum ether. (The first solvent is the stationary phase; the second, the mobile phase.) In the "normal" chromatographic system, the polar (or the more polar) phase functions as the stationary one. In the "reversed" system (system with reversed phases), the stationary phase constitutes the more lipophilic component (e.g., paraffin oil–methanol/water). Hence it is important to know the character of the chromatographic system being employed. In the reversed system, the order of the eluted species is reversed as compared to the normal system. The normal systems are preferred in the separation of hydrophilic compounds. Contrastingly, the reversed systems are employed for lipophilic compounds. However, lipophilic compounds are in general more easily separated by means of adsorption chromatography. Partition chromatography of this type is therefore most suitable for the separation of polar substances such as amino acids [180], oligopeptides [181], and carbohydrates [182]. Furthermore, partition chromatography is specially suited for separating homologous series, for which it is more responsive than adsorption chromatography. Thus Ali and Laurence [183] have described a technique for the quantitative separation of alkane mono- from di- and polysulfonic acids on cellulose powder over a wide carbon–number range (C_9–C_{20}). Recently, fundamental progress has been made in partition chromatography by preparing stationary phases that are chemically bonded to the support. These were specially developed for modern high-resolution liquid chromatography (e.g., Permaphase by DuPont).

Owing to its limited capacity with respect to the quantity of the sample to be separated, partition chromatography is more suitable for analytical than for preparative purposes. Nevertheless, milligram amounts of working material can be separated in a partition column without technical difficulties [180].

2. Preparation of the Column

In preparing a Celite column [184], the solid support is first treated with the polar component of the solvent system (e.g., water saturated with butanol). The column is packed portionwise in 3-mm layers, using a glass rod to compact the layer and to eliminate air bubbles. Then the organic phase (e.g., butanol saturated with water) is applied until the effluent is clear (i.e., free from droplets of the polar phase). The vertical homogeneity of the column is tested with azobenzene as follows: A little of the compound, dissolved in butanol, is placed on top of the packed column, which has been equilibrated with the liquid phases. At several time intervals the distances travelled by the azobenzene (a) and by the drop in the solvent level (b) are recorded. Using these data, the R value is calculated as the ratio a/b. The R values remain constant for a homogeneously packed column. The homogeneity of the column usually improves on additional washing with the mobile phase. The flow rate of the Celite column is slow; it can be increased by applying slight pressure.

The cellulose column [180] is packed with a slurry of cellulose in acetone. After sedimentation, the excess of acetone is drained off and replaced by the aqueous phase of the equilibrated mixture of solvents (e.g., water saturated with butanol). Finally, the aqueous phase is replaced by the organic phase (e.g., butanol saturated with water), and the washing is continued until no more droplets of water are observed in the effluent. The vertical homogeneity of the column is checked with methyl violet for constant R value as described above.

3. Elution and Recovery

The elution technique, the detection of individual components in the effluent, the collection of fractions, and their working up for preparative purposes are analogous to the procedures described for adsorption LC (Section IV.B) and DCC (Section III). The chief difference between partition and adsorption chromatography is that in partition chromatography a single mobile phase is employed throughout the separation process. This principle does not hold for the stationary phase, which is chemically bonded to the support and therefore suitable for gradient elution. In classical partition, the sample to be transferred to the column must be dissolved in the same solvent used for elution. Since the composition of mutually saturated liquid phases is temperature dependent, it is advantageous to use a partition column surrounded by a thermostatically controlled jacket (see Fig. 4.27). In contrast with adsorption chromatography, the columns for partition chromatography, such as Celite columns, are reusable.

D. MODERN HIGH-RESOLUTION LIQUID CHROMATOGRAPHY

1. Special Features

Modern high-resolution liquid chromatography [185, 186] (HRLC) differs from classical liquid chromatography in efficiency, speed, and control. These features are attained by optimizing the parameters of the column (length and diameter), packing (particle size and geometry), and solvent flow rate, and by continuous analysis of the effluent. The separation is achieved by utilizing adsorption, partition, gel-permeation, and ion-exchange principles. HRLC can be used for both analytical and preparative separations.

In principle, any material used in classical LC can be used for packing HRLC columns if the particle size is appropriately reduced. The most common size of particles for HRLC is about 30 μm in diameter, compared to 150 μm for classical LC. However, fundamental advances have been made in finding new materials [187]. At present, the best packing material consists of spherical glass microbeads of controlled surface porosity (e.g., Corasil and Parasil by Waters Associates; Zipax by Du Pont). These microbeads are used as adsorbents in liquid-solid chromatography, or as support in partition chromatography after being coated with an efficient stationary liquid phase (e.g., β,β-oxydipropionitrile, trimethylene glycol, cyanoethylsilicone, Carbowax).

For partition chromatography, in order to avoid stripping the column, the mobile phase should be saturated with the stationary phase as described in Section IV.C. It is particularly objectionable if the stationary phase is "eluted" in preparative chromatography; in such a case, the separation of the stationary liquid phase from the sample in the effluent would be difficult. Furthermore, the original composition of the mobile phase should be maintained throughout the procedure. Microbeads coated with polymers (e.g., polyamide, ion-exchange resins) and chemically bonded stationary phases [188] (e.g., octadecylsilane and ether-type packing materials such as Permaphase by DuPont) are resistant to stripping. The solvent used for elution with such stationary phases does not require presaturation. The gradient elution technique is applicable to these colmns which are durable and reusable.

2. Operation

Understandably, the small size of the particles used for packing the column increases the resistance and hence decreases the flow rate. This difficulty is circumvented by applying pressure from a pump or from a cylinder of compressed gas. In the first case a pulseless pump is employed. Pressures from several hundred up to several thousand psi are applied. In the second case a special cylinder valve with pressure adjustable up to 3000 psi is needed. Scott and co-workers [189] have constructed a sample-injection valve that is satisfactory for pressures up to 5000 psi without leakage during 4000 hr of working. Halasz and co-workers [190] have studied the problems of rapid liquid chromatography with high inlet pressures, including the need to smooth out pulsations, difficulties with the permeability of the column, and the influence of peak broadening.

Continuous monitoring is performed by a detecting device coupled to the chromatograph at the outlet of the column. These detectors operate on various principles, such as absorption of ultraviolet light, changes in refractive index, flame ionization, and polarography. The ultraviolet detector is very sensitive, but its use is restricted to compounds that absorb light in the region of 254 to 280 nm. The signals from the detector are recorded continuously and automatically.

The solution containing the mixture to be separated is transferred to the column by means of a syringe inserted into the injection port of the chromatograph or by means of the sample valve. When the injection port is used, the septum should be changed frequently to prevent leakage resulting from the high pressure applied to the system. In most HRLC applications thermostatization of the column is not necessary.

3. Uses

HRLC applied to analytical purposes usually yields excellent results. The best use for HRLC is the separation of nonvolatile, temperature-sensitive, or polar compounds for which gas chromatography is not suitable. For obvious reasons, preparative applications of HRLC are less common, although suitable techniques have been developed.

The quantities of working material required for HRLC analytical separation are extremely small (<1 mg). In order to maintain high resolution in separations of large samples, the analytical procedure needs modification with respect to both equipment and experimental conditions. The capacity of the column has to be increased; otherwise the column is overloaded. Increasing the diameter of the column, however, results in decreased column efficiency, as predicted by the LC theory [185]. Therefore the length of the column should be increased with its diameter, and also the flow rate. The same effect of lengthening the column can be achieved by the recycling technique; this involves special equipment. Low flow rates cause broadening of zones by diffusion, while increased flow rate reduces the separation efficiency if the equilibrium conditions are not satisfied. These three parameters must be balanced for a particular problem with respect to sample size and difficulty of separation. Table 4.8 gives typical experimental conditions on Parasil and Corasil columns.

In an analytical HRLC experiment the results are presented in the form of a record (chromatogram) in which chromatographic peaks are evaluated qualitatively and quantitatively. For the identification of a compound, the time delay necessary for eluting the compound from the column is

TABLE 4.8. **Experimental Conditions for Preparative Separations
on Porasil and Corasil Columns**

Column o.d. (in.)	⅛	⅜	1
Packed void volume per 2 ft length			
for Porasil (ml)	2.2	25	250
for Corasil (ml)	1.3	14.5	145
Typical column length (ft)	2	8	8
Typical flow rates (ml/min)	3	10	10
Sample load for packings			
for easy separation	20 mg	500 mg	5 g
for difficult separation	2 mg	50 mg	500 mg
Typical injection volume	5–100 μl	0.5–4 ml	4–10 ml

determined by measuring the elution time (from injection of sample to maximum of peak). This elution time is a physical constant characteristic of each chemical species under the experimental conditions used, such as column specifications (diameter, length, packing material), solvent system, and flow rate. The quantification of the chromatographic peak is obtained by integrating the peak area, and evaluated with a calibration graph.

For preparative separations the chromatogram is used primarily to evaluate the resolution of peaks in order to determine the efficiency of separation of the individual components in the collected fractions. If the peaks overlap, the separation is incomplete. In collecting fractions it is more convenient to change vials when the flow is stopped. Apparatus that is most suitable for sharp separation of fractions has minimal retention volume in the tubing between the detector and the outlet.

E. MODIFICATIONS

Fox and co-workers [191] have described the construction and operation of a continuous-chromatography apparatus. Hills and Payne [192] have designed a refillable constant-head device for continuous operation, while Davis [193] has described a simple eluent reservoir that can maintain head pressure to within ±1 mm. Flynn and Michl [193a] have performed column chromatography at the temperature of liquid nitrogen.

The apparatus for use in incremental gradient elution has been reviewed by Scott and Kucera [194). Hegenauer and co-workers [195] have described a mixing device for generating single gradients, while Schmidtmann [196] has constructed one for the preparation of pH gradients. Hirsch and co-workers [197] have separated petroleum distillates using gradient elution through dual-packed (silica-gel–alumina) adsorption columns.

Popl and co-workers [198] have described a method of calculating retention volumes in gradient adsorption LC. Farrell [199] has published a computer-generated table for the selection of column systems to separate organic bases. Kochen and Lauer [200] have constructed a simple electronically controlled apparatus for the automation of column chromatography.

F. APPLICATIONS

1. Classical Methods

Applications of classical liquid chromatography are well documented in the manuals [2–5] and handbooks [201]. This technique has been used most extensively in biochemistry, the separation of natural products, and organic syntheses. Practically all classes of compounds have been tested. For example, Schwartz and co-workers [202] have published a series of papers on the separation of phenylhydrazones and osazones, and on the separation of alcohols, amines, diols, and thiols through their derivatives of pyruvic acid 2,4-dinitrophenylhydrazones. Column chromatography has been employed to isolate and identify new constituents in milk fat [203]. Recently dimethylsulfoxide has been used as a solvent [204]. Unstable compounds have been separated; for example, hydroperoxides have been loaded in columns packed with zirconium dioxide and eluted with o-dichlorobenzene [205].

An interesting application of elution chromatography developed by Schwartz and co-workers involves performing chemical reactions in the packed column [206]. An inert support is impregnated with one of the reactants dissolved in an aqueous phase. The other reactant, dissolved in a water-immiscible solvent, is then passed over the column, the reaction taking place on the surface of the coated support. These two-phase column reactions have been found to be extremely efficient for reacting micro quantities of specific classes of compounds present in a very dilute solution. The impregnated solid particles can be packed into capillary tubes as shown in Fig. 4.21. Besides requiring minimal manipulation, the chemical reactions are usually faster to run; they also tend to go to completion because of high molar ratio of one reactant to the other and the continuous removal of the products from the reaction medium. Reactions that have been tested by this technique include oxidation by chromic acid or periodic acid, catalytic hydrogenation, and sodium bisulfite addition [207]. These workers also have found that cholesterol is rapidly and quantitatively removed from butterfat under very mild conditions by passage of the fat in benzene over a column of Celite impregnated with an aqueous solution of digitonin [208].

2. Recent Developments in High-Resolution Liquid Chromatography

Although high-resolution (high-pressure, high-speed) liquid chromatography is a relatively new development [185], its use is already widespread. Since this technique is particularly well suited for microsamples, it has been recommended for trace, residue, and pollution studies. Many papers have been published about its application to the analysis of pharmaceuticals [209], vitamins [210], steroids [211], carbonyl compounds and glycols [212], complex biological mixtures [213], natural products, dyes, polymers, and so on. Schmidt and co-workers [214] have investigated the effect of temperature on reversed-phase column efficiency and reported that an increase in column temperature increases the efficiency owing to decreased viscosity of the mobile phase and increased sample capacity. For every 30°C increase there is a 50% decrease in retention time. Modification of the mobile phase with varying proportions of certain water-immiscible organic solvents has been found to have a great effect on the capacity values of the solutes, so that it should be possible to adjust the capacity to any desired value by changing the modifier.

The instrumentation of high-resolution liquid chromatography is more sophisticated than for classical LC. A relatively simple and inexpensive apparatus has been constructed, using a microdetector based on polarography [215]. Chandler and McNair [216] have compiled a list of commercial equipment currently available. Done and Knox [217] have reviewed the recent developments with reference to components of the apparatus, packing materials, and choice of mobile phase for applications in biochemical analysis.

Kirkland [218] has investigated the use of high inlet pressure (500 to 5000 psi) and long, small-diameter columns packed with materials of particle size 20 to 60 μm, and of microspheres [219] (about 5-μm diameter; 35-nm pores), which are packed as a slurry in ammonium hydroxide into a stainless-steel column (250×3 mm) under high pressure and then impregnated with 30% of 3,3'-oxydipropionitrile by an *in situ* coating technique. Pearce and Thomas [220] have discussed the methods for introducing small samples of liquids into the high-pressure column. Scott and co-workers [189] have frabricated a six-port sample-injection valve capable of operation up to 5000 psi. Machin and co-workers [221] have constructed a simple pneumatic pump that provides a pulseless flow of solvent at pressures of 2000 to 3000 psi. Cassidy and Frei [222] have described a fluorescence detector that comprises a flow cell (7.5-μl capacity, made by drawing out a glass tube) attached to the column by means of Teflon tubing and a fluorimeter. Since capillary-tube connections between column and detector can adversely affect performance, Scott and Kucera [223] have studied the relationships between the variance of the solute band and the length of the connecting tube, the radius of the tube, and the flow rate through the tube. Equations have been derived to calculate the optimum dimensions of a connecting tube that would restrict the increase of the bandwidth of eluted peaks to 5%.

V. GAS CHROMATOGRAPHY

A. PRINCIPLE

In gas chromatography [224–226] (GC), the gaseous or vaporized sample is transported through the column by means of a carrier gas. As a result of their different affinities for the stationary phase, various compounds in the mixture are separated into individual components. The interaction of the molecules of the sample with the stationary phase is based on either partition or adsorption. In the partition process, molecules are distributed between the carrier gas and a liquid that is used for coating the solid support; this version of GC is known as gas-liquid chromatography (GLC). In the adsorption process, molecules are distributed between the carirer gas and a solid adsorbent; hence this technique is called gas-solid chromatography (GSC). In general, GLC is preferred, because its experimental conditions can be widely varied owing to the large choice of stationary liquids available. Thus, for each case of separation the most selective stationary liquid phase can be found, on the basis of some theoretical assumptions and certain empirical rules and experience. Besides partition and adsorption, exclusion principles are involved in gas chromatography when the stationary solid is furnished with uniform pores of molecular dimensions (molecular sieves).

B. EQUIPMENT AND MATERIALS

The popularity and extensive use of GC is a result of its advanced experimental and instrumental techniques. Basically the GC apparatus consists of a gas cylinder equipped with a fine-adjustment valve, a chromatographic column with injection port, a detector, and a recorder (see Fig.

4.32). Simple and moderately priced gas chromatographs are commercially available [227, 228]. McNair and Chandler [229] have compiled a list of current gas-chromatography equipment.

The carrier gas plays roles in two independent operations [230]. First, it provides the transportation for the sample through the column and participates in the separation process. Secondly, it also participates in the detection process. As a transportation medium, carrier gases of high molecular weight are preferred in order to minimize vertical diffusion and zone broadening. Thus nitrogen has an advantage over hydrogen or helium. With respect to its role in detection, the choice of the carrier gas depends on the detector used in the apparatus. For instance, hydrogen or helium is preferred to nitrogen when a catharometer is employed. On the other hand, nitrogen is generally used with the flame ionization detector, which does not sense any of the common carrier gases.

GC columns are usually fabricated from glass, copper, or stainless tubing [231]. The packing material of the column for GLC, which consists of a solid support and a liquid phase, is dependent on the characteristics of the sample to be separated. Generally the support has uniform size in the range from 40 to 100 mesh and is of uniform porosity. The common material is diatomaceous earth (synonyms: diatomite, Chromosorb, Kieselguhr), usually treated chemically to achieve high adsorptive and chemical indifference. Such indifference is recognized by the appearance of sym-

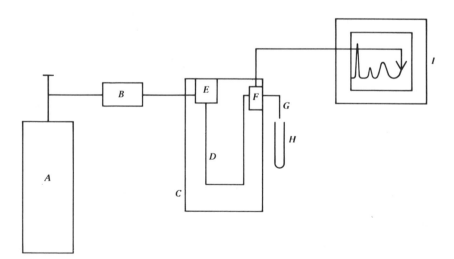

Fig. 4.32. Diagrammatic sketch of the gas chromatograph: *A*, carrier-gas cylinder; *B*, flow controller; *C*, thermostat (oven); *D*, chromatographic column; *E*, injector pot; *F*, detector; *G*, gas exit; *H*, sample trap; *I*, recorder.

metrical peaks with no tailing when the column is tested with polar compounds such as methanol.

C. OPERATION

1. Injection of the Sample

The solution containing the mixture to be separated is introduced into the column through the injection port [232, 233]. A microsyringe (with capacity 10 μl or 50 μl) is used for injection. The injection port communicates directly with the column. While both the column and injection port are thermostatized, the port is maintained at a higher temperature to ensure fast vaporization of the sample. Either isothermal conditions or temperature gradients [234] are imposed, depending on the nature of the sample. The temperature-gradient technique allows the separation of a mixture with widely spread chromatographic characteristics, which would not be separable in one-run operation under isothermal conditions.

2. Selection of the Stationary Phase

The choice of the stationary liquid [235, 236] is of great importance for efficient separation, particularly of compounds with similar structures. Specific characteristics of the stationary phase [237], along with the boiling points of the compounds, the flow rate of the carrier gas, and the column temperature, are the chief factors controlling the separations. The stationary liquids are classified as polar and nonpolar. The nonpolar liquids are represented by the high-boiling fractions of crude petroleum oil (e.g., Apiezon), squalane, and silicone polymers. The polar stationary liquids are either high-boiling and thermally stable monomeric species such as esters (e.g., di-n-octyl phthalate), nitriles (e.g., β,β-oxydipropionitrile, tetracyanoethylpentaerythritol), and polyhydric alcohols (e.g., mannitol, diglycerol), or polymeric species such as polyesters (e.g., polyethyleneglycol adipate), polyethers (e.g., polyethyleneglycols, Carbowax) and polyamides (e.g., Versamid). For separation with nonpolar stationary phases, the order of elution follows the same sequence as the boiling points. In contrast, polar stationary phases interact specifically with the polar groups of the molecules. As a result of these interactions, the contribution of the boiling point to the migratory aptitude of the molecules may be reduced. It is due to these interactions that the separation of complex mixtures and of compounds with very similar structures (e.g., isomers) can be achieved [238]. The amount of stationary phase in the column affects the separation (it ranges between 1 and 20% of the support). The larger amounts make the elution slower, peaks broader, and capacity higher. On the other hand, the use of an insufficient amount of the stationary liquid may result in interference by the solid surface due to adsorption.

3. Detection and Recovery

Continuous monitoring of the outflow is performed by a detector [239, 240]. Various types of detectors have been used, among which the most common devices are based on thermoconductibility (catharometer) or flame ionization. The catharometer operates on the principle of changes in the electrical conductivity of a hot filament as a result of differences between the heat transferred by the carrier gas and by the individual compounds eluted from the column. Two such cells are built into a Wheatstone bridge. The resistance of the cell that monitors the outflow from the column is compared with the resistance of the reference cell flushed with the carrier gas only. The sensitivity of the detection depends on the difference in thermoconductivity λ of the carrier gas and the sample. The highest sensitivity is achieved with hydrogen and helium, both being considerably more thermoconductive than other substances ($\lambda \times 10^5$ for H_2, 41.6; He, 34.8; N_2, 5.8; hexane, 3.0).

The flame ionization detector (FID) is a microburner with two electrodes. In this device the outflow from the column is mixed with hydrogen and combusted. The ions and electrons produced in the combustion cause a change in the potential between the electrodes as result of their electroconductivity. No reference cell is necessary. The FID is particularly suited to the detection of organic molecules. Since many substances are not detected by the FID, they can serve as carrier gas or solvent without interfering with the detection of the analyzed products. These substances include the inert gases, N_2, O_2, CS_2, H_2S, SO_2, oxides of nitrogen, NH_3, H_2O, CO, and CO_2.

The proper choice of the detector is important in preparative GC. Whereas the separated compounds are usually not changed with the catharometer, the FID is destructive. Consequently, when the FID is used, the major portion of the separated compounds should bypass the detector. This is achieved by dividing the outflowing gas from the column in a certain ratio, for example, 1:100. For this purpose a T tube with arms of different inside diameters is connected to the column, the narrow arm being joined to the detector and the wide arm to the collecting trap. In the trap the outflowing products are separated from the carrier gas and condensed by means of efficient cooling. Various types of traps have been used [241–243]. Detailed descriptions are given in Chapter 7, Sections III.B and V.C. If the trap is composed of straight and spiral parts, more efficient condensation is achieved if the gases enter at the spiral part. Thus the droplets of the mist deposit on the cold walls by centrifugal forces, and the path to the bottom of the collecting trap is longer.

D. USES AND GUIDELINES

Qualitative and quantitative information can be obtained from a GC experiment. Qualitatively, the retention time is a physical constant that characterizes a particular compound under the operational conditions of a particular gas chromatograph. It indicates the time delay that results from interactions of the molecules with the stationary phase. The retention time is measured from the injection time (or from the air peak used as a marker) to the peak maximum. Quantitatively, the result is obtained either by comparing the peak height against a standard, or by integrating the peak area. The peak area is usually linearly proportional to the amount of the particular compound. For accurate quantification, however, the calibration graph for each component of the mixture should be established.

The sample to be separated by GC is injected as received (for solutions and gas mixtures) or dissolved in a suitable solvent (for solids and liquids) before injection. The removal of nonvolatile components from the sample extends the life of the column; preliminary distillation, selective extraction, or use of a precolumn serves this purpose. "On-column injection" is the common technique for introducing the working material for preparative GC. When large samples are introduced, the gas flow should be stopped during the injection period in order to minimize broadening of the migrating zones.

The theoretical aspects of gas chromatography have been under extensive investigation [244]. They are important in finding the optimum conditions for the separations. These conditions constitute a multiparameter problem involving factors such as temperature, flow rate, efficiency of the column (in theoretical plates), selectivity of the packing material, length and width of the column, and properties of the carrier gas. For example, Ettre [245] has studied the factors affecting the speed of gas-chromatographic separations. Oberholtzer and Rogers [246] have reported on the effects of micropores on peak shape and retention volume in GSC. Bierl and co-workers [247] have studied the effect of functional group position on the retention time of several classes of compounds on four stationary phases; as the stationary phase becomes more polar, the change in retention index with increasing temperature becomes more marked. Martynyuk and Vigdergauz [248] have proposed some rules relating to the retention of organic compounds on columns containing colloidal stationary phases. Ecking and Lenz [249] have advocated the classification of stationary phases and the calculation of retention data by means of infrared spectrometry. Walker and Wolf [250] have observed the operating characteristics for the separation of similar compounds, using the interrupted elution technique. Gröenendijk and Kemenade [251] have evaluated the linearity of the logarithmic plot of adjusted retention times versus carbon number in a homologous series.

An ideal GC separation is characterized by maximum resolution (no overlapping) of peaks. This is achieved by keeping the peaks narrow and far apart. The first objective can be realized by minimizing the longitudinal diffusion; the second, by proper selection of the stationary phase. It should be recognized, however, that to optimize all of the parameters is complicated, since the same parameter may control different phenomena simultaneously. Moreover, the change in one parameter may influence different phenomena in opposite ways. Therefore, in order to reach the highest resolution, the values for all parameters must represent a compromise. Generally speaking, the efficiency of the column will increase with decreasing particle size and column temperature at optimum flow rate. A plot of the diffusion versus flow rate yields a distorted U-shaped curve with the highest efficiency at the lowest point of the graph.

E. MODIFICATIONS

While GC columns made of metals have the advantage of being durable and pliable, they cannot be used to separate corrosive materials. For instance, in the procedure for separating alkyl iodides [252], the column consists of a nine-coil Pyrex glass spiral packed with glass spheres, which are coated with tricresyl phosphate as the liquid stationary phase. Currently various types of glass GC columns are commercially available [236]. Crossley [253] has described an 18-in. column made from Teflon tubing of 0.04-in. i.d. and packed with 1% XE-60 on Chromosorb G. The injection block and part of the CsBr thermionic detector block are also lined with Teflon. Fielder and Williams [254] have constructed a Teflon injection port for the introduction of corrosive liquids and gases.

The literature on GC solid supports and liquid phases is legion; over 2500 references [236] were indexed up to 1968. Recently Cadogan and Sawyer [255] have studied various thermally activated and chemically modified silicas for GSC. The variation of the functional-group contributions with the extent of thermal modification has been used to establish the optimum column activation temperatures for difficult separations. Ross and Jefferson [256] have advocated the use of formed-in-place polyurethane for GC support, citing the following advantages: (1) the porosity and surface characteristics of the support can be controlled by adjusting reaction conditions; (2) liquid phases can be incorporated by adding them to the precursor solutions; (3) the polyurethane adheres tightly to the column walls, thus preventing channeling and eliminating the need of baffles for preparative GC. With porous polymer-bead packings, Ackman [257] has found that adding formic acid to the carrier gas helps to suppress tailing and other adsorption effects without affecting the column life. Jowitt [258] has described a cold-packing method for filling GC columns with Teflon supports. It should be noted that Teflon is not an inert support [259]. Similarly, reaction may occur between the substrate and molecular sieves [260] that are commonly used as packing material for the separation of hydrocarbons [261].

GC experiments have been performed on solid samples directly. These experiments are usually undertaken to study the thermal decomposition of solid materials, using GC to separate the resulting products. Pyrolysis attachments for the GC of solids are commercially available [262]; or easily made [262a]; they consist of a glass tube with a built-in heating element and fit into the injection port of the gas chromatograph. According to DePietro [263], the crystalline solid sample (or powder) should be placed in a quartz capillary as shown in Fig. 4.33. Pease [264] has described a solid-sample injection system that is used to introduce mixtures containing low-boiling liquids and nonvolatile components into the gas chromatograph. The sample, cooled in Dry Ice, is packed into the sample holder, also cooled in Dry Ice. As illustrated in Fig. 4.34, the steel rod carrying the holder is inserted through the O-ring seal on the outer chamber. The ball valve is open and the rod advanced until it encounters the column inlet. Karayannis and co-workers [265] have constructed a high-pressure chromatograph for the GLC of normally nonvolatile compounds like porphyrins and metals chelates; dichlorodifluoromethane above the critical temperature is used as carrier gas, and detection is by a flow-through cell scanned by a spectrophotometer.

Nikelly [266] has evaluated porous-layer open-tubular columns made by the dynamic method; such columns have an i.d. of 0.02 in. and lengths up to 100 ft. A new trapping technique has been described by Willis [267]. Kirsten and Mattsson [268] have constructed a heated precolumn coupled to a capillary column for high-efficiency trace determination using an

Direction of He ⟶

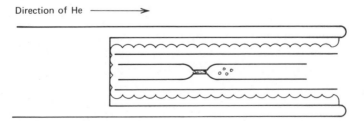

Fig. 4.33. Gas chromatography of solids, showing the position of the quartz capillary in the pyrolysis attachment [263].

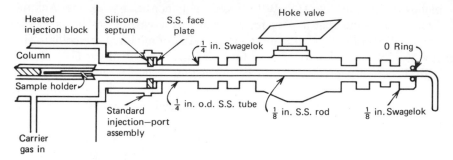

Fig. 4.34. Solid-sample injection system [264]. (Courtesy *Anal. Chem.*)

electron-capture detector. Jeltes and Veldink [269] have utilized the solid support as a subtractor for the trace analysis of polar compounds. Levins and Ikeda [270] have described a technique for separating volatile material from solids in the injection port. Walker and Wolf [270a] have reported on the characteristics of interrupted elution GC, whereby the separation process in a column is stopped for a specific period of time while chemical analysis is performed upon a particular eluate.

F. APPLICATIONS

Since extensive bibliographies [201, 201a] have been compiled on the applications of GC, only selected recent publications related to microsamples will be mentioned in this section. Understandably this technique is most widely used for the separation of thermally stable organic compounds such as alcohols [271], esters, amines [272], phenols [273], and hydrocarbons. Elaborate chromatographs are found in the petroleum industry; a 450-ft capillary column has been constructed, and 180 components in gasoline are separated in 2 hours on a 200-ft column [7]. GC is a convenient tool for investigating volatile substances such as perfumes and flavors [274]. Recently GC has been employed extensively in environmental studies such as those of air pollution [275] and pesticide residues [276, 277]; over 900 papers have been published on the latter subject [278].

GC has been applied to the analysis of pharmaceuticals [279–281] and narcotic drugs [282, 283]. It has also been used to separate biochemical compounds that are themselves nonvolatile, such as amino acids [284, 285] vitamins [286], steroids [287, 288], and metabolites [289]. Although uncommonly, GC has been applied to inorganic materials through the formation of volatile metal complexes [290, 291] and by anion-exchange techniques [292].

Another application of GC to organic analysis involves inducing a chemical reaction (e.g., esterification, silylation, etc.) on the sample, followed by the detection or determination of the product in a gas chromatograph [293]. One method is called "carbon-skeleton chromatography," in which the functional groups of the molecule are stripped off, resulting in a hydrocarbon in which to identify the carbon skeleton [294]; it has been employed in the analysis of a wide range of compounds. For quantitative analysis, a specific functional group in the sample is reacted with a suitable reagent to produce a known gaseous compound, which is subsequently separated and measured by GC. Organic functional groups that have been determined by this technique include the amino [295], carboxyl [296], alkoxyl [252], azo, and nitro [297] groups, as well as sulfonates [297] and carbamates [298]. A pyrolysis and reaction chamber for GC has been patented [299]. A carbon molecular sieve has been described that can be used in GC to retain molecules that contain the methylene group [300].

Preparative GC has been investigated by a number of workers [301–304]. It is a useful tool for the separation of small amounts of volatile liquid mixtures. For large quantities, however, fractional distillation in an efficient column (see Chapter 5) is preferred. For instance, milliliter amounts of mixtures of hydrocarbons, aromatics, halocarbons, ketones, esters, or alcohols have been separated by GC on very long columns, but the chromatograms show typical distortions [305].

Since the gas chromatograph is frequently used for repetitive operations, automation [306] and computerization are logical developments in GC. The use of computers for the optimization of temperature or pressure programming [307], the separation of overlapping peaks [308], and the planning of new experiments [309] have been reported. In view of the vast number of GC data, computer storage and retrieval [310] are indispensable in large laboratories.

The coupling of the gas chromatograph with other analytical instruments further extends the applications of GC. For example, it is possible to combine GC with TLC [311], thermogravimetry [312], and infrared [313–314a], nuclear-magnetic-resonance [315], and mass spectrometry [228, 315a, 316, 316a].

VI. GEL CHROMATOGRAPHY

A. PRINCIPLE

Inorganic and organic gels characterized by pores of uniform size are suitable for separating compounds according to their molecular size (i.e., molecular weight) and shape. This separation technique [317–319], called gel chromatography, is also known as gel filtration, gel exclusion, and gel permeation chromatography. In the dynamic process of separation, molecules with diameters larger than that of the gel pores simply bypass the gel particles. In contrast, the "matching" molecules (molecules that fit the pores) enter the porous structure of the gel. Consequently, these molecules are delayed in the elution process owing to their extended pathway through the gel. The critical value of molecular weight above which the molecules are excluded from the porous structure is called "molecular-weight exclusion limit." It should be noted, however, that the exclusion limit is influenced by the experimental conditions (e.g., the solvent). The exclusion principle operates not only in gel chromatography, but also in adsorption, ion-exchange, and affinity chromatography, as a result of the special characteristics of the particular macromolecular matrix (e.g., the cross-linked polystyrene matrix of an ion-exchanger, or the agarose used in affinity chromatography). Furthermore, the chemical contribution to affinity is superimposed on the common tendency of gels to fractionate on the basis of steric exclusion [320]. A mathematical treatment of the principles of gel chromatography has been presented by Vink [321].

Generally speaking, molecules with large molecular weights are eluted in gel chromatography before the small molecules. About 25% difference in molecular weight is usually required for the distinct separation of two species. The effect of molecular weight on the rate of migration in gel chromatography is reversed in comparison with adsorption. In view of the nature of the separation, it is understandable that gel chromatography has become especially popular in the field of macromolecular chemistry [322].

B. PACKING MATERIALS—GELS

Various types of gels are available [323]. They are manufactured by different firms [324], and classified according to chemical properties (e.g., hydrophilic, lipophilic) or mechanical properties (e.g., rigid, semirigid, soft). The values of the molecular-weight exclusion limit range from 10^2 to 10^6 or higher. For the best performance, the gel particles should have uniform size and shape, and the pores should have uniform diameter. Organic gels are mostly polymerized in emulsions, from which uniform spherical beads are obtained. Rigid gels are required for packing columns of high-pressure liquid chromatographs, while soft gels can be used with slow-flow columns only.

The most popular inorganic gels are silica gels and porous glasses; they are rigid and operate well in both aqueous and organic solvents. These gels are produced in various pore sizes (from 75 to 2500 Å) and are characterized by molecular-weight exclusion limits from 28,000 to 2,000,000.

One group of organic gels is represented by cross-linked polystyrene-divinylbenzene gels (PS gels—e.g., Styragel by Waters Associates) and polyvinylacetate gels (PVA gels—e.g., Merck-o-gel by Merck). These gels are semirigid, permitting high flow rates through the column. PS gels are used exclusively with organic solvents such as toluene, dichlorobenzene, tetrahydrofuran, and dimethylformamide. PVA gels can also be used with alcohols that are not compatible with PS gels.

Another type of organic gels is represented by those based on dextrane, starch, and agarose structures (e.g., Sephadex by Pharmacia). These are hydrophilic soft gels and are used exclusively with water and its mixtures. Synthetic hydrophilic gels (e.g., Bio-Gels by Bio-Rad) are obtained by polymerizing hydrophilic acrylamides.

The commercial organic gels have molecular-weight exclusion limits between 10^3 and 10^8. Gels for particular purposes can be made in the laboratory. For example, the procedures for preparing polyamide [325], agarose-acrylamide [326], and lipophilic-hydrophobic polysaccharide [327] gels have been reported.

C. SELECTION OF EXPERIMENTAL CONDITIONS

In a gel chromatography experiment three factors are of interest [328]: (1) the resolution (i.e., the completeness of the separation), (2) the time required, and (3) the amount of material that can be separated. For microsamples, resolution is of chief concern. As indicated in the equation

$$R_s = \frac{\Delta X}{W} = \frac{\text{distance between zone centers}}{\text{zone width}}$$

the resolution R_s is increased either by increasing the distance between the peaks, ΔX, or by increasing their sharpness. For a given gel type, the distance between the peaks depends on the bed volume. Thus, for any fixed bed diameter, the distance between the peaks is directly proportional to the bed height L. However, the width of the peaks, W, is also related to the bed height: $W \propto \sqrt{L}$. Therefore, the resolution is proportional to the square root of the bed height. Since long gel columns are difficult to pack, improved resolution is achieved either by recycling chromatography (see Fig. 4.35) or by connecting several columns in series. Johnson and co-workers [329] have discussed methods for optimizing the resolution in gel chromatography.

D. EQUIPMENT AND OPERATION

Ellis [330] and Williams [331] have reviewed the equipment for gel chromatography, including pumping systems, columns and detectors, the

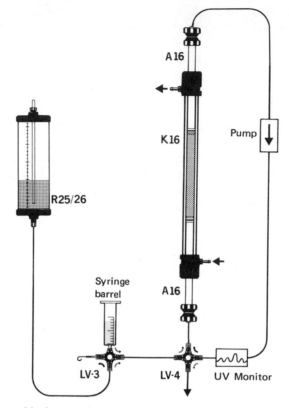

Fig. 4.35. Assembly for recycling chromatography [328]. (Courtesy Pharmacia Fine Chemicals, Inc.)

measurement of the elution volume, and fraction collection. In general, the techniques described previously for column chromatography (Section IV.B) and TLC (Section II) can be adapted to gel chromatography.

Gullberg and Brostroem [332] have constructed an apparatus for simultaneous gel filtration through 10 small columns. Nichilas and Fox [333] have designed a continuous chromatograph for use with Sephadex gels. Cassera [334] has described a simple tube for gel filtration. Jaworek [335] has performed thin-layer gel chromatography as follows: The swollen gel is spread on a 20-× 20-cm plate, which is then placed in a sandwich tank that is capable of tilting the plate to any desired angle. The developing solvent is transferred to the plate through filter-paper strips. Berret and co-workers [336] have used aminoethylcellulose paper for separation. Emneus [337] has described a procedure for the filtration of viscous solutions. Junowicz and co-workers [338] have studied gel filtration with vibration; vibrations perpendicular to the axis of the column improve the fractionation pattern of the high-molecular-weight components.

E. APPLICATIONS

Most of the applications of gel chromatography are in biochemistry and polymer chemistry. Macromolecular substances such as proteins can be desalted, or the buffers can be exchanged by means of gel filtration, using a gel with low exclusion limit; in both cases the original salt components are retarded while the proteins are eluted from the column. For buffer exchange, the column before use is equilibrated with the new buffer, which also constitutes the elution liquid.

Critical experimental conditions may be necessary for the separation of molecular species with only a slight difference in molecular weights. Sometimes gel chromatography is coupled with adsorption or partition chromatography [339]. The molecular-weight exclusion principle may be combined with ion exchange (see Section VIII) by employing resins of uniform porosity. Polymeric substances that have been separated by gel chromatography include oligostyrenes [340], polystyrenes [341], polyethylenes [342], cellulose nitrates [343], phenol-formaldehydes [344], dextrines [341], and hydroxymethylmelamines [345]. Preparative gel chromatography has been reported [346]. Gladen [347] has made extensive studies on gel chromatography in organic media.

Gel chromatography also has been utilized to separate monomeric substances such as fats [348], fatty acids [349], adenine and other purines [350], iodo compounds [351], and flavonoids [352]. Aromatic hydrocarbons having five to eight benzene nuclei are isolated from petroleum distillates by gel chromatography [353]. A method to determine the glyceride dimer content of corn oil by means of gel permeation chromatography has been described [354]. Thin-layer gel chromatography has been employed for the analysis of proteins [355, 356], and the determination of their molecular weights [357].

VII. ELECTROPHORESIS

A. CHARACTERISTICS OF ELECTROPHORETIC SEPARATION

Electrophoresis [358] is a separation technique that operates through differences in the migratory velocity of the electrically charged particles in a liquid medium where they are exposed to an electric field. The efficiency of the method is dependent on the molecular electrical properties and the molecular weights of the migrating species, and on the properties of the meduim, such as its pH and the electric field. This technique, especially its two-dimensional version, has certain aspects in common with chromatography, especially TLC and paper chromatography. If the electric field operates simultaneously with some of the principles involved in chromatographic separation, a selectivity higher than that of chromatography alone is achieved [359]. This becomes more evident when the electric field operates perpendicular to, or against, the flow of a mobile liquid phase.

The characteristics of electrophoresis make it more useful for separating macromolecules than small molecules. The requirement of electrically charged particles limits electrophoretic separation to acids and bases in a medium where they exist in ionic forms (ionophoresis). Other classes of organic compounds, however, may become amenable to electrophoresis through suitable chemical transformations. For example, carbohydrates form strong anionic complexes [360] in a borex buffer, which can be subjected to electrophoretic migration. A reason for using electrophoresis to separate small molecules is their high polarity and hydrophilicity, and resulting poor affinity for lipophilic organic solvents. It should be recognized, however, that chromatography on ion exchangers (Section VIII) may be more convenient for this type of small molecules.

B. SUPPORTING MATERIALS FOR ELECTROPHORESIS

The electrophoresis of microsamples is performed on solid supports such as silica gel, cellulose, and gels. The reason for using the support is to decrease diffusion and facilitate manipulation. The considerations in selecting the stationary phase are several including (1) the molecular weights of the sample, (2) the ease of isolation of the species from individual zones of the electrophorogram, and (3) the stability of the stationary phase in the pH employed. For instance, silica gel cannot be used for compounds of high molecular weight, and it is chemically unstable in alkaline media. Agar gel [361] is useful for large molecules and is stable in any pH region, but the isolation of the separated substances is difficult. The recovery of the compounds after electrophoresis is most convenient from porous materials and powders that are saturated with the buffer solution during the experiment. Electrophoresis on filter paper is recommended for routine work, because cellulose is a compact material and possesses superb chemical and mechanical properties. Filter paper of high purity, chemically modified cellulose membranes (e.g., cellulose nitrate [362], cellulose acetate), and synthetic polymers of specific characteristics (e.g., polyacrylamide gel) are commercially available [324]. Flexible agarose gel support strips [363] have been advocated for immunoelectrophoresis.

C. SELECTING THE ELECTROPHORETIC SYSTEM

The electrophoretic system, which involves the applied voltage, solid support, and supporting electrolyte, is rather complex. It is necessary to consider all factors in order to select the correct experimental conditions. The choice of solid support is discussed above. The supporting electrolyte can be selected rationally by considering the buffer, pH, and ionic strength, while recognizing that the most efficient separation occurs between ions and neutral molecules, or between ions with opposite charges. Nevertheless, the efficient separation of compounds with small differences in dissociation constants and molecular weights has been achieved by electrophoresis under properly selected conditions. Isoelectric focusing [364, 365] is an improvement in the electrophoretic technique; the separation is performed in a medium characterized by a pH gradient. The combination of several factors operating simultaneously in such a method results in a substantial increase in separation power.

D. APPARATUS

The main components of an electrophoresis apparatus using supporting materials are a source of direct current, the electrodes, the anodic and cathodic chambers, and the supporting structure for the electrophoretic medium (e.g., glass plates). The assembly usually includes a cooling syssystem. When working with medium flow, there is also an arrangement for dispensing the electrolyte.

In low-voltage electrophoresis, the voltage ranges between 100 and 300 V dc. However, more efficient separation is achieved with 3000 to 10,000 V dc, which is required for separating mixtures of compounds with closely related structure. Apparatus operating at high voltage requires an efficient cooling device and safety precautions.

In the most common types of electrophoretic apparatus, a strip of paper, a film of gel, or a slurry in a buffer is sandwiched between two horizontal glass plates (see Fig. 4.36). Cooling is provided from top and bottom. The mechanical strength of filter paper is utilized by hanging the paper in the two chambers and holding the ends with glass rods.

Variations on the electrophoretic apparatus can be made in order to combine electrophoresis with some principles of chromatography. Thus, in countercurrent electrophoresis, the migrating species move against the gravity flow of the electrolyte. In the apparatus for continuous electrophoresis [366], the electric field is oriented perpendicular to the flow of the electrolyte. As illustrated in Fig. 4.37, the basic assembly is similar to that used in descending paper chromatography. The upper edge of the

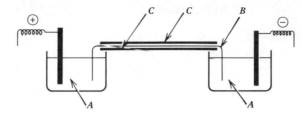

Fig. 4.36. Electrophoresis on paper strip: *A*, electrode compartment; *B*, paper strip; *C*, glass plates (support).

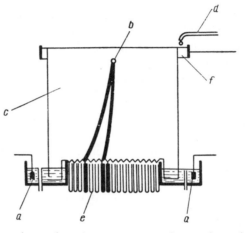

Fig. 4.37. Continuous electrophoresis on paper curtain: *a*, electrode compartments; *b*, spot for introduction of sample; *c*, paper with wedge-shaped cuts for collecting individual fractions; *d*, device for supplying trough with solvent; *e*, test tubes for collection of individual fractions; *f*, trough containing the solvent.

paper reaches into a trough containing the electrolyte. The lower end of the paper is cut in such manner so that the two narrow strips on the two sides are immersed in the electrolyte in the electrode chambers, while there are a number of V-shaped terminals to provide for the collection of separated fractions. The sample is introduced continuously at the center of the upper part of the paper curtain, and migrates down with the flow of the electrolyte. The electric field causes additional migration horizontally. As a result, the downward motion is not rectilinear.

E. OPERATION PROCEDURES

1. On Paper Strips

A strip of filter paper (moistened with the selected buffer) is placed in the operative position of the electrophoresis apparatus. After selecting the voltage, the system is equilibrated; $20\mu l$ of the solution containing the sample is applied at the starting line (25 mm long), which is equidistant from the two electrodes. Performing the experiment in a cold chamber (about 5°C) provides sufficient cooling in the low voltage system; this is recommended especially for working with polypeptides. An internal marker (e.g., caffeine, dyes [367]) can be spotted on the strip, and its migration indicates the movement of the liquid medium in the electric field. After electrophoresis, sections are cut out to detect the separated components by means of suitable reagents. Then the electrophorogram is cut into segments to recover each species.

2. On Starch Gel

The trough of the electrophoresis apparatus (designed for gel operation) is filled with a slurry of starch in the appropriate buffer. The sample (dissolved in water or buffer) is placed midway between the two electrodes. Owing to the high capacity of starch gel, volumes of about 1 ml can be introduced. The selected voltage is applied, and after a certain period, the progress of separation is observed as follows: The electrophoresis is temporarily stopped by disconnecting the power supply. The cover of the apparatus is opened, and a strip of filter paper is pressed lightly against the surface of the gel and removed to detect the positions of the separated species. The electrophoresis is resumed if the separation is insufficient. After the experiment, the starch slab is sliced and the respective segments are eluted.

F. APPLICATIONS AND MODIFICATIONS

The applications of electrophoresis have been compiled in monographs [368–370]. This technique is widely employed for analytical and preparative separation of proteins and polypeptides. It has also been used for separating simple compounds such as alkaloids [371], purines [372], and sugars [373].

Simple electrophoresis techniques recently published include thin-layer [374, 375] and sandwich disc [376] methods, and vertical electrophoresis on microscope slides [377]. Buchanan and Corfield [378] have designed an easily constructed device for two-dimensional paper electrophoresis. Vecerek and co-workers [379] have described a quantitative electrophoretic apparatus to separate up to five components in a 15-μl sample, and a multi-purpose cooling device [379a].

Fingerhut and Ortiz [380] have evaluated a system which produces 8 electrophorograms per strip of cellulose acetate and can be increased to 16 samples. Krawczyk [381] has studied ionophoresis of organic compounds in nonaqueous solvents. Strohl and Dunlop [382] have proposed an electrosorption method to separate quinones that is based on adsorption of the quinones on a column of graphite particles (to which a potential is applied) after passage of the sample through the column; thus the technique is related to electrophoresis.

VIII. ION-EXCHANGE METHODS

A. PRINCIPLE

Unlike the techniques discussed in the previous sections, separation by ion exchange [383, 384] methods is based primarily on reversible chemical reactions which occur between the species in the liquid phase and the ions associated with the solid phase. It involves the exchange, equivalent for equivalent, of mobile solvated ions at a macromolecular matrix (solid) for ions of the same charge in solution. The solid material can be organic or inorganic; it may be a synthetic polymer, natural substance (e.g., zeolite) or modifications thereof (e.g., modified cellulose [385]). The character of the macromolecular matrix is important since it controls the accessibility of the ions to the inner portions of the macromolecules (see Section VI.A). In general, when the number of cross-linkages in the polymer is large, the matrix becomes less accessible to large ions in the solution. This "sieving" (i.e., exclusion) phenomenon contributes to the selectivity of the separation process. Furthermore, the macromolecular structure is related to the capacity, swelling, and mechanical stability of the ion exchanger.

B. TYPES OF ION EXCHANGERS

The chemical properties of ion exchangers are governed by their functional groups. The common acidic groups are $-SO_3H$, $-COOH$, and $-OH$ (phenolic); basic groups are $-\overset{+}{N}R_3$, and $-NR_2$. The strongest acidic ion exchangers carry $-SO_3H$; strongest basic, $-\overset{+}{N}R_3$ (R = alkyl). Amphoteric [386] ion exchangers are those containing both acidic and basic groups. If the acidic and basic groups are so situated in the matrix that they can reach one another, then no additional factor is needed for desorption besides the liquid medium.

There is another type of ion exchangers that carries functional groups suitable for complexation [387]; most of them are polystyrene-iminodiacetate resins. Their selectivity is generally superior to that of the acid-base ion exchangers [388]. Resins with functionalities based on the well-established chelating agents such as 8-hydroxyquinoline [389] and resacetophenone oxime [390] are similar in nature. Polystyrene-polyhydric alcohol resin is selective for boric acid [391].

In general, the function of an ion exchanger is dependent on the formation of insoluble (or slightly ionizable) species. The particular ion is attached to the functional group which is connected to the macromolecular matrix through a covalent bond. However, the bonding of the ions to the ionic sites of the macromolecule are weak so as to make the replacement of one ion by another readily possible. In the competition of ions for the functional group of the ion exchanger, the number of charges and specific affinities of the ions, as well as the mass-action law, are involved. In common cases, the ions carrying more charges have greater affinity than those with fewer charges. In special cases, however, the specific affinities are the controlling factors. For example, filter paper impregnated with chlorophyll retains silver selectively [392]. Proposals have been made to characterize ion-exchange resins by their ir [393] or nmr [394] spectra, and by pyrolysis gas chromatography [395].

C. LIQUID MEDIA IN ION EXCHANGE

The common media for working with ion exchangers are water and its mixture with polar liquids (e.g., alcohols). However, water can be partly or completely replaced by organic solvents [396]. In such cases, water is replaced stepwise if the organic liquid is immiscible with water. For instance, water cannot be directly replaced by benzene, but the medium can be first replaced with *iso*-propyl alcohol. It should be noted that both the ion exchanger and ions change characters in organic media. In nonpolar liquids of low dielectric constants, the acid ion exchanger operates with a base primarily through hydrogen bonding instead of direct protonation [396a]. Thus, dialkyl sulfoxides are retained by sulfonic resin in benzene but can be desorbed quantitatively by ethanol (see Section F.2).

The composition of the liquid medium can be varied to effect separation of ionic species by shifting the dissociation equilbrium. In an aqueous-organic mixture (e.g., aq. alcohol, acetone, tetrahydrofuran), the composition reflects its dielectric constant and hence its ionization power. Utilizing this principle, $Fe(III)$, $Ni(II)$, and $Co(II)$ are separated on an anion exchanger in a medium of 80% acetone-20% $6M$ HCl. $Fe(III)$ runs through the column as $H^+FeCl_4^-$ aggregate with zero net charge; $Ni(II)$ and $Co(II)$ are trapped on the column as $NiCl_4^{2-}$ and $CoCl_4^{2-}$ respectively, and then eluted in this order by passing water through the column [397].

Oleophilic ion exchangers are designed for use in nonpolar solvents;

the special choice of functional groups permits proper swelling in organic liquids [398, 399]. Separation by ion exchange has been achieved in fused salts [400]; the preference of ions in these media is reverse of the order in water [401].

D. PHYSICAL PROPERTIES OF ION-EXCHANGE RESINS

Most commercial ion exchangers are well defined in chemical characteristics, uniform in size and shape (usually spherical), and kept in a swollen state. These conditions are essential for the efficiency and reproducibility of the separation. For instance, the diffusion into small resin particles has different characteristics as compared to the diffusion into large particles. This factor influences particularly the sharpness of the separation. The best separation is achieved by using extremely fine spherical particles (5–20 μm in diameter) for which high pressure is required to move the liquid medium [402]. The uniformity of size should be watched carefully when the ion exchangers are recycled many times, particularly for those which are not mechanically stable. Swelling changes may break the spherical particles into fragments. Freeman and co-workers [403] have studied the encapsulation of Na^+ or Ca^{2+} for the counter-ion mass range below 1 ng, and described procedures for preparing beads in pure counter-ion form and also for preventing subsequent ion exchange so that a virtually uncontaminated single bead is obtained.

The swelling of ion-exchanger particles is an important factor controlling its functions. It should be emphasized that complete loss of water (liquid) irreversibly alters the characteristics of most ion exchangers, and the internal portion of such macromolecular structure can never operate to the extent of the original state. Moreover, these particles become fragile and their life is shortened. Hence the ion exchangers are always kept in a liquid medium and the ion-exchange column should never be allowed to dry.

E. USES AND MODIFICATIONS

The uses of ion-exchange separation may be divided into two categories. In the first category, the species to be separated exhibit extreme difference in characteristics. This category includes the separation of anions from cations, of ionic species from neutral molecules (e.g., desalting), and the substitution of ions present in the solution with those fixed on the column under conditions of complete adsorption and desorption. These operations are known as nonchromatographic ion exchange. In contrast, the second category involves separation of ionic species of similar character (e.g., bases with close pK values) and is known as chromatographic ion exchange. For the latter purpose, the liquid medium should be selected so that its desorbing power is capable of distinguishing small differences, or the desorbing power of the eluent should be increased gradually. The pK of the individual species in the sample and pH of the liquid medium cooperate intimately in the separation. Sometimes ion exchange and sieving effects are combined, such as the action of macroporous gels [404, 405] and artificial zeolites that are inaccessible to large inorganic ions [406].

Ion-exchange methods of the nonchromatographic category are used extensively in microscale separations, such as purification of natural products [407], analysis of drugs [408], and isolation of polar constituents from petroleum [409]. It should be recognized that the ion exchangers are not inert solid supports, and unwanted chemical reactions may occur in the liquid medium. For instance, the strongly acidic or basic resins can cause ester hydrolysis [410] and aldol condensation [411]; sugars are bound irreversibly to ion exchangers having primary amino groups due to formation of Schiff's bases [412]. These side effects can be minimized by selecting the proper resins and liquid medium. Thus the hydrolysis of esters is prevented by using alcohol instead of water. Nauman [413] has reported that esterification occurs with certain amino acids in methanol medium but not in *tert*-butyl alcohol.

Ion exchangers have been employed as supporting material for performing spot tests. The sensitivity of the test is increased about tenfold owing to the accumulation of the sample and reagent on the resin bead [414]. Similarly, filter paper loaded with ion exchanger can extend the limit of detection [415]. Whereas ion exchange methods are generally used with solid or liquid samples, Chester and Juvet [292] have studied the anion-exchange GC for the elution of metal halides.

Chemical modification of the molecules to be separated can make them amenable to ion exchange. For example, glycols do not interact with acid-base ion exchangers. However, by converting 1,2-glycol into boric acid complexes which are strong acids, they are easily separated. This technique has been utilized for the separation of carbohydrates [416]. Another approach is to modify the resins chemically. Thus, by esterifying sulfonated resins in order to impart hydrophobic characteristics, Gorski and Moszcynska [417] were able to separate Zr and Hf as thiocyanates. Even the low solubility of certain salts does not restrict the use of ion exchange. For instance, calcium can be isolated by mixing its insoluble oxalate with chelating resins in the sodium form [418].

The lower limit of microsamples separable by ion exchange can be extended by the use of pellicular ion exchangers [419] which are made by depositing a film of the resin on glass beads packed in capillary columns

(1 mm×150 cm). The liquid medium is forced through under pressure (up to 200 atm.); as little as 10^{-12} mole of nucleotides has been processed by this technique [420]. Liquid ion exchangers (e.g., tri-caprylamine [421], (2-ethylhexyl)phosphoric acid [422]) with highly lipophilic properties also can be coated on inert stationary phase.

Automated ion-exchange chromatography is useful for routine separation of multicomponent mixtures. Leich and Langanke [423] have described a simple apparatus for analytical as well as preparative ion exchange. Among the commercial devices, those for amino acid separations are well established [424]. The important part of these machines is the detector of the fractions eluting from the column. Most detectors use u.v. spectrometry or refractive index, but monitors employing flame ionization are also available. Peterson and co-workers [425] have tested some simple introduction valves for the amino acid analyzer. Lange and Hempel [426] have used this instrument to fractionate phenolic acid, aldehydes, and alcohols.

F. OPERATION

1. Preparation of the Ion-Exchange Bed

Ion exchange can be performed batchwise (in a flask or test tube) or in a column. Columns of various designs are commercially available; some simple columns are shown in Fig. 4.23. For microsamples, two or more columns may be conveniently connected by interchangeable joints, as illustrated in Fig. 4.38. Thus, for desalting, one column is packed with cation exchanger and the other with anion exchanger. Long and narrow columns are recommended for chromatographic ion-exchange, with sintered glass disc at the bottom and capillary inlet and outlet. Smith [427] has described a plastic column to be used with solutions containing HF. As mentioned above, uniform particle size is essential for effective chromatographic ion exchange. If the resin is nonhomogeneous, the best way to differentiate the particles is by sedimentation (see Chapter 3, Section VI.B).

Conversion of the ion-exchange resin into the proper form (e.g., H^+ or Na^+ form for cation exchangers, OH^- or Cl^- form for anion exchangers) can be done before or after the resin is placed into the vessel or column. This operation involves complete saturation of the ion exchanger with the solution of the modifying reagent that is used in excess. This excess of reagent is removed by thorough washing with the pure solvent. Distilled water should be freshly boiled to expel carbon dioxide. The resin suspen-

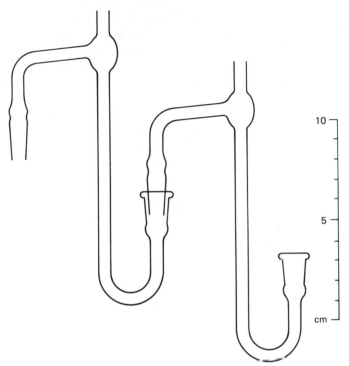

Fig. 4.38. Two ion-exchange columns connected by ground-glass joints.

sion should be deaerated by mild heating and evacuating just before being transferred into the column.

Operation on resin-loaded filter paper or modified cellulose paper is similar to that described for TLC and paper chromatography (see Section II), although the separation method is governed by the laws of ion exchange. While this technique is used mainly for analytical purpose, preparative separation of micro amounts of inorganic [428, 429] and organic [430, 431] compounds have been demonstrated.

2. Procedures For Non-Chromatographic Ion-Exchange

An example of separating basic substance from neutral or acidic compounds is given by the removal of 2,4-dinitrophenylhydrazine (slightly basic) from its corresponding hydrazone (slightly acidic). The procedure is as follows [432]: AG 50W-X4 cation exchanger (from Bio-Rad Laboratories) is slurried in water and transferred into the column, then treated sequentially with (a) 2 column volumes of M NaOH, (b) water until the effluent is neutral, (c) two-column volumes of M HCl, (d) water until the effluent is neutral, (e) four-column volumes of methanol, and (f) two-column volumes of purified methanol (carbonyl-free) : benzene (1:1). The hydrazine-hydrazone mixture is dissolved in the minimum of methanol : benzene (1:1) and transferred to the column. The resin is washed with the solvent until the effluent is colorless. One gram of resin retains about 100 mg of 2,4-dinitrophenylhydrazine. The resin cannot be regenerated.

A procedure in nonaqueous medium is illustrated by the separation of aliphatic sulfoxides from sulfides and sulfones as follows [396a]. A slurry of Dowex 50 (a strongly acidic cation exchanger) is transferred into the column and washed with (a) M NaOH, (b) water until the effluent is neutral, (c) M HCl, (d) water until the effluent is neutral, (e) ethanol, and (f) benzene. The crude sulfoxide is dissolved in a small volume of benzene and added to the column up to one-fifth of its capacity (recognized by the disappearance of transparency of the resin). The column is washed with benzene to remove sulfides and sulfones. The sulfoxides are recovered by elution with ethanol.

Nonchromatographic ion exchange is a powerful tool for separating trace metals from a large volume of liquid by filtration through highly selective ion exchangers. Thus using short columns of chelating resins, cobalt [433], molybdenum and vanadium [434], zinc and cadmium [435], thalium [436], gold [437] and silver [438] have been collected from sea water and other natural waters.

3. Procedures For Chromatographic Ion Exchange

Chromatographic ion exchange requires strict experimental conditions. In complicated cases, the operation is carried out with solvents of varying compositions. Gradient solvent system (pH gradient, salt gradient, solvent gradient) are frequently employed; the gradient can be discontinuous or continuous. For example, continuous pH gradient is prepared by continuously mixing two buffers with different pH values. Simple as well as sophisticated devices are available [324].

Separation of the rare earth elements has been carried out by ion exchange. For instance, scandium and ytterbium are separated by means of a strongly basic anion exchanger in acidic medium in a solvent of low dielectric constant which controls the ion-pair formation. The procedure follows [439]: The column (1-cm i.d.) is packed with Dowex 1-X8 (Cl⁻ form, 100–200 mesh) and washed with (a) $6M$ HNO₃ until test for Cl⁻ is negative, (b) water, and (c) methanol. The solution containing 0.5 mg of Sc and 0.6 mg of Yb is transferred to the column, which is eluted with a solvent mixture of tetrahydrofuran: $1.2M$ HNO$_3$ (90:10 v/v). Aliquots of 10 ml are collected. The result is shown in Fig. 4.39.

Geometric isomers of cobalt(III) iminodiacetate are separated on anion exchanger in the following manner [440]: Amerlite CG-400 (Cl⁻ form, 100–200 mesh) is transferred into a glass tubing (16-mm i.d., 10 cm in height) with a drip tip, and washed with water. A solution containing 25 mg of the purple isomer and 40 mg of the brown isomer in 5 ml of water is added to the column. After washing with 5 ml of water, the system is eluted with 0.1 M NaCl at the rate of 1 drop/20 sec. Visible zones migrate through the column within 2 hr.

Separation of 10-μg amounts of aromatic sulfonic acids has been described by Stehl [441]. The anion exchanger used is the polyalkylencamine type, with which the ion exchange process is not interfered by adsorption that occurs in styrene resins. For elution a mixture of water and organic liquids is employed, and the separation is enhanced by means of a salt gradient. The chromatography is performed in a sophisticated liquid chromatograph equipped with a u.v. detector and recorder. The Bio-Rex 5 resin (Cl⁻ form, 270–325 mesh) is washed with 4 HCl, water, and methanol. The column is 2 mm × 50 cm, and the flow rate is 1.0 ml/min. Fig. 4.40 shows the graph of a 10-μl solution containing 10 μg of orthanilic acid and 20 μg of dimethylorthanilic acid separated in the liquid medium of water:acetonitrile:methanol (1:1:1).

Additional examples of ion-exchange chromatography can be found in

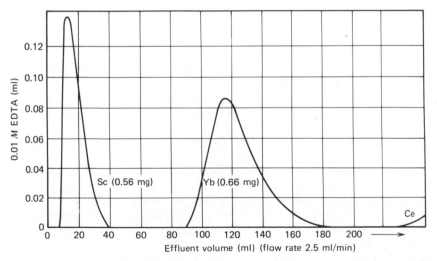

Fig. 4.39. Separation of scandium and ytterbium by ion-exchange chromatography [439]. (Courtesy *Mikrochim. Acta.*)

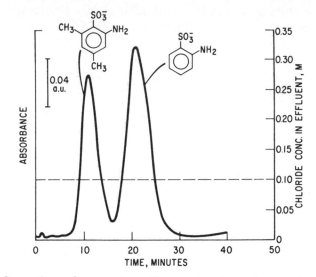

Fig. 4.40. Separation of aromatic sulfonates by ion-exchange chromatography [441]. (Courtesy *Anal. Chem.*)

the separation of carbohydrates [442, 443], nucleotides [444], alkali metals [445], dicarboxylic and tricarboxylic acids [446], benzene derivatives by anion exchanger in butanol [447], alkyl sulfates by salting-out technique [448], and alcohols on iron (III)-loaded cation exchanger [449].

REFERENCES

1. F. G. Helferich and G. Klein, *Multicomponent Chromatography.* Dekker, New York, 1970.
2. E. Heftmann, *Chromatography,* 2nd ed. Reinhold, New York, 1967.
3. E. Lederer and M. Lederer, *Chromatography,* 2nd ed., Elsevier, Amsterdam, 1957.
4. O. Mikes (Ed.), *Laboratory Handbook of Chromatographic Methods.* Van Nostrand, London, 1966.
5. R. Stock and C. B. F. Rice, *Chromatographic Methods*, 2nd ed., Chapman and Hall, London, 1967.
6. W. E. Harris, *J.Chromatogr. Sci.,* **11**, 184 (1973).
7. D. L. Camin and A. J. Raymond, reported at the Anniversary Symposium on Chromatography, 164th ACS National Meeting, New York, 1972.
8. E. Heftmann, *J. Chromatogr. Sci.,* **11**, 295 (1973).
9. E. Lederer, *Chromatogr. Rev.,* **16**, 361 (1972).
10. A. J. P. Martin, reported at the Anniversary Symposium on Chromatography, 164th ACS National Meeting, New York, 1972.

11. Y. Ito and R. L. Bowman, *J. Chromatogr. Sci.*, **11**, 284 (1973); *Anal Chem.*, **43**, 69A (1972).

12. R. L. Anackler, J. H. Simmons, and E. Ribi, *J. Chromatogr.*, **62**, 93 (1971).

13. A. C. Lanser, J. O. Ernest, W. F. Kwolek, and H. J. Dutton, *Anal. Chem.*, **45**, 2344 (1973).

14. F. W. Karasek, and D. M. Kane, *J. Chromatogr. Sci.*, **10**, 673 (1972).

15. P. Cuatrecasas and I. Parikh, *Biochemistry*, **11**, 2291 (1972).

15a. H. W. Weetall, *Sep. Purif. Methods*, **2**, 199 (1973).

16. W. Lautenschlager, S. Pahlke, and G. Tolg, *Z. Anal. Chem.*, **260**, 203 1972).

17. R. Kaiser, *Chromatographia*, **3**, 38 (1970).

18. S. Turina and V. Jamnicki, *Anal. Chem.*, **44**, 1892 (1972).

19. R. A. Rimerman and G. W. Hatfield, *Science*, **182**, 1268 (1973).

19a. H. P. Lenk and H. Gruber, *Mikrochim. Acta*, **1972**, 646.

20. N. Haber, *Chem. Eng. News*, Oct. 11, 1971, p. 23.

21. R. A. Keller, *J. Chromatogr. Sci.*, **11**, 49 (1973).

22. A. J. P. Martin, *Endeavor*, **6**, 21 (1947).

23. G. Rouser, *J. Chromatogr. Sci.*, **11**, **60** (1973).

24. R. E. Majors, *J. Chromatogr. Sci.*, **11**, 88 (1973).

25. L. S. Ettre, *Am. Lab.*, Dec., **1970**, p. 28.

26. J. K. Haken, *J. Chromatogr. Sci.*, **11**, 144 (1973).

27. V. Maynard and E. Grushka, *Anal Chem.*, **44**, 1427 (1972).

28. R. S. Swingle and L. B. Rogers, *Anal Chem.*, **44**, 1415 (1972).

29. D. MacNaughtan, Jr., L. B. Rogers and G. Wernimont, *Anal. Chem.*, **44**, 1421 (1972).

30. T. J. Farrell and R. L. Pecsok, *J. Assoc. Off. Anal. Chem.*, **52**, 999 (1969).

31. D. W. Underhill, J. A. Reeds, and R. Bogen, *Anal. Chem.*, **45**, 2314 (1973).

32. B. E. Bowen, S. P. Cram, J. E. Leitner, and R. L. Wade, *Anal. Chem.*, **45**, 2185 (1973).

33. Twelfth Annual Meeting on The Practice of Chromatography, presented by Committee E-19 of the American Society for Testing and Materials, Oct. 21–24, 1973, San Diego, Calif.

34. J. Polesuk and D. G. Howery, *J. Chromatogr. Sci.*, **11**, 226 (1973).

35. A. T. Blades, *J. Chromatogr. Sci.*, **11**, 251 (1973).

36. W. A. Aue and S. Kapila, *J. Chromatogr. Sci.*, **11**, 255 (1973).

37. R. J. Hesselberg and J. L. Johnson, *Bull. Environ. Contam. Toxicol.*, **7**, 115 (1972).

38. C. R. Brewington, D. P. Schwartz, and M. J. Pollansch, *Proc. Soc. Exp. Biol. Med.*, **139**, 745 (1972).

39. J. S. Fritz and G. L. Latwesen, *Talanta*, **17**, 81 (1970).

40. S. P. Cram and S. N. Chester, *J. Chromatogr. Sci.* **11**, 391, (1973).

41. R. Coneden, A. H. Gordon, and A. J. P. Martin, *Biochem. J.*, **38**, 224 (1944).

42. A. J. P. Martin and R. L. M. Synge, *Biochem. J.,* **35,** 1358 (1941).
42a. T. Smith, and J. G. Feinberg, *Paper and Thin-Layer Chromatography and Electrophoresis,* 2nd ed., Longman, London, 1972.
43. J. Petit, J. A. Berger, J. L. Chabard, G. Besse, and V. Voissier, *Bull. Soc. Chim. Fr.,* **1969,** 1027.
44. J. Petit, J. A. Berger, G. Gaillard, and G. Meguiel, *J. Chromatogr.,* **39,** 167 (1969).
45. E. A. Morris, *Lab. Prac.,* **16,** 37 (1967).
46. A. Grüne, *Fette Seifen Anstrichm.,* **69,** 916 (1967).
47. H. Jork, *Pharm. Int., Engl. Edn.,* **1968,** 11.
48. C. G. Honneger, *Helv. Chim. Acta,* **45,** 1409 (1962); **46,** 1772 (1963).
48a. Kontes Glass Co., Vineland, N. J., cat. Nos. K–416000 and K–416050.
49. Scientific Manufacturing Industries, Emeryville, Calif.
50. The Lab Apparatus Co., Cranwood, Cleveland, Ohio.
50a. D. Herbert, R. E. A. Gadd, and S. Clayman, *Lab. Pract.,* **32,** 22 (1973).
51. R. R. Lowry, *Chem.-Anal.,* **55,** 119 (1966).
52. T. S. Ma and R. F. Sweeney, *Mikrochim. Acta,* **1956,** 191.
53. S. Prydz, *Anal. Chem.,* **45,** 2317 (1973).
54. K. Dohmann, *Lab Prac.,* **14,** 808 (1965).
55. P. Wollenweber, *Lab. Pract.,* **13,** 1194 (1964).
56. M. L. Wolfrom, D. L. Patin, and R. M. de Ledercremer, *J., Chromatogr.,* **17,** 488 (1968).
57. R. D. Houghton, *J. Chromatogr.,* **24,** 494 (1965).
58. R. Schaad, *Microchem. J.,* **15,** 208 (1970).
59. I. S. Taylor, *J. Chromatogr.,* **37,** 120 (1968).
60. T. Okumura, T. Kadone, and M. Nakatani, *J. Chromatogr.,* **74,** 73 (1972).
61. H. Endres and Hörmann, *Angew, Chem., Int. Ed.,* **2,** 254 (1953).
62. L. Hörhammer, *Polyamide Chromatography in Polyphenal Chemistry,* Pergamon Oxford, 1964.
63. G. Hesse, H. Engelhardt, and R. Kaltwasser, *Chromatographia,* **1,** 302 (1968).
64. K. Randenrath, *Angrew. Chem.,* **73,** 674 (1961).
65. G. Duncan, L. Kitching, and R. J. T. Graham, *J. Chromatogr.,* **47,** 232 (1970).
66. B. J. Radola, *J. Chromatogr.,* **38,** 61, 78 (1968).
66a. H. Thielemann, Mikrochim. Acta, **1972,** 672.
67. E. Stahl, *Pharm. Rundsch.,* **1,** (2), 1 (1959).
67a. D. L. Massart and H. De Clercq, *Anal. Chem.* **46,** 1988 (1974).
68. J. Jacques and J. P. Mathieu, *Bull. Soc. Chim. F.,* **1946,** 94.
69. S. Hermanek, V. Schwartz, and Z Cekan, *Collect. Czech. Chem. Commun.,* **28,** 2031 (1963).
70. M. Stuchlik, I. Csiba, and L. Krasnec, *Czech. Farm.,* **18,** 91 (1969); *J. Chromatogr.,* **30,** 543 (1967); **36,** 522 (1968).
71. E. Röder, E. Mutschler, and H. Rochelmeyer, *Arch. Pharm. Ber.,* **301,** 624 (1968).

72. S. Hermanek, V. Schwartz, and Z. Cekan, *Collect. Czech. Chem. Commun.*, **26**, 3170 (1961).

73. A. Waksmundzki and J. K. Rozylo, *Chem. Anal.*, **17**, 1079 (1972).

74. Boleb Inc., Derry, N. H.

75. A. K. Munson, J. R. Mueller, and M. E. Yannone, *Microchem. J.*, **15**, 95 (1970).

76. G. Kasang and H. Rembold, *J. Chromatogr.*, **71**, 101 (1972).

77. M. H. Coleman, *Lab. Pract.*, **13**, 1200 (1964).

78. F. Musil and E. Fossilien, *J. Chromatogr.*, **47**, 116 (1970).

79. E. Stahl and E. Dumont, *J. Chromatogr.*, **39**, 157 (1969).

80. E. Stahl, *J. Chromatogr.*, **37**, 99 (1968); *Analyst*, **94**, 723 (1969).

81. M. F. Bacon, *Chem. Ind.*, **1965**, 1692.

82. B. A. Bessonov and Y. S. Zvenigorodskii, *Zavod, Lub.*, **34**, 157 (1968).

83. C. B. Mankinen and R. M. Sachs, *J. Chromatogr.*, **47**, 509 (1970).

84. Applied Science Laboratories, Inc., State College, Pa.

85. Cordis Laboratories, Miami, Fla.

86. S. Samuels, *Chem.-Anal.*, **54**, 122 (1965).

87. Anti-oxidation spotting chamber, Kensco, Emeryville, Calif.

88. H. Halpaap, *Chem. Ing. Tech.*, **35**, 488 (1963).

89. A. Pittoni and P. L. Scessi, *J. Chromatogr.*, **32**, 422 (1968).

89a. R. A. deZeeuw, *Anal. Chem.*, **40**, 2134 (1968); *J. Chromatogr.*, **32**, 43 (1968); **33**, 222 (1968).

90. M. H. Stutz, *Anal. Chem.*, **40**, 258 (1968).

90a. J. R. Perry, K. W. Haag, and L. J. Glunz, *J. Chromatogr., Sci.*, **11**, (1973); apparatus by Regis Chemical Co., Morton Grove. Ill.

91. Laboratory Supplies Co., Hicksville, N. Y.

92. P. M. Houpt, *X-Ray Spectrom.*, **1**, 37 (1972).

93. M. Bounias, *Chim. Analyt.*, **51**, 76 (1969).

94. Auto-zonal plate scraper, Isolab Inc., Akron, Ohio.

95. M. S. Shraiber, *Chromatogr., Rev.*, **16**, 367 (1972).

96. J. G. Kirschner, *J. Chromatogr. Sci.*, **11**, 180 (1973).

97. E. Stahl (Ed.), *Dünsichicht-Chromatographie, Ein Laboratoriums–Handbuch*, 2nd ed., Springer, Berlin, 1969. *Thin-layer Chromatography* (translated from German). Springer, New York, 1970.

98. J. G. Kirschner, *Thin-layer Chromatography*. Wiley-Interscience, New York, 1967.

99. G. Vernin, *Thin-layer Chromatography; Techniques and Application in Organic Chemistry* (translated from French). Dunod, Paris, 1970.

100. K. Randerath, *Thin-layer Chromatography* (translated from German), 2nd ed., Academic, New York, 1966.

101. A. Niederwieser and G. Pataki (Eds.), *Progress in Thin-Layer Chromatography and Related Methods*. Humphrey, Ann Arbor, Mich., Vol. 1, 1970; Vol. 2, 1972.

102. B. J. Haywood, *Thin-layer Chromatography, and Annotated Bibliography: 1964–1968*. Ann Arbor Science Publishing, Ann Arbor, Mich., 1968.

102a. *Camac's The Cumulative Bibliography,* Vol. III (1970–1973). Camac Inc., New Berlin, Wis., 1974.

103. D. F. G. Pusey, *Chem. Br.,* **5**, 408 (1969).

104. R. A. DeZeeuw, *Crit. Rev. Anal. Chem.,* Vol. 1, p. 119. Chemical Rubber Co., Cleveland, Ohio, 1970.

105. R. M. Scott, *J. Chromatogr. Sci.,* **11**, 129 (1973).

106. A. Waksmundzki and J. K. Rozylo, *Chem. Anal.,* **16**, 277, 283, 291 (1971).

107. M. Vanhaelen, *Ann. Pharm. Fr.,* **26**, 565 (1968).

108. R. J. Hurtubise, P. F. Lott, and J. D. Dias, *J. Chromatogr.,* **11**, 476 (1973).

109. G. Pataki, *Dünschicht-Chromatographie in der Aminosäure– und Peptid-Chemie.* Walter de Gruyter, Berlin, 1966.

110. G. Steinke and F. Schmidt, *Dtsch. Apothekztg.* **110**, 1787 (1970); **109**, 1489 (1969).

111. R. A. Eigl, *Dtsch. Apotheztg.,* **111**, 987 (1970).

112. S. Goenechea, *Z. Anal. Chem.,* **225**, 30 (1967).

113. J. Bujna and F. Machovicova, *Farm. Obz.,* **41**, 349 (1972).

114. P. E. Haywood and M. S. Moss, *Analyst,* **93**, 737 (1968).

115. J. Polesuk, J. M. Amodeo, and T. S. Ma., *Mikrochim. Acta,* **1973**, 507.

116. D. P. Schwartz and C. R. Brewington, *Michrochem. J.,* **12**, 1 (1967); **13**, 30 (1968); **14**, 556 (1969).

117. P. E. Kolattukudy, *Analabs, Inc., Res. Notes,* **13**, (2), 1, (1973).

118. D. P. Schwartz and A. I. Virtanen, *Acta Chem. Scand.,* **22**, 1717 (1968).

119. T. S. Ma et al, *Mikrochim, Acta,* **1965**, 1098; **1967**, 960; **1968**, 436; **1969**, 352 815; **1970**, 677; **1971**, 267, 662; **1972**, 313.

120. A. Fono, A Sapse, and T. S. Ma., *Mikrochim, Acta,* **1965**, 1100.

121. J. Polesuk and T. S. Ma, *J. Chromatogr.,* **57**, 315 (1971).

122. B. J. Marcus, A. Fono and T. S. Ma., *Mikrochim, Acta,* **1967**, 962.

123. V. Novacek, *Am. Lab.,* Dec., **1969**.

124. J. S. Chahi and C. C. Kratzing, *Clin. Chim. Acta,* **26**, 177 (1969).

125. S. Samuels and C. Fisher, *J. Chromatogr.,* **71**, 297 (1972)

126. E. C. Kyriakides and J. A. Balint, *J. Lapid Res.,* **9**, 142 (1968).

127. R. de Medeiros and W. K. Simmons, *Microchem. J.,* **18**, 449 (1973).

127a. M. H. Stutz, W. D. Ludemann, and S. Sass, *Anal. Chem.,* **40**, 258 (1968).

127b. O. Lerch and J. G. Moffatt, *J. Orig. Chem.,* **36**, 3396 (1971).

128. G. Zweig and J. Sherma, *J. Chromatogr.,* **11**, 279 (1973), *Paper Chromatography,* Vol. 2. Academic, New York, 1971.

129. R. J. Block, E. L. Durrum, and G. Zweig, *Paper Chromatography and Paper Electrophoresis,* 2nd ed., Academic, New York, 1958.

130. I. M. Hais and K. Macek, *Handbuch der Papierchromatographie* (3 vols.). VEB Gustav Fischer, Jena, 1963.

130a. K. Macek, *Pharmaceutical Applications of Thin-Layer and Paper Chromatography.* Elsevier, Amsterdam, 1972.

131. L. Reio, *J. Chromatogr.,* **68**, 183 (1972).

132. E. Pertot and M. Blinc, *Staerke,* **24**, 260 (1972).

133. J. L. Hirshman and T. S. Ma,*Mikrochim. Acta,* **1968**, 262.

134. P. S. Gupta, A. Mondal, and S. N. Mitra, *J. Inst. Chem. India,* **44**, 49 (1972).

135. J. A. Levine, *J. Chromatogr.,* **67**, 129 (1972).

136. C. L. de Ligny and N. G. v. d. Veen, *J. Chromatogr.,* **71**, 307 (1972).

137. C. B. Mankinen and D. Fischer, *J. Chromatogr.,* **36**, 112 (1968).

138. B. V. Milborrow, *J. Chromatogr.,* **20**, 180 (1965).

139. R. de Wachter, *J. Chromatogr.,* **36**, 109 (1968).

140. I. E. Bush and K. Crowshaw, *J. Chromatogr.,* **19**, 114 (1965).

141. T. S. Ma and R. Roper, *Mikrochim. Acta,* **1968**, 169.

142. P. Jeroschewski, T. X. Gian, and H. Berge, *Chem. Anal.,* **17**, 503 (1972).

143. I. E. Bush, *J. Chromatogr.,* **29**, 157 (1967).

144. L. B. Rockland, *J. Chromatogr.,* **16**, 547 (1964).

145. Spin thimbles, Reeves Angel, Clifton, N. J.

146. Z. A. Shevchenko and I. A. Favorskaya, *Vestn. Leningr. Gos. Univ., Ser. Fiz. Khim.,* **1964**, 148.

147. J. H. P. Tyman and A. Higdon, *Lab. Pract.* **21**, 559 (1972).

148. A. V. DeThomas and M. Zdankiewicz, *Microchem. J.,* **15**, 74 (1970).

149. A. V. DeThomas, M. Zdankiewicz, and D. Witkovic, *Microchem. J.,* **15**, 71 (1970).

150. A. V. DeThomas, C. R. DeThomas, R. Lazar, and D. Verrastro, *Microchem. J.,* **16**, 52 (1971).

151. T. T. Martin, *Lab. Pract.* **21**, 732 (1972).

152. D. Edgar, *J. Chromatogr.,* **43**, 271 (1969).

153. H. P. Lenk and H. Gruber, *Mikrochim. Acta,* **1972**, 646.

154. A. Lepoivre, *Bull. Soc. Chim. Belg.,* **81**, 213 (1972).

155. M. H. Hashmi, M. A. Chaudhry, N. A. Chugtai, and R. U. Rehman, *Mikrochim. Acta,* **1970**, 200.

156. M. E. Hutzul and G. F. Wright, *Can. J. Pharm. Sci.,* **3**, 4 (1958).

157. E.A. Mistryukov, *Collect. Czech. Chem. Commun.,* **26**, 2072 (1961).

158. J. A. Berger, G. Meymil, and J. Petit, *Bull. Soc. Chim. Fr.,* **1964**, 3179.

159. A. Niederwieser, *Chromatographia,* **2**, 23 (1969).

160. A. L. Misra, R. B. Pontami, and S. J. Mule, *J. Chromatogr.,* **71**, 554 1972).

161. S. Turina, V. Marjanovic-Krajovan, and Z. Soljic, *Anal. Chem.,* **40**, 71, (1968).

162. S. Sandroni and H. Schlitt, *J. Chromatogr.,* **52**, 169 (1970).

163. Vario-KS-chamber, Camac Inc., New Berlin, Wis.

164. M. Miroslawa, *Chem. Anal.,* **17**, 1139 (1972).

165. J. A. Perry, K. W. Haag, and L. J. Gwunz, *J. Chromatogr. Sci.,* **11**, 44 (1973); *Chem. Eng. News,* Aug. 13, 1973, p. 13.

166. *Chem. Eng. News,* Mar. 20, 1972, p. 42.

167. B. Loev and K. M. Snader, *Chem. Ind.,* **1965**, 15.

168. B. Loev and M. G. Goodman, *Chem. Ind.,* **1967**, 2026; *Intra-Sci. Chem. Rep.* **4**, 283 (1970).

169. A. C. Casey, *J. Lipid Res.,* **10**, 456 (1969).

170. D. P. Schwartz, private communication.
171. Walter Coles Co., London, England; "Unibar," Analtech, Inc., Wilmington, Del.
172. H. E. Hodd and E. Caspi, *J. Chromatogr.,* **71**, 353 (1972).
173. J. L. Waters, *J. Chromatogr. Sci.,* **9**, 428 (1971).
173a. Z. Dayl, J. Rosmus, M. Juricova, and J. Kopecky, *Bibliography of Column Chromatography 1967–1970,* Elsevier, New York, 1973.
174. J. L. Waters and L. James, *J. Chromatogr.,* **55**, 213 (1971).
175. Laboratory Data Cnotrol, Riviera Beach, Fla.
176. Rapid fraction evaporator, RHO Scientific, Inc., Commack, N. Y.
177. Fisher Scientific Co., Pittsburgh, Pa.
178. Pierce Chemical Co., Rockford, Ill.
179. L. Schutte, *J. Chromatogr.,* **72**, 303 (1972).
179a. C. Liteanu and T. Hodisan, *Stud. Cercet. Chim.,* **21**, 857 (1973).
180. V. Holeysovsky and F. Sorm, *Collect. Czech. Chem. Commun.,* **20**, 586 1955).
181. R. L. M. Synge, *Biochem. J.,* **38**, 285 (1944).
182. L. A. Boggs, L. S. Cuendet, T. Higuchi, and N. C. Hill, *Anal. Chem.,* **24**, 491 (1952).
183. W. R. Ali and P. T. Laurence, *Anal. Chem.,* **45**, 2426 (1973).
184. O. Knessl and B. Keil, *Chem. Listy,* **45**, 145 (1951).
185. J. J. Kirkland, *Modern Practice of Liquid Chromatography.* Wiley-Interscience, New York, 1971.
185a. P. M. Rajcsanji and L. Otvos, *Sep. Purif. Methods,* **2**, 361 (1973).
186. S. G. Perry, R. Amos, and P. I. Brewer, *Practical Liquid Chromatography,* Plenum, New York, 1972.
187. J. J. Kirkland, *Anal Chem.,* **43**, (12), 37A (1971).
188. D. C. Locke, *J. Chromatogr. Sci.,* **11**, 120 (1973).
189. C. D. Scott, W. F. Johnson, and V. E. Walker, *Anal. Biochem.,* **32**, 182 (1969).
190. I. Halasz, A. Kroneisen, H. O. Gerlach, and P. Walking, *Z. Anal. Chem.,* **234**, 81 (1968).
191. J. B. Fox, Jr., R. C. Calhoun, and W. J. Eglinton, *J. Chromatgr.,* **43**, 48, 55 (1969).
192. B. A. Hills and R.B. Payne, *J. Chromatogr.,* **40**, 171 (1969).
193. J. G. Davis, *J. Chromatogr.,* **40**, 169 (1969).
193a. C. R. Flynn and J. Michl, *J. Amer. Chem. Soc.,* **36**, 3287 (1974).
194. R. P. W. Scott and P. Kucera, *J. Chromatogr.,* **11**, 83 (1973).
195. J. C. Hegenauer, K. D. Tartof, and G. W. Nace, *Anal. Biochem.,* **13**, 6 (1965).
196. W. Schmidtmann, *Chem. Ztg.—Chem. Appar.,* **89**, 231 (1965).
197. D. E. Hirsch, R. L. Hopkins, H. L. Coleman, F. O. Cotton, and C. J. Thompson, *Anal. Chem.,* **44**, 915 (1972).
198. M. Popl, V. Dolansky, and J. Mostecky, *Anal. Chem.,* **44**, 2082 (1972).
199. T. J. Farrel, *J. Assoc. Off. Anal. Chem.,* **52**, 999 (1969).
200. W. Kochen and G. Laver, *Chromatographia,* **2**, 213 (1969).

201. G. Zweig and J. Sherma (Eds.), *CRC Handbook of Chromatography*, (2 vols.).Chemical Rubber Co., Cleveland, Ohio, 1972.

201a. C. E. H. Knapman and R. J. Maggs (Eds.), *Gas and Liquid Chromatography Abstracts*. Applied Science Publishers, Essex, England, 1973.

202. D. P. Schwartz, C. R. Brewington, J. Shamey, J. L. Weilhrauch, and O. W. Parks, *Microchem. J.*, **12**, 186, 547 (1967); **13**, 310, 407 (1968); **17**, 302 (1972).

203. C. R. Brewington, E. A. Caress, and D. P. Schwartz, *J. Lipid Res.* **11**, 355 (1970).

204. E. Smith, *J. Assoc. Anal. Chem.*, **53**, 603 (1970).

205. I. A. Sokolova, E. S. Boichinova, A. M. Grebzde, V. M. Potekhin, and V. A. Proskuryakov, *Zh. Prikl. Khim. Leningr.*, 45, 1985 (1972).

206. D. P. Schwartz, *Agric. Sci. Rev.*, **8**, 41 (1970).

207. D. P. Schwartz, J. L. Weihrauch, and C. R. Brewington, *Microchem. J.*, **14**, 597 (1969); **17**, 63, 234, 677 (1972); **18**, 249 (1973).

208. D. P. Schwartz, C. R. Brewington, and L. H. Burgwald, *J. Lipid Res.* **8**, 54 (1967).

209. J. E. Evans, *Anal. Chem.*, **45**, 2428 (1973).

210. R. C. Williams, J. A. Schmidt, and R. A. Henry, *J. Chromatogr. Sci.*, **10**, 494 (1972).

211. F. A. Fitzpatrick and S. Siggia, *Anal. Chem.*, **45**, 2310 (1973).

212. M. A. Carey and H. E. Persinger, *J. Chromatogr. Sci.*, **10**, 537 (1972).

213. C. D. Scott, D. D. Chilcote, S. Katz, and W. W. Pitt, Jr., *J. Chromatgr. Sci.*, **11**, 96 (1973).

214. J. A. Schmidt, R. A. Henry, R. C. Williams, and J. F. Dieckman, *J. Chromatogr. Sci.*, 9, 645 (1971).

215. R. Stillman and T. S. Ma., *Mikrochim. Acta*, **1973**, 491.

216. C. D. Chandler and H. M. McNair, *J. Chromatogr. Sci.*, **11**, 468 (1973).

217. J. N. Done and J. H. Knox, *Process Biochem.*, **7**, (9), 11 (1972).

218. J. J. Kirkland, *Anal. Chem.*, **43**, (12), 36A (1971).

219. J. J. Kirkland, *J. Chromatogr. Sci.*, **10**, 593 (1972).

220. B. Pearce and W. L. Thomas, *Anal. Chem.*, **44**, 1107 (1972).

221. A. F. Machin, C. R. Morris, and M. P. Quick, *J. Chromatogr.*, **72**, 388 (1972).

222. R. M. Cassidy and R. W. Frei, *J. Chromatogr.*, **72**, 293 (1972).

223. R. P. W. Scott and P. Kucera, *J. Chromatogr. Sci.*, 9, 641 (1971).

224. A. Littlewood, *Gas Chromatography*. 2nd ed., Academic, N. Y., 1970.

225. J. H. Purnell (Ed.), *Progress in Gas Chromatography*. Wiley, New York, 1968.

226. H. M. McNair and E. J. Bonelli, *Basic Gas Chromatography*. Consolidated Printers, Oakland, 1965.

227. Carle Instrument, Inc., Fullerton, Calif.

228. Hewlett-Packard, Avondale, Pa.

229. H. M. McNair and C. D. Chandler, *J. Chromatogr.*, **11**, 454 (1973).

230. M. Verzele, *J. Chromatogr.*, **15**, 482 (1964).

231. Chemical Research Services, Inc., Addison, Ill.

232. D. J. Malcolme-Lawes and D. S. Ureh, *J. Chromatogr.*, **44**, 609 (1969).
233. M. Singliar and J. Bricha, *Chem. Ind.*, **1960**, 225.
234. M. Verzele, *J. Chromatogr.*, **9**, 116 (1962).
235. T. R. Lyn, C. L. Hoffman, and M. M. Austin, *Guide to Stationary Phases for Gas Chromatography*. Analabs, Inc., Hamden, Conn., 1968.
236. *A Guide to Selected Liquid Phases and Adsorbents used in Gas Chromatography*, Chromatogr. Assoc., New Castle, Del. 1972.
237. D. M. Ottenstein, *J. Chromatogr. Sci.*, **11**, 136 (1973).
238. R. Stern and E. R. Atkinson, *Chem. Ind.*, **1962**, 1758.
239. C. H. Hartman, *Anal. Chem.*, **43**, (2), 113A (1971).
240. T. Todd and D. DeBord, *Am. Lab.*, Dec., 1970, p. 56.
241. Y. I. Kholkin and G. S. Gridynshka, *J. Chromatogr.*, **53**, 354 (1970).
242. D. A. Cronin, *J. Chromatogr.*, **52**, 375 (1970).
243. H. Gopier and L. Schutte, *J. Chromatogr.*, **47**, 464 (1970).
244. J. C. Giddings, *Dynamics of Chromatography*. Dekker, New York, 1965.
245. L. S. Ette, *Am. Lab.*, Dec., 1970, p. 28.
246. J. E. Oberholtzer and L. B. Rogers, *Anal. Chem.*, **41**, 1590 (1969).
247. B. A. Bierl, M. Beroza, and M. H. Aldridge, *J. Chromatogr. Sci.*, **10**, 712 (1972).
248. R. N. Martynuk and M. S. Vigdergauz, *Izv. Akad. Nauk S.S.S.R., Ser. Khim.*, **1972**, 1173.
249. W. Ecking and E. Lanz, *J. Chromatogr.*, **64**, 7 (1972).
250. J. Q. Walker and C. J. Wolf., *Anal. Chem.*, **45**, 2263 (1973).
251. H. Groenendijk and A. W. C. van Kemenade, *Chromatographia*, **1**, 472 (1968).
252. M. M. Schachter and T. S. Ma, *Mikrochim. Acta*, **1966**, 65.
253. J. Crossley, *J. Chromatogr. Sci.*, **8**, 426 (1970).
254. D. Fielder and D. L. Williams, *Anal. Chem.*, **45**, 2304 (1973).
255. D. F. Cadogan and D. T. Sawyer, *Anal. Chem.*, **42**, 190 (1970).
256. W. D. Ross and R. T. Jefferson, reported at the 6th International Symposium on Advances in Chromatography, Miami Beach, Fla., June, 1970.
257. R. G. Ackman, *J. Chromatogr., Sci.*, **10**, 560 (1970).
258. J. Jowitt, *Chem. Ind.*, **1968**, 683.
259. J. R. Conder, *Anal. Chem.*, **43**, 367 (1971).
260. J. E. Connett, *Lab. Pract.* **21**, 545 (1972).
261. T. A. Washall, S. Blittman, and R. S. Mascieri, *J. Chromatogr. Sci.*, **8**, 663 (1970).
262. Chemical Data System, Oxford, Penn.
262a. J. R. Parrish, *Anal. Chem.*, **45**, 1659 (1973).
263. C. DePetro, personal communication, 1972.
264. E. C. Pesse, *Anal. Chem.*, **45**, 1584 (1973).
265. N. M. Karayannis, A. H. Corwin, E. W. Baker, E. Klesper, and J. A. Walter, *Anal. Chem.*, **40**, 1736 (1968).
266. J. G. Nikelly, *Anal. Chem.*, **44**, 623 (1972).
267. D. E. Willis, *Anal. Chem.*, **40**, 1597 (1968).
268. W. J. Kirsten and P. E. Mattson, *Anal. Lett.*, **4**, 235 (1971).

269. R. Jeltes and R. Veldink, *J. Chromatogr.*, **32**, 413 (1968).

270. R. J. Levins and R. M. Ikeda, *J. Gas Chromatogr.*, **6**, 331 (1968).

270a. J. Q. Walker and C. J. Wolf, *Anal. Chem.*, **45**, 2263 (1973).

271. R. Bassette, C. R. Brewington, and D. P. Schwartz, *Microchem. J.*, **13**, 297 (1968).

272. W. Biernacki, *J. Chromatogr.*, **50**, 135 (1970).

273. T. S. Ma and D. Spiegel, *Microchem. J.*, **10**, 61 (1966).

274. O. W. Parks, N. P. Wong, C. A. Allen, and D. P. Schwartz, *J. Dairy Sci.*, **52**, 953 (1969); **46**, 295 (1963).

275. L. A. Shadoff, G. J. Kellos, and J. S. Woods, *Anal. Chem.*, **45**, 2341 (1973).

276. W. Ebing, *J. Gas Chromatogr.*, **6**, 79 (1968).

277. C. C. Cassil, R. P. Stanovick, and R. F. Cook, *Residue Rev.*, **26**, 63 (1969).

278. *Gas Chromatography of Plant Protection Agents*, Vol. 1. Chromatog. Assoc., New Castle, Del. 1972.

279. K. H. Kubeczka, *Dtch. Pharm. Ges.*, **41**, 278 (1971).

280. H. Ehrsson, T. Walle, and H. Broetell, *Acta Pharm. Suec.*, **8**, 319 (1971).

281. S. J. Romano, J. A. Renner, and P. M. Leitner, *Anal. Chem.*, **45**, 2327 (1973).

282. D. C. Fenimore, R. R. Freeman, and P. R. Loy, *Anal. Chem.*, **45**, 2331 (1973).

283. I. M. Ryabtseva, M. Kaleshova, B. A. Rudenko, and V. F. Kucherov, *Izv. Akad. Nauk S.S.S.R., Ser. Khim.*, **1970**, 2676.

284. J. P. Hardy and S. L. Kerrin, *Anal. Chem.*, **44**, 1497 (1972).

285. W. A. Bonner, *J. Chromatogr. Sci.*, **11**, 101 (1973).

286. P. W. Wilson, D. E. M. Lawson, and E. Kodicek, *J. Chromatogr.*, **39**, 75 (1969).

287. A. L. German and E. C. Horning, *J. Chromatogr. Sci.*, **11**, 76 (1973).

288. P. M. Adhikary and R. A. Harkness, *J. Chromatogr.*, **42**, 29 (1969).

289. A. Zlatkis, H. A. Lichtenstein, A. Tishbee, W. Bertsch, F. Shunbo, and H. M. Liebich, *J. Chromatogr. Sci.*, **11**, 299 (1973).

290. R. E. Sievers, K. J. Eisentraut, M. S. Black, J. J. Brooks, and F. D. Hileman, reported at 164th ACS National Meeting, New York, August, 1972.

291. L. C. Hansen, W. G. Scribner, T. W. Gilbert, and R. E. Sievers, *Anal. Chem.*, **43**, 349 (1971).

292. S. N. Chester and R. S. Juvet, Jr., reported at 164th ACS National Meeting, New York, August, 1972.

293. V. G. Berezkin, *Analytical Reaction Chromatography* (translated from Russian). Plenum, New York, 1968.

294. M. Beroza and M. N. Inscie, in L. S. Ettre and W. H. McFadden (Eds.), *Ancillary Techniques of Gas Chromatography*, p. 89. Wiley-Interscience, New York, 1969.

295. E. R. Hoffman and L. Lysyz, *Microchem. J.*, **6**, 45 (1962).

296. T. S. Ma., C. T. Shang, and E. Manche, *Mikrochim. Acta,* **1964,** 571.
297. P. C. Rahn and S. Siggia, *Anal. Chem.,* **45,** 2336 (1973).
298. A. S. Ladas and T. S. Ma, *Mikrochim. Acta,* **1973,** 853.
299. A. Reichle, M. Wandle, H. Borger, and H. Henger, British Patent 1,086,442 (1966).
300. R. Kaiser, *Chromatographia,* **3,** 38 (1970).
301. E. J. Debrecht, "Some Practical Applications of Small Scale Preparative Gas Chromatography," in I. I. Donski and J. A. Perry (Eds.), *Recent Advances in Gas Chromatography,* p. 231. Dekker, New York, 1971.
302. K. P. Hupe, *Chromatographia,* 1, 462 (1968).
303. M. Verzele, J. Bouche, A. DeBruyne, and M. Verstrappe, *J. Chromatogr.,* **18,** 253 (1965); **26,** 485 (1967).
304. B. M. Mitzner and W. V. Jones, *J. Gas Chromatogr.,* **3,** 294 (1965).
305. M. Verzele, *J. Gas Chromatogr.,* **3,** 186 (1965).
306. P. B. Stockwell and R. Sawyer, *Lab. Pract.* **19,** 277 (1970).
307. E. M. Sibley, C. Eon, and B. L. Karger, *J. Chromatogr. Sci.,* **11,** 309 (1973).
308. H. D. Metzer, *Chromatographia,* 3, 64 (1970).
309. S. P. Cram, S. N. Chester, J. E. Leitner, and B. E. Bowen, reported at the 164th ACS National Meeting, New York, August, 1972.
310. R. W. McKinney, G. R. Garst, R. E. Raver, and W. O. Harris, *J. Gas Chromatogr.,* **6,** 115 (1968).
311. C. J. Cleumett, *Anal. Chem.,* **43,** 490 (1971).
312. G. Blandenet, *Chromatographia,* **2,** 184 (1969).
313. R. P. W. Scott, I. A. Fowlis, D. Welti, and T. Wilkins, reported at the International Symposium on Gas Chromatography, Rome, 1966.
314. J. E. Crooks, D. L. Gerrard, and W. F. Maddams, *Anal. Chem.,* **45,** 1823 (1973).
314a. A. Jalnik, *Chem.Anal.,* **18,** 29 (1973).
315. T. Tsuda, M. Ojika, I. Fujishima, and D. Ishii, *J. Chromatogr.,* **69,** 194 (1972).
315a. C. Fenselau, *App. Spectrosc.,* **28,** 305 (1974).
316. Finnigan Corp., Sunnyvale, Calif.
316a. I. Otvos, S. Iglewski, D. H. Hunneman, B. Bartha, and G. Palyi, *J. Chromatogr.,* **78,** 309 (1973).
317. L. Fischer, *An Introduction to Gel Chromatography.* North-Holland, London, 1969.
318. W. Heitz, *Angew. Chem., Int. Ed.,* **9,** 689 (1970).
319. J. Cazes, *J. Chem. Educ.,* **47,** A 461, 505 (1970).
320. D. H. Freeman, *J. Chromatogr.,* **11,** 175 (1973).
321. H. Vink, *J. Chromatogr.,* **52,** 205 (1970).
322. W. V. Smith, *Rubber Chem. Technol.,* **45,** 667 (1972).
323. A. Lambert, *Br. Polym. J.,* **3,** 13 (1971).
324. Bio-Rad Laboratories, Richmond, Calif.; Pharmacia Fine Chemicals, Piscataway, N. J.
325. K. Urbanek and J. Lachman, *Chem. Listy,* **66,** 1094 (1972).

326. E. Boschetti, R. Tixier, and J. Uriel, *Biochimie*, **54**, 439 (1972).

327. E. James, A. Bjoern, and J. Sjoevall, *Acta Chem. Scand.*, **24**, 463 (1970).

328. *Sep. News*, May, 1973.

329. J. F. Johnson, A. R. Cooper, and R. S. Porter, *J. Chromatogr. Sci.*, **11**, 292 (1973).

330. R. A. Ellis, *Pigm. Resin Technol.*, **1**, 4 (1972).

331. T. Williams, *J. Mater. Sci.*, **5**, 811 (1970).

332. R. Gullberg and H. Brostroem, *Clin. Chim. Acta.*, **39**, 475 (1972).

333. R. A. Nicholas and J. B. Fox, Jr., *J. Chromatogr.*, **43**, 61 (1969).

334. A. Cassera, *Lab. Pract.*, **17**, 54 (1968).

335. D. Jaworek, *Chromatographia*, **2**, 289 (1969).

336. R. Berret, A. Gavauden, and J. Hirtz, *Ann. Pharm. Fr.*, **25**, 365 (1967).

337. N.I.A. Emmeus, *J. Chromatogr.*, **32**, 243 (1968).

338. E. Junowicz, S. E. Charm, and H. E. Blair, *Anal. Biochem.*, **47**, 193 (1972).

339. H. Bende, *Fette Seifen Anstrichm.*, **70**, 937 (1968).

340. W. Heitz, B. Bömer, and H. Ullner, *Makromol. Chem.*, **121**, 102 (1968).

341. M. LePage, R. Beau, and A. J. DeVries, *J. Polym. Sci.*, *(C)*, **21**, 119 (1968).

342. L. Wild and G. Guiliana, *J. Polym. Sci.*, *(A-2)*, **5**, 1087 (1967).

343. L. Segal, *J. Polym. Sci. (C)*, **21**, 267 (1968).

344. E. J. Quinn, H. W. Osterhandt, J. S. Heckles, and D. C. Ziegler, *Anal. Chem.*, **40**, 547 (1968).

345. D. Braun and V. Legradic, *Angew. Markomol. Chem.*, **25**, 193 (1972).

346. J. L. Mulder and F. A. Buytenuhuts, *J. Chromatgr.*, **51**, 459 (1970).

347. R. Gladen, *G.I.T. Fachz. Lab.*, **16**, 1159 (1972).

348. K. Aitzetmueller, *J. Chromatogr.*, **71**, 355 (1972).

349. L. Winkler, T. Heim, H. Schenk, B. Schlag, and E. Goetze, *J. Chromatogr.*, **70**, 164 (1972).

350. G. Ghilardelli and R. Pergo, *Farmacol., Ed. Prat.*, **27**, 467 (1972).

351. W. L. Green, *J. Chromatogr.*, **72**, 83 (1972).

352. F. Thomas, P. Gomez, J. J. Mataix, and O. Carpena, *Rev. Agroquim. Tecnol. Aliment.*, **12**, 106 (1972).

353. J. F. McKay and D. R. Latham, *Anal. Chem.*, **45**, 1050 (1973).

354. W. P. Ferren and W. E. Seery, *Anal. Chem.*, **45**, 2278 (1973).

355. H. Determan, *Experientia*, **18**, 430 (1968).

356. B. J. Radola, *J. Chromatogr.*, **38**, 61, 78, (1968).

357. D. Jawarek, *Chromatographia*, **3**, 414 (1970).

358. D. J. Shaw, *Electrophoresis.* Academic, New York, 1969.

359. R. J. Wiene, *Agar Gel Electrophoresis.* Elsevier, Amsterdam, 1965.

360. J. Kellen and A. Bellog, *Z. Physiol. Chem.*, **311**, 283 (1958).

361. W. Ghidalia, R. Vendrely, and Y. Coirault, *Bull. Soc. Chim. Biol.*, **52**, 110 (1970).

362. T. I. Pristoupil, *Chromatogr. Rev.*, **12**, 109 (1970).

363. Millipore Immuno Agaro Slide, Millipore Corp., Bedford, Mass.

364. H. Haglund, *Methods Biochem. Anal.*, **19**, 1 (1971).

365. D. Wellner, *Anal. Chem.,* **43**, 59A (1971).
366. A. Karler, C. L. Brown, and P. L. Kirk, *Mikrochim. Acta,* **1956**, 1585.
367. K. J. Stevenson, *Anal. Biochem.,* **40**, 29 (1971).
368. B. J. Haywood, *Electrophoresis—Technical Application.* Humphrey Science, Ann Arbor, Mich., 1969.
369. M. Bier (Ed.), *Electrophoresis,* Vol. 2. Academic, New York, 1967.
370. J. R. Whitaker, *Electrophoresis on Stabilizing Media.* Academic, New York, 1967.
371. S. N. Tewari, *Microchem. J.,* **1968**, 390.
372. M. Stuchlik, I. Csiba, and L. Krasnac, *Cesk. Farm.,* **16**, 187 (1967).
373. L. B. Jaques, R. F. Ballieux, C. P. Detrich, and L. W. Kavanagh, *J. Physiol. Pharmacol,* **46**, 351 (1968).
374. A.S.C. Wan, *J. Chromatogr.,* **60**, 371 (1971).
375. H. Hazama and H. Uchimura, *Microchem. J.,* **17**, 318 (1972).
376. S. V. Koppikar, P. Fatterpaker, and A. Sreenivasion, *Anal. Biochem.,* **33**, 366 (1970).
377. A. Chalvaodjian, *Clin. Chim. Acta,* **26**, 174 (1969).
378. J. H. Buchanan and M. C. Corfield, *J. Chromatogr.,* **31**, 274 (1967).
379. B. Vecerek, J. Stephan, I. Hynie, and K. Kacl., *Collect. Czech. Chem. Commun.,* **33**, 141 (1968).
379a. J. Stepan and B. Vecerek, *Chem. Listg,* **67**, 646 (1973).
380. B. Fingerhut and A. Ortiz, *Clin. Chem.,* **17**, 34 (1971).
381. A. Krawczyk., *Chem. Anal.,* **16**, 657 (1971).
382. J. H. Strohl and K. L. Dunlop, *Anal. Chem.,* **44**, 2166 (1972).
383. O. Samuelson, *Ion Exchange Separations in Analytical Chemistry.* Wiley, New York, 1963.
384. J. Inczedy, *Analytical Applications of Ion Exchangers* (translated from Russian). Pergamon, London, 1966.
385. R. A. A. Muzzarelle, in J. C. Giddings and R. A. Keller (Eds.), *Advances in Chromatography,* Vol. 5, p. 127. Dekker, New York, 1968.
386. J. Bosholm, *J. Chromatgr.,* **21**, 286 (1966).
387. Y. Marcus and A. S. Kertes, *Ion Exchange and Solvent Extraction of Metal Complexes.* Wiley, New York, 1969.
388. R. Rosset, *Bull. Soc. Chim. Fr.,* **1966**, 59.
389. H. Bernard, and F. Grass, *Mikrochim. Acta,* **1966**, 426.
390. V. Sykora and F. Duvski, *Collect. Czech. Chem. Commun.,* **32**, 3342 (1967).
391. J. Deson and R. Rosset, *Bull. Soc. Chim. Fr.,* **1968**, 4307.
392. I. N. Ermolenko and G. A. Kamalyan, *Arm. Khim. Zh.,* **21**, 264 (1968).
393. D. Whittington and J. R. Miller, *J. Appl. Chem.,* **18**, 122 (1968).
394. J. P. de Villiers and J. R. Parrish, *J. Polym. Sci. (A),* **2**, 1331 (1964).
395. J. R. Parish, *Anal. Chem.,* **45**, 1659 (1973).
396. W. R. Heuman, *Crit. Rev. Anal. Chem.,* **2**, 425 (1971).
396a. V. Horak and J. Pecka, *J. Chromatogr.,* **14**, 97 (1964).
397. H. F. Walton, *Anal. Chem.,* **40**, 57R (1968).
398. M. Shida and H. P. Grepor, *J. Polym. Sci., (A),* **4**, 1113 (1966).

399. A. F. Tsuk and H. P. Grepor, *J. Am. Chem. Soc.,* **87**, 5538 (1965).
400. G. Alberti, and S. J. Allulli, *J. Chromatogr.,* **32**, 379 (1969).
401. G. H. Naneollas and B. V. Tilak, *Inorg. Nucl. Chem.,* **31**, 213 (1969).
402. G. Aubouin and J. Laverlochue, *J. Radioanal. Chem.,* **1**, 123 (1968).
403. D. H. Freeman, L. A. Currie, E. C. Kuehner, H. D. Dixon, and R. A. Paulson, *Anal. Chem.,* **42**, 203 (1970).
404. D. Cozzi, P.G. Desideri, L. Lepri, and V. Coas, *J. Chromatogr.,* **43**, 463 (1969).
405. B. Z. Egan, *J. Chromatogr.,* **34**, 382 (1968).
406. H. S. Sherry, *J. Phys. Chem.,* **72**, 4086 (1968).
407. D. P. Schwartz, C. R. Brewington, and O. W. Parks, *Microchem. J.,* **13**, 125 (1968); *Agric. Food Chem.,* **21**, 38 (1973).
408. P. Larson, E. Murgia, T. J. Hsu, and H. F. Walton, *Anal. Chem.,* **45**, *Chem.,* **45**, 2306 (1973).
409. G. J. Moody and J. D. R. Thomas, *Lab. Pract.* **19**, 387 (1970).
410. S. A. Bernhard and L. P. Hammett, *J. Am. Chem. Soc.,* **75**, 1798 (1953).
411. G. Durr, *Compt. Rend.,* **235**, 1314 (1952).
412. D. T. Murphy, G. N. Richards, and E. Senogles, *Carbohydr. Res.,* **7**, 460 (1968).
413. L. W. Nauman, *J. Chromator.,* **36**, 398 (1968).
414. A. Tsuji and H. Kakihana, *Mikrochim. Acta,* **1962**, 475.
415. K. A. Hooton and M. L. Parsons, *Appl. Spectrosc.,* **27**, 480 (1973).
416. J. A. Marilnsky (Ed.), *Ion Exchange, A Series of Advances,* Vol. 2. Dekker, New York, 1964.
417. A. Gorski and J. Moszczynska, *Chem. Anal.,* **9**, 1071 (1964).
418. E. Segall and G. Schmukler, *Talanta,* **14**, 1253 (1967).
419. C. G. Horvoth, B. A. Preias, and S. R. Lipsky, *Anal. Chem.,* **339**, 1422 (1967).
420. C. G. Horvoth and S. R. Lipsky, *Anal. Chem.,* **41**, 1227 (1969).
421. J. S. Fritz and R. K. Gillette, *Anal. Chem.,* **40**, 1777 (1968).
422. E. Herrmann, *J. Chromatogr.,* **38**, 498 (1968).
423. V. T. Leich and A. Langanke, *Chem. Ztg—Chem. Appar.,* **90**, 540 (1966).
424. Technicon Instruments Corps., Tarrytown, N. Y.; Sanda, Inc., Philadelphia, Pa., Beckman Instruments, Fullerton, Calif.
425. J. I. Peterson, F. Wagner, F. Anderson, G. M. Thomas, Jr., *Anal. Biochem.,* **32**, 128 (1969).
426. H. W. Lange and R. Hempel, *Beckman Rep.,* **1971**, 18.
427. J. D. Smith, *Lab. Pract.* **20**, 496 (1971).
428. J. Sherma, *Sep. Sci.,* **2**, 177 (1967).
429. D. I. Ryabchikov and M. P. Volynets, *Zh. Anal. Khim.,* **21**, 1348 (1966).
430. R. D. Brown and C. E. Holt, *Anal. Biochem.,* **20**, 358 (1967).
431. S. Zagrodzki and A. Kurkowska, *Chem. Anal.,* **12**, 159 (1967).
432. D. P. Schwartz, A. R. Johnson, and O. W. Parks, *Microchem. J.,* **6**, 37 (1962).
433. M. G. Lai and H. A. Goya, U.S. C.F.S.T.I., AD 648,485 (1968).

434. J. P. Riley and D. Taylor, *Anal. Chim. Acta,* **41**, 175 (1968).
435. J. P. Riley and D. Taylor, *Anal. Chim. Acta,* **40**, 479 (1968).
436. D. A. Mathews and J. P. Riley, *Anal. Chim. Acta,* **48**, 25 (1969).
437. T. T. Chao, *Econ. Geol.,* **64**, 287 (1969).
438. T. T. Chao, M. J. Fishman, and J. W. Ball, *Anal. Chim, Acta,* **47**, 189 (1969).
439. C. W. Walter and J. Korkisch, *Mikrochim. Acta,* **1971**, 162.
440. J. A. Weyh, *J. Chem. Educ.,* **47**, 715 (1970).
441. R. H. Stehl, *Anal. Chem.,* **42**, 1802 (1970).
442. E. F. Walborg, Jr., and L. E. Kondo, *Anal. Biochem.,* **37**, 320 (1970).
443. J. S. Hobbs and J. G. Lawrence, *J. Chromatogr.,* **72**, 311 (1973).
444. R. H. Lindsay, M. Y. Wong, C. J. Romaine, and J. B. Hill, *Anal. Biochem.,* **24**, 506 (1968).
445. F. W. E. Strelow, C. J. Liebenberg, and F. von S. Toerien, *Anal. Chim. Acta,* **43**, 465 (1968).
446. L. Bengtsson and O. Samuelson, *Anal. Chim. Acta,* **44**, 217 (1969).
447. W. Funasaka, T. Hanoi, K. Kujimura, and T. Ando, *J. Chromatogr.,* **72**, 187 (1972).
448. S. Fudano and K. Konishi, *J. Chromatogr.,* **71**, 93 (1972).
449. V. Shaw and H. F. Walton, *J. Chromatogr.,* **68**, 267 (1972).

FRACTIONATION OF LIQUID MATERIALS

I. GENERAL

A. METHODS FOR FRACTIONATION OF LIQUID MIXTURES

In this chapter we consider the methods that may be employed for the microscale fractionation of liquid mixtures. It is assumed that the working material as received is a homogeneous liquid at ambient temperature. The common technique for the separation of liquid mixtures is based on the differences of the vapor pressures of the components, and is known as distillation (see Sections II, III, IV). Less frequently, fractionation is carried out by the principle of solubility, depending either on a temperature change in the system (Section V) or on the partition coefficient between two solvents (Section VI). It should be noted that chromatographic methods have been discussed in Chapter 4. Chromatography is most valuable for fractionating liquid mixtures below the milligram region.

B. THE DISTILLATION PROCESS

Distillation [1, 2] involves heating the liquid sample at its boiling point (except in the case of molecular distillation) followed by condensing the vapors at a lower temperature. This process can be carried out either under atmospheric conditions or at reduced pressures. Thermodynamically speaking, the boiling point of a liquid is the temperature at which the vapor pressure of the liquid is the same as the gas pressure above the liquid surface. When the liquid sample is a homogeneous mixture, its vapor pressure represents the sum of all vapor pressures contributed by its components. Thus, fractionation by distillation is possible if the composition of the vapor phase is considerably different from that of the liquid phase. In general, two liquids whose boiling points are more than $50°C$ apart can be separated by means of relatively simple distillation apparatus, while liquids whose boiling points differ by less than $30°C$ require elaborate fractionation devices.

The effectiveness of separation using a specific distillation apparatus is measured by the number of theoretical plates. One theoretical plate corresponds to the effect of separation that is achieved in an equilibrium state at the boiling temperature in a simple distilling flask. In an apparatus where liquid vapor equilibrium is not established (e.g., in molecular distillation), the separation effect is less than one theoretical plate.

Fractional distillation under reduced pressure is required if the compounds in the liquid mixture are unstable at high temperatures. The vacuum that can be obtained with a water aspirator is about 20 torr, and with a mechanical oil pump about 10^{-2} torr. Compared with distillation under atmospheric pressure, the boiling point of a compound is lowered approximately 100°C at 20 torr and 200°C at 10^{-2} torr. Since fluctuation of the vacuum has great influence on the performance of fractionation, the pressure in the system should be kept fairly constant by means of a manostat.

The course of fractionation is usually followed by plotting the volume of distillate collected against the boiling temperature. This graph is known as the distillation curve, an example being shown in Fig. 5.1. The horizontal portions of the graph indicate the respective fractions containing "pure" compounds, while the steep portions indicate "mixed" fractions containing two adjacent components. Obviously, the smaller the "mixed"

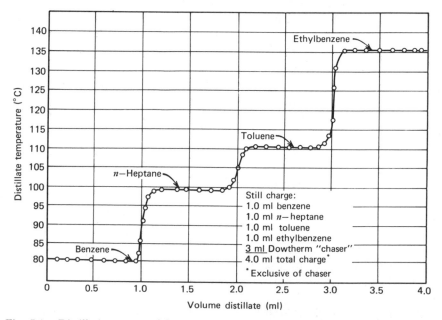

Fig. 5.1. Distillation curve of four component mixture containing equal volumes of benzene, n-heptane, toluene, and ethylbenzene. (Courtesy Thomas Y. Crowell Co.)

fractions, the more effective is the separation process. Paulik and co-workers [3] have proposed to measure the weight change against temperature in microdistillation, and reported that the automatic registration of thermogravimetric curves gives reliable results for the fractionation of 1 g or smaller amounts of ternary mixtures of essential oils.

When the objective of fractionation is to obtain one or more pure components for subsequent analytical experiments, micro amounts of the distillate may be adequate. Thus Colichman and co-workers [4] have separated radiation-damaged samples of terphenyls by vacuum distillation on a micro scale and determined the fractions by infrared spectrometry. On the other hand, the analytical distillation of petroleum mixtures requires the collection of larger volumes of distillate. Javes and co-workers [5] have found that distillates and residues from 5-ml portions of crude oil cut in a microstill with a packed column have properties agreeing with those prepared from several liters of crude oil using a 14-plate column. Tests with microstills on 0.8-ml samples give results agreeing with those obtained by standard methods, but the experimental error is 3 times as great. An apparatus and a method have been described by the British Standards Institution for the determination of the distillation yield and distillation range of liquids [5a].

There are many difficulties in adapting macro distillation methods to microscale operations. A serious problem is the loss of the micro sample, which adheres to the packing material and the walls of the distilling apparatus. Unless a special technique is utilized, this loss due to holdup may amount to a large proportion of the working material. One technique is to add a heat-stable high-boiling liquid, known as the "chaser," to the mixture. The boiling point of the chaser should be at least 30°C higher than that of the last fraction, and the amount of chaser should be a little more than the estimated holdup. Thus, in the example shown in Fig. 5.1, all components of the original mixture are recovered before any of the chaser reaches the receiving vessel. Needless to say, the chaser should neither react nor form azeotropes with the compounds present in the liquid to be fractionated. Among the chasers commonly employed are cymene (boiling point 175°C), diphenyl (254°C), acenaphthene (277°C), and phenanthrene (340°C).

C. COMPARISON BETWEEN FRACTIONAL DISTILLATION AND CHROMATOGRAPHY

Whereas in chromatographic processes (see Chapter 4) the individual components of the original mixture are usually completely separated by means of a solvent or gas placed between the fractions, in fractional distillation the individual components follow each other immediately (as in the displacement technique in chromatography). Consequently, while the occurrence of overlapping in chromatography is often prevented by reducing the sample size, the opposite is true for fractional distillation. As the volume of the distillate becomes smaller, the difficulty of quantitative collection tends to increase, and the ratio between the volume of the pure fraction and that of the mixed fraction decreases.

To some extent, chromatography—particularly gas chromatography—and fractional distillation complement each other for the separation of liquid mixtures. In general, fractionation can be performed more easily by distillation when the volume exceeds 1000 μl, while chromatography is the preferred technique for sample sizes below 100 μl.

II. FRACTIONAL DISTILLATION WITH SIMPLE APPARATUS

A. FOR VOLUMES LARGER THAN 1 ML

A number of simple devices have been described [6] for the fractional distillation of 10 ml of liquid sample or less. The distilling tube shown in Chapter 2, Fig. 2.1f can be employed for mixtures that contain compounds with boiling points far apart. The vessel is heated in a metal block (Figs. 2.16, 2.17) or in an oil bath. Since there is no thermometer inside the distilling tube, the temperature of the heating bath (usually about 20°C higher than the vapor temperature inside) serves as an indication of the boiling temperature. The temperature of the heating bath is gradually raised. When the distillate begins to condense in the elbow receiver, the heating bath is kept at that temperature until the particular fraction has been collected. The distilling tube is now detached from the heating bath, and the fraction is withdrawn. The apparatus is reassembled and again placed in the heating bath. The next fraction is obtained as the temperature is raised.

Various designs for the distilling apparatus have been descirbed whereby several fractions can be collected without disconnecting the assembly. Such a device is advantageous for fractionation under reduced pressure, since it permits continuous operation without breaking the vacuum system. The assembly proposed by Nemec [7] is shown in Fig. 5.2. Schrader and Ritzer [8] have described an apparatus that has a flat-bottom distilling vessel of 4-ml capacity and a column 12 cm long (Fig. 5.3). The inside of the column has indentations like a Vigreux column and is surrounded by an evacuated jacket.

Dobrowsky [9] has described a distillation vessel that prevents bumping or boiling over of the distilling liquid. Pinkava [10] has constructed

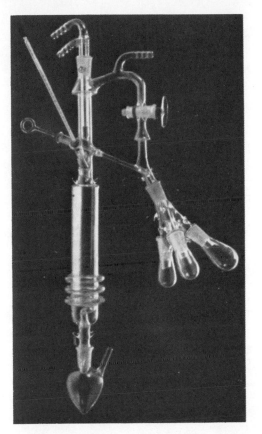

Fig. 5.2. Fractional distillation apparatus of Nemec [7]. (Courtesy Kavalier Glassworks.)

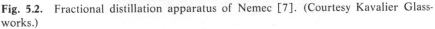

a distilling flask especially suitable for the atmospheric or vacuum distillation of foaming liquids. The liquid is boiled in an electrically heated side arm, from which it flows tangentially into the flask, Jacobson and Miller [11] have fabricated an apparatus particularly designed for distilling substances that are viscous liquids or solids at normal temperature. Sherwood [11a] has described an apparatus with internal condenser and receiver for distilling 3 ml or less of material without losing any distillate by trapping.

Another type of distillation apparatus that works well in semimicro operations is the Hickman [12] flask. As depicted in Fig. 5.4, the collar, which is situated above the distilling bulb, serves as a trap to collect the condensate. The main advantages of the Hickman design are: (1) a short

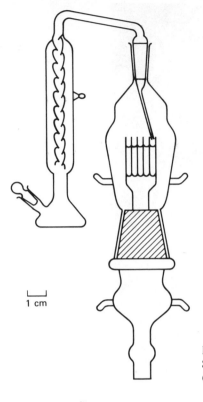

1 cm

Fig. 5.3. Fractional distilling apparatus of Schrader and Ritzer [8]. (Courtesy *Anal. Chem.*)

Fig. 5.4. Hickman distillation apparatus [12]: *a*, glass wool; *b*, thermometer.

pathway for vapors to travel from the evaporation to the condensation compartment, (2) simple construction, (3) convenient handling, and (4) easy cleaning. It can be seen from the illustration that only one fraction can be collected at a time. After the distillation of one fraction is com-

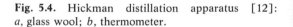

pleted, the process has to be discontinued and the distillate withdrawn from the collar by means of a long capillary pipet. In order to prevent contamination of the next fraction, the collar is washed with a suitable solvent (e.g., ether) and the washings pipetted off and discarded. The next fraction is then distilled at higher temperature. The modified apparatus shown in Fig. 5.5, which is provided with an exit tube for the distillate, can be used for continuous distillation.

Riley [13] has described a semimicro multipurpose distillation unit that is suitable for fractionation, steam distillation, or continuous extraction. Sevick [14] has constructed an azeotropic-distillation apparatus that can be used with solvents lighter or heavier than water. A similar apparatus was proposed by Martin [15]. Lee and Lynch [16] have described an improved distillation head for organic aqueous mixtures that is a modified form of the fractional-distillation head designed by Chin [17].

Instead of the conventional round-bottomed or pear-shaped distilling vessel, Henderson and Kamphausen [18] have proposed the use of an inverted conical flask, which is heated by means of an internal electric element. This device removes any obstruction of the escaping vapor due to the shape of the vessel. Schrecker [19] has described a bulb-tube assembly for vacuum distillation that consists of three or more glass bulbs connected one by one through standard tapered joints. The bulb containing the sample is heated in a box oven. (The apparatus is commercially available [19a].) Round vessels are also utilized by Enzell and co-workers [20] for the low-temperature vacuum distillation of volatile matter from natural products. The apparatus is illustrated in Fig. 5.6. Carbon dioxide of high purity is passed through a filter of active carbon

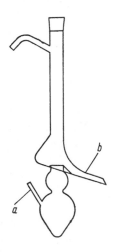

Fig. 5.5. Modified Hickman apparatus for continuous distillation: *a*, connecting tube to vacuum line; *b*, exit tube for distillate.

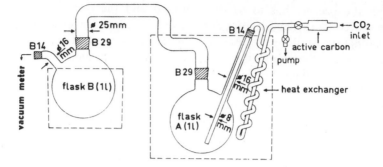

Fig. 5.6. Assembly for low-temperature, closed-vacuum-system distillation [20]. (Courtesy *Acta Chem. Scand.*)

and a heat exchanger into flask *A*, and finally condensed in flask *B*, which is kept at $-180°C$ by means of a liquid-nitrogen bath. The pressure, which is measured with the aid of a vacuum meter attached to flask *B*, is mainly regulated by the carbon dioxide flow. Before starting the distillation, the air is removed from the system to prevent the condensation of oxygen. The operation is performed in the following manner: (1) The sample, preferably presaturated with carbon dioxide, is transferred into flask *A*. (2) Flask *A* is connected to the system and cooled in a Dry Ice–acetone bath. (3) The system is evacuated to 1 torr and filled with carbon dioxide up to 700 torr. This procedure is repeated four times. (4) The Dry Ice–acetone bath is removed, and flask *B* is cooled with liquid nitrogen; the resulting pressure is about 0.1 torr. Carbon dioxide is led via the active carbon filter and heat exchanger into flask *A*, which is slowly heated to the desired temperature in a thermostatted bath. The tube connecting flasks *A* and *B* is kept at the same temperature by means of a heating tape. (5) The distillation is interrupted when on removal of the heating tape no condensation is observed. (6) The cooling and heating baths are removed. When the pressure of the system has increased somewhat, carbon dioxide is introduced until atmospheric pressure is attained. The receiver is disconnected and the solidified carbon dioxide is allowed to evaporate, leaving the distillate in flask *B*. In test with a 2.5 g of a mixture containing equal amounts of C_6 to C_{28} *n*-alkanes distilled for 30 min at 50°C with 100 g of carbon dioxide, the yields were >99%, and the only overlap between distillate and residue occurred with the C_{14} to C_{17} compounds.

B. FOR VOLUMES SMALLER THAN 1 ML

When the liquid sample to be fractionated is less than 1 ml, it is sometimes difficult to utilize the distillation apparatus described in the previous section. Distillation in capillary fractionating tubes is the preferred technique. Since the capillary tube can be easily prepared, it is advisable to construct a new distillation vessel each time. This prevents contamination and also avoids the tedium of cleaning the used capillary. The distillation vessel is fabricated from clean glass tubing of 2- to 6-mm i.d. After one end of the tubing has been sealed, clean, inert, fibrous material (e.g., asbestos, steel wool) is pushed into the bottom. Then the tube is constricted at one or two points about 10 mm apart, forming the lower chamber of 3 to 6 cm in length. The constrictions are simply made by thickening the walls (see Fig. 5.7) instead of drawing out the tubing. The liquid is delivered into the lower chamber with the aid of a syringe needle or fine glass capillary, and then centrifuged down to the bottom of the fractionating tube. The tube is wrapped with asbestos paper, which serves as insulator, and placed in the well of the block heater (Chapter 2, Figs. 2.16, 2.17), as illustrated in Fig. 5.8. Depending on the boiling points of the fractions, the condensation compartment of the distillation vessel can be cooled by air, water (by using wet paper), or Dry Ice. As the temperature of the block heater is gradully raised, a fraction of the liquid distils

Fig. 5.7.   Fractionating tube for distilling less than 1 ml of liquid material.

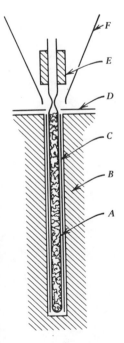

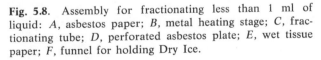

Fig. 5.8. Assembly for fractionating less than 1 ml of liquid: *A*, asbestos paper; *B*, metal heating stage; *C*, fractionating tube; *D*, perforated asbestos plate; *E*, wet tissue paper; *F*, funnel for holding Dry Ice.

and condenses at the constriction area. This fraction is collected by means of a capillary and can be identified by boiling-point determination [21], gas chromatography, or spectral analysis. Collection of the distillate is facilitated by constructing a bent fractionating tube, as shown in Fig. 5.9, which also illustrates the arrangement for vacuum distillation [22].

Using capillary tubes of 1.5- to 2-mm bore and 120-mm length, Morton and Mahoney [23] were able to obtain as many as 70 fractions from initial samples of about 25 µl. The beginner may test this microtechnique by distilling 50 µl of a 1:1 mixture of toluene (boiling point 111°C) and o-xylene (144°C), and then 30 µl of a 1:1 mixture of toluene and aniline (183°C).

Other devices for the fractional distillation of less than 1 ml of liquid material have been proposed. The apparatus described by Babcock [24] is shown in Fig. 5.10. The distilling tube *ABC* is fabricated from 6-mm glass tubing and has a length of about 23 cm. Nemec [25] has described a device that is suitable for distilling volumes up to 120 µl. The condenser and absorption part of the apparatus are situated inside the distillation vessel, thereby avoiding loss of distillate. Walls [26] has constructed a distillation apparatus shown in Fig. 5.11 which comprises a distilling tube, heating gun, and Dry Ice jacket. A planchet-to-planchet microstill for handling 100 µl of liquid has been designed by Larrabee [26a].

A comprehensive review of microscale distillation equipment has been published by Stage [27]. It discusses apparatus for the distillation of samples of weight down to 1 mg. Horizontal distillation apparatus, various

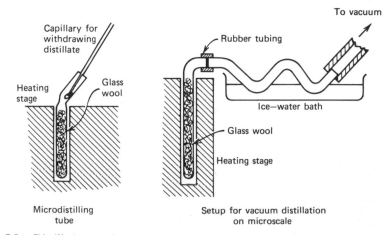

Fig. 5.9. Distillation technique at the microliter level [22]. (Courtesy *Microchem. J.)

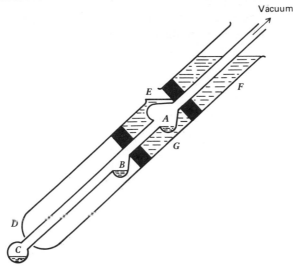

Fig. 5.10. Microdistillation apparatus of Babcock [24]. (Courtesy *Anal. Chem.*)

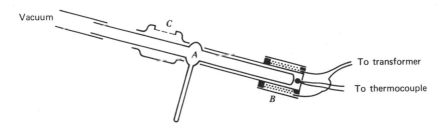

Fig. 5.11. Microdistillation apparatus of Walls [26]: *A*, distilling tube; *B*, heating gum; *C*, Dry Ice jacket. (Courtesy Microchem. J.)

split-tube columns, fractional-distillation analysis, and automatic control devices are included. Huwyler [28] has described a method for the vacuum distillation of microgram amounts of material as follows. A solution of the sample (e.g., 10 μg of octanoic acid) in a volatile solvent is introduced into a region around a bend at about 125° in a capillary tube of 1-mm i.d.; the two arms of the tube are 50 and 160 mm long. The solvent is removed in a vacuum desiccator, and the filling process is repeated until the required amount of sample has been introduced. The end of the shorter arm of the tube is sealed and immersed in a heating bath, and the open end of the tube is connected to a source of high vacuum. Two paper cones filled with powdered Dry Ice are suitably positioned on the upper arm of the tube to provide cooling zones. The tube is broken to recover the distillate. It should be noted that volumes less than 1 ml (e.g., 50 μl) can be successfully distilled in miniature Hickman flasks (see Section A above).

III. USE OF FRACTIONATING COLUMNS

The efficiency of separation by distillation can be substantially improved by the use of fractionating columns. Components of a liquid mixture with boiling point differences less than 30°C in general require column distillation for complete separation. By means of superior columns, the fractionation of liquids whose boiling points are 1°C apart has been achieved. Column distillation is a common technique in industrial operations and ordinary laboratory experiments. Its application in microscale manipulation, however, has certain limitations. Generally speaking, in the best cases it can be adapted to the separation of quantities in the region of 1 ml, although the column fractionation of a 10-μl sample has been reported by Gould and co-workers [29].

It should be recognized that the incapacity of the column distillation for separating very small volumes of liquid results directly from the principle of this technique. The increased efficiency of column distillation, as compared with simple fractional distillation, is based on the multiple equilibration between the liquid and vapor phases established on the large surface of the column. In order to achieve this state, the whole inner surface of the column has to be coated with a liquid film while the inner space is filled with vapors. Even with columns of the smallest dimensions (e.g., the capillary columns) the quantity of the retained liquid is considerable. Furthermore, for high performance of the column, it is necessary to regulate carefully the ratio of the reflux from the top of the column to the uptake of the distillate. Such regulation in the column head is difficult to maintain with small volumes.

In evaluating the usefulness of a particular distillation column, two criteria should be considered: (1) the number of theoretical plates and (2) the holdup of the column. The number of theoretical plates is a measure of the efficiency of the column; columns for large-scale distillation with efficiencies of 100 theoretical plates are commercially available. The holdup is a measure of the volume of the liquid that permanently occupies the column during the entire distillation procedure (dynamic holdup) as well as what is retained by the column after the distillation has been completed (static holdup). The holdup factor is of paramount importance in judging the column for microfractionation. Obviously, only columns with small holdups can be successfully employed for such a purpose. The problem is not mainly in the static holdup, which can be ameliorated by adding a "chaser" to the sample (see section I.B). The real problem is the dynamic holdup. If the volume of one component in the liquid mixture passing through the column is smaller than the dynamic holdup, the equilibrium for this component is not established in the entire length of the column, and the separation efficiency decreases rapidly. One of the basic rules of column distillation demands that the volume of the smallest component in the mixture should be 5 to 10 times larger than the dynamic holdup of the column [30]. If the increase of the number of theoretical plates is obtained by increasing the length of the column, it may also increase the holdup of the column. Hence lengthening the fractionating column does not necessarily improve the separation of small quantities of liquid mixtures.

The optimal operation of a fractionating column is dependent on several factors, one of which being the thermal insulation of the column. In order to reduce losses of energy and of efficiency, the column is sealed into a mantle made from glass tubing, which is evacuated and coated with silver mirror. Another factor is the distillation process. The column operates efficiently under certain conditions only, known as the equilibrated state. Under these conditions there exists true equilibrium between the liquid phase and the gas phase in the entire column at the boiling temperature. It should be recognized that the column is not in its equilibrated state at the beginning of the experiment; this state is attained through an equilibration process. That process can be followed by reading the temperature in the distillation head under full operation of the reflux. The temperature reaches a maximum at the beginning of the equilibration process and then falls slowly until it becomes constant at the equilibrated state. The observed temperature changes indicate the variation of the composition of the vapor phase in the distillation head, since more and more high-boiling components are removed from the column head in the equilibration process. Therefore, columns that cannot be equilibrated owing to lack of regulation of the reflux operate with poor efficiency, particularly at the beginning of the distillation. The equilibration of a column may take several hours.

Column distillation at reduced pressure requires that the vacuum be kept reasonably constant. The fluctuation of the pressure in the system causes disruption of the equilibrated state and hence rapid deterioration of the separation efficiency. A steady vacuum is maintained by using a good pump and manostat, or by incorporating a large reservoir in the vacuum line. It should be noted that all columns lose some efficiency when operated at reduced pressure.

Several types of columns suitable for fractionating small volumes are discussed below.

A. COLUMNS WITH PACKING

A wide range of fractionating columns with packing are commercially available. The Podbielniak columns [31] are probably the most commonly employed for liquid volumes larger than 1 ml. Brezina [32] has described columns with glass-fiber packing, which have the advantage that their characteristics are not changed during vacuum distillations.

Juchheim [33] has constructed columns (length 20, 30, 50, and 100 cm) that are packed with rolls of stainless-steel wire net. The columns are electrically heated, and the temperature is controlled by means of a contact thermometer connected to a relay. It is claimed that equilibrium is established far more rapidly than with equivalent vacuum-jacketed columns. Previously, Gould and co-workers [29] had described a capillary column of 2.5-mm i.d. packed with stainless-steel wire spiral (Fig. 5.12). The column has a holdup of 20 to 30 μl and operates without reflux. At a distillation rate of 5 to 20 μl/min, the efficiency of the column is 9 theoretical plates.

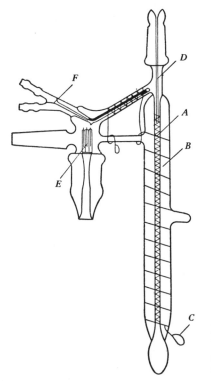

Fig. 5.12. Fractionating column for distillation of 10 to 100 μl of liquid [29]: *A*, packing; *B*, vacuum jacket; *C*, heating wire; *D*, tube for thermistor; *E*, receiver; *F*, condenser.

B. EMPTY CAPILLARY COLUMNS

Levin [34] has described an empty capillary column made from glass tubing of 2-mm i.d. and 100-cm length (Fig. 5.13). Since organic liquids do not form liquid films of sufficient capacity on a smooth glass surface, the inner wall of the capillary column is made rough by grinding with sand. This column is designed for distillation under atmospheric pressure. The maximum efficiency is 33 theoretical plates with total reflux, and 25 theoretical plates under operational conditions when 0.5 ml of distillate is collected per hour.

C. CONCENTRIC-TUBE COLUMNS

In a fractionating column containing two concentric tubes, the inner surface is increased and the empty volume is diminished in comparison with the empty capillary column. A simple model designed by Craig [35] is shown in Fig. 5.14. This column is provided with some reflux that cannot be regulated; its holdup is about 100 μl. It is suitable for distilling 0.5 to 2 ml of material, and the efficiently is 8 theoretical plates at a distillation rate of 0.3 ml/hr.

The technique of fabricating columns from concentric tubes has been

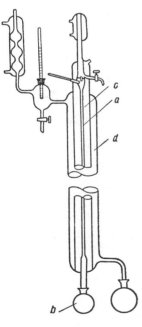

Fig. 5.13. Empty capillary column [34]: *a*, capillary tubing; *b*, distilling flask; *c*, evacuated jacket; *d*, jacket heated with a boiling liquid.

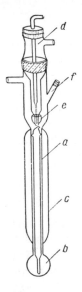

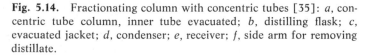

Fig. 5.14. Fractionating column with concentric tubes [35]: *a*, concentric tube column, inner tube evacuated; *b*, distilling flask; *c*, evacuated jacket; *d*, condenser; *e*, receiver; *f*, side arm for removing distillate.

described by Rozengart [36]. Recently Fischer [37–39] has improved this type of column by making the condensate descent spirally in the gap between two concentric, grooved glass tubes. Such columns are now commerically available [40] under the trade name Micro Spaltrohr-Columns. Their characteristics are given in Table 5.1. The schematic drawing of the column and accessories is shown in Fig. 5.15.

D. SPINNING-BAND COLUMNS

Spinning-band columns are the most convenient tools for the fractional distillation of liquid mixtures in the range from several milliliters up. One of the advantages of a spinning-band column is fast equilibration, which can be attained within 10 to 20 min. The basic component of this type of column is a spinning metal [41] or Teflon [42] band, or glass cylinder [43], inside a glass tube. The rotor speed ranges from about 2000 to 6000 rev/min. A commercial apparatus [44] is shown in Fig. 5.16; it can handle 2 to 70 ml of liquid. The theoretical-plate values range from 20 to 150, depending on boilup rate, rotor speed, band material, and sample characteristics. Teflon-band columns have larger holdup and pressure drop than metal-band columns.

Boivin [45] has demonstrated the effect of rate of rotation on separation by distilling a mixture of benzene and toluene using a nickel-chromium ribbon about 550 mm long. There is marked separation of pure benzene when the band is spinning at 2000 rev/min, and the separation is im-

TABLE 5.1. Characteristics of the Spaltrohr Columns[a]

Type	MMS 150	MMS 200	MS 300	MS 500	HMS 500
Number of theoretical plates (approx.)	10	35	40	65	75
Lowest pressure in distillation flask (torr)	1.0	0.5	0.05	0.1	0.1
Hold-up (ml)	0.1	0.5	1.3	2.0	2.3
Average quantity for distillation (ml)	0.5–5	1–10	3–50	5–50	5–50
Column length, distillation flask + column head (mm)	200	300	450	650	800
Distillation flask	fused on		(choice)	separate	

[a] Supplied by Brinkmann Instruments, Inc.

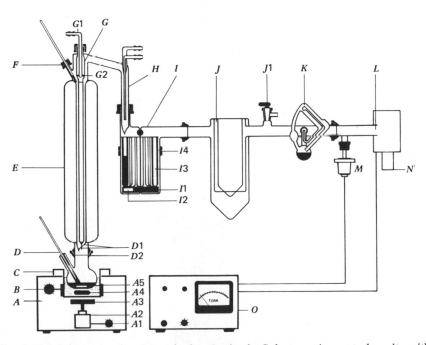

Fig. 5.15. Schematic drawing of the Spaltrohr-Column: *A*, control unit with built-in oil bath (*A*1, speed regulator; *A*2, stirrer motor, magnetic; *A*3, magnet, rotating; *A*4, magnetic follower in oil bath; *A*5, magnetic follower in distillation flask); *B*, electronic temperature regulator; *C*, oil bath; *D*, thermometer pocket-distillation flask; *E*, Spaltrohr-Column; *F*, thermometer entry—column head; *G*, dephlegmator (*G*1, cold finger; *G*2, O-ring); *H*, cold finger; *I*, fraction collector (*I*1, holder; *I*2, rotary magnet; *I*3, receiving tubes; *I*4, flange); *J*, cooling trap; *K*, vacuum gauge; *L*, magnetic valve; *M*, vacuum probe; *N*, vacuum connection; *O*, vacuum regulator with built-in electric indicator, range 50 to 10^{-3} torr. (Courtesy Brinkmann Instruments.)

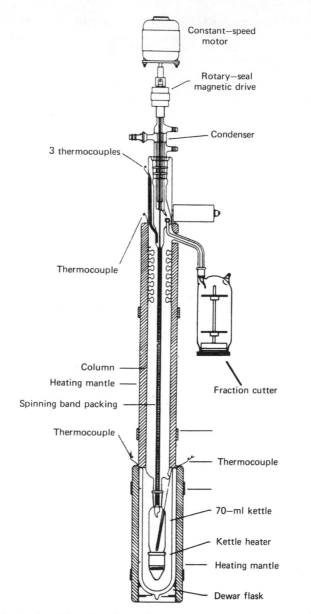

Fig. 5.16. Spinning-band fractionating column. (Courtesy Kontes Glass Co.)

proved slightly by increasing the speed to 5000 rev/min. By contrast, no pure benzene is obtained when the band is stationary. Nernheim [42] has studied the effects of band design with the aim of increasing the efficiency of miniature spinning-band columns. A four-bladed Teflon band seems to offer the most advantages. Tests at both total and partial reflux show that this band gives the high separating power of packed columns as well as the low pressure drop and small holdup inherent in band columns. The increased efficiency is probably caused by increased vapor-liquid contact and lower frictional heat at high band speed. Pease and co-workers [46] have described an improved stirrer and bearing, which provides effective stirring even when the volume of liquid has diminished to a few milliliters. Jentoft and co-workers [47] have designed a still head, which accommodates the rotating shaft, in such manner that the condensate returns to the system via a metal valve, with no holdup. A timing device activates the valve to allow condensate to enter the outlet tube. Nester and Nester [48] have constructed an apparatus (Fig. 5.17) in which the annular band is extended into the pot, and a stirrer is incorporated for agitation and high boilup. This apparatus is available in various sizes [49]; the micro model has a column length of 200 cm and a bore of 6 mm, a pot range of 5 to 10 ml, a holdup of 0.1 ml and theoretical-plate value of 10.

IV. MOLECULAR DISTILLATION

A. PRINCIPLE

In the molecular distillation process [50, 51], molecules of a liquid migrate in an appropriate vacuum from the surface of the liquid to the cooled surface of a condenser; hence it is particularly suited for microscale manipulations. It should be noted, however, that this process can be economically utilized only under certain conditions. The most important factor is the vacuum. The pressure at which the system operates should be low enough to prevent collision of the molecules of the distilling liquid with the gas molecules in the apparatus. Specifically, the mean free path of the gas molecules in the atmosphere inside the apparatus should be larger than the distance between the surface of the liquid and the condensing surface. For this reason, molecular stills are constructed in such manner that the distance between the surface of the liquid sample and the condenser is small, usually about 10 mm. The conditions for molecular distillation are usually achieved in a vacuum below 1×10^{-3} torr, at which pressure the mean free path for air molecules is 56 mm. The molecules of the liquid released from the surface shoot through the inner space of the apparatus, and some of them reach the cooled surface of the condenser.

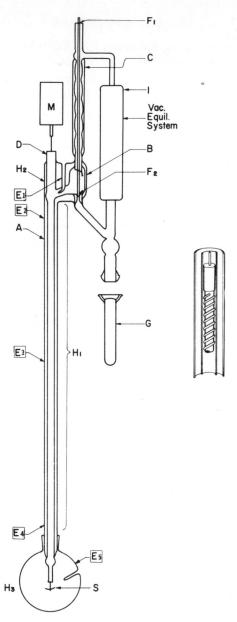

Fig. 5.17. Spinning-band distillation apparatus of Nester [48]. (Courtesy Perkin-Elmer Corp.)

Here the thermal energy of the molecules is absorbed, and condensation takes place. Like sublimation, this phenomenon does not occur at specific temperatures, such as the boiling point of a specific liquid sample. In general, the higher the temperature of the liquid, the faster is the molecular distillation process. The rate of molecular distillation is also increased by (1) reducing the distance between the surface of the liquid and the condensing surface, or (2) increasing these surfaces, or (3) increasing the temperature difference between the liquid and condenser.

B. APPARATUS

Several types of molecular distillation apparatus are available. A simple static type is illustrated in Fig. 5.18. The apparatus shown in Fig. 5.19 is designed for the distillation of 50 to 150 μl of material. Paschke and co-workers [52] have described a molecular still made of 3-cm-bore glass tubing; the upper chamber houses a spring-balance device, which consists of a quartz helix carrying a cross wire and a scale. The sample (about 500

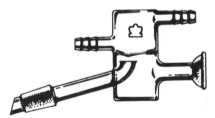

Fig. 5.18. Simple molecular-distillation apparatus. (Courtesy Kontes Glass Co.)

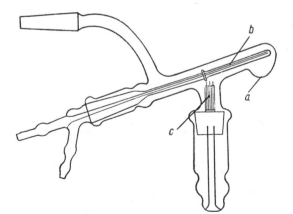

Fig. 5.19. Molecular-distillation apparatus for 50 to 150 μl of liquid [29]. *a*, distilling flask; *b*, condenser; *c*, receiver.

mg) is distributed on a roll of sheet glass wool, carried by a wire holder, which is suspended from the quartz helix by means of a glass fiber. The part of the tube enclosing the sample and its carrier is surrounded by an aluminum-block heater. At 2 to 3 cm below the heater, the tube is constricted to take either a device for collecting distilled fractions in ampuls or a Dry Ice–cooled trap. This apparatus was used for the fractionation of normal polymers. Rushman and Simpson [53] have proposed a molecular still for the quantitative analysis of 100-mg samples; the apparatus is essentially a micro hotplate in a high vacuum.

Sanders and Helwig [54] have constructed a molecular distillation chamber for handling 1 to 5 mg of material. As shown in Fig. 5.20, the chamber is exhausted from the bottom, thus preventing any of the distillate from getting entrained in the vacuum system. The apparatus is easily dismantled and cleaned; visibility is unobstructed. Fraction collecting is simple and precise, and temperature and pressure are measured in an absolute manner. This device is employed to analyze the organic particulate portions of airborne pollutants following solvent extraction from the air filter (see Chapter 7, Section II.B). The cold-finger condenser, refrigerated with Dry Ice–acetone, is 5 cm above the sample holder. Initially the pressure measurements must be the same on both the McLeod and the

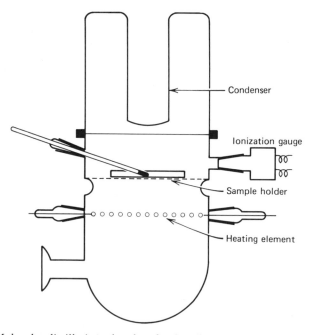

Fig. 5.20. Molecular-distillation chamber for handling 1 to 5 mg of material [54]. (Courtesy *Anal. Chem.*)

ionization gage. The ionization gage measures condensable as well as non-condensable gas pressures, while the McLeod gage measures only the noncondensable gas pressure.

A fluid-film-type molecular-distillation apparatus has been described by Loev and co-workers [55]. As illustrated in Fig. 5.21, the assembly consists of two round flasks connected to a rotating device used in rotary evaporators. This apparatus eliminates spattering due to residual traces of solvent or to superheating, and minimizes the exposure of the material to high temperatures, since the heat is applied only to a thin film of the liquid. The smallest volume of liquid that can be handled is 0.5 ml, using 25-ml flasks. The degree of vacuum attainable is limited by the seal at the rotating joint. Employing Apiezon Grease N, these workers have been able to evacuate the apparatus to about 1×10^{-2} torr.

Fischer [56] has designed a split-tube molecular-distillation device. It is similar to the concentric-tube fractionating columns (see Section III.C). A commercial molecular-distillation apparatus [49] using this principle is shown in Fig. 5.22. The sample is placed in the upper reservoir C and allowed to enter the still at a rate preset by the needle valve. The liquid material is swept down the heated wall in a thin descending film by the action of the rotating spiral band. Vaporization occurs on the inside of the heated wall, and the vaporized molecules are collected on the center cold finger. Condensed distillate flows down along the cold finger into the fraction collector A. Unvaporized sample is forced through the still by the rotating band and collected in the reservoir B. The mechanical recycling pump D returns this undistilled material to the sample reservoir C, where it is again admitted to the still.

C. OPERATION TECHNIQUES

The operation technique for molecular distillation varies with the type of the apparatus. Generally speaking, the molecular still is connected to a

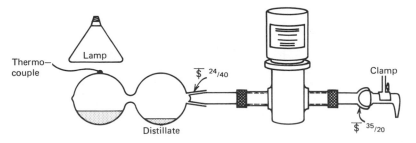
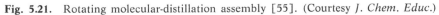

Fig. 5.21. Rotating molecular-distillation assembly [55]. (Courtesy *J. Chem. Educ.*)

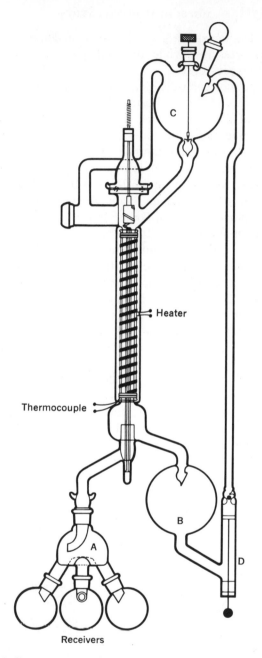

Heater

Thermocouple

C

B

A

D

Receivers

Fig. 5.22. Falling-film molecular-distillation apparatus [49]. (Courtesy Perkin-Elmer Corp.)

vacuum line, which consists of a mechanical forepump, a mercury diffusion pump, a manometer, and a system of cold traps. Prior to the distillation of the liquid material, the dissolved gases and other volatile components such as the solvent should be expelled. In the case of micro samples, this operation is carried out directly in the molecular distillation assembly by heating the apparatus while evacuating with a water aspirator.

It should be recognized that the effect of separation by molecular distillation is less than one theoretical plate. Therefore this method is not suitable for the fractionation of liquid mixtures whose components have boiling points close to one another. The chief advantage of molecular distillation lies in the fact that it permits vaporization and condensation at low temperature. It is the best procedure when the liquid sample is thermally unstable or cannot be boiled up in a conventional distillation apparatus. It is frequently employed for separating high-boiling and heat-sensitive liquids (e.g., vitamin A) from nonvolatile contaminents such as polymeric and inorganic materials.

V. DEEP-FREEZE CRYSTALLIZATION

Deep-freeze crystallization is a process in which the temperature of the liquid mixture is gradually lowered, whereby the individual components solidify and are separated in a very cold environment. For instance, McCully and McKinley [57] have described a method to freeze out lipids and waxes from benzene-acetone solutions. The semimicro freeze-out apparatus designed by Grussendorf and co-workers [58] is shown in Fig. 5.23. It is used to freeze out lipids and waxes at $-70°C$ from acetone solution in the following manner: Place acetone in tube D up to about 1 cm over a glass-wool plug. Cool the apparatus in a Dry Ice–acetone bath in a Dewar flask for 5 to 10 min; then raise the apparatus so that only that part which contains acetone is submerged in the freezing mixture. With a 1-ml syringe, transfer the liquid sample containing lipids and waxes to tube D and lay it gently on the acetone seal. Now insert pressure heads AB into joint C, lower the whole assembly into the Dewar flask, and add enough crushed Dry Ice so that level of freezing mixture is just below C. Place flask N in position and add 10 ml of acetone to reservoir A. Thus, the lipids and waxes which solidify are collected in tube D, while the acetone solution is recovered in flask N.

Gorbach [59] has constructed a cooling device to separate fatty acid glycerides from natural fats. Microcooling is achieved by means of thermoelectric cooling elements, and a cooling range from $+12$ to $-22°C$ can be progressively established. Friedrich [60] has described a low-temperature crystallization apparatus and claims that it reduces the problems of tem-

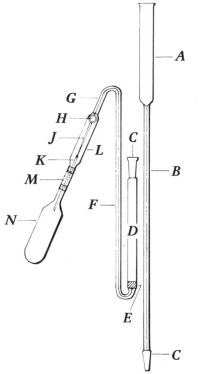

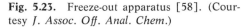

Fig. 5.23. Freeze-out apparatus [58]. (Courtesy *J. Assoc. Off. Anal. Chem.*)

perature and moisture control, as well as transference troubles, to a minimum and is suitable for semimicro quantitative work. Shapiro [61] has utilized the principle of freezing out for the quantitative concentration of small volume of dilute solutions. Mechanical stirring during freezing is an essential part of the process.

VI. COUNTERCURRENT DISTRIBUTION

Fractionation by countercurrent distribution [62] is based on repetitive equilibration of the individual components in a mixture between two liquid phases in an assembly consisting of several to hundreds of extraction units. The process involves multiple operations of liquid-liquid extraction (see Chapter 6, Section II.A). Hence the compounds separated may be solids or liquids at room temperature, and they are obtained by evaporation of the solutions in the respective extraction units at the end of the experiment. Fig. 5.24 shows a single unit of the extraction tube designed

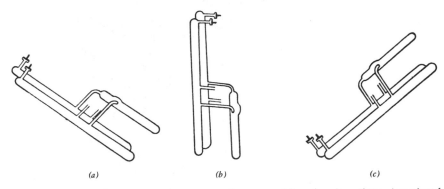

(a)   (b)   (c)

Fig. 5.24. Unit of countercurrent extraction assembly, showing three functional positions: (a) extraction; (b) transfer; (c) extraction.

by Craig [62]. Equilibration takes place when the tube is shaken in the oblique position (a). Subsequently, the upper liquid phase is automatically transferred from one unit through the storage compartment (b) to the next tube (position c). A simple apparatus (Fig. 5.25) for manual counter current distribution has been described by Bush and Post [63]; the units are made to hold 7 ml of the lower phase and up to 25 ml of the upper phase. Wall [64] has designed a micro mixer-settler that gives a 100% stage efficiency for phase volume samples of 4 ml.

During the countercurrent distribution process, the compounds are actually separated as a result of the stepwise migration of two liquid phases with respect to each other. Hence the similarity between countercurrent distribution and partition chromatography (see Chapter 4, Section I) is apparent. The separation of the components of a mixture depends on their respective partition coefficients in the migrating solvents. Since equilibrium is fully established in each extraction unit, the process can be subjected to mathematical evaluation. For successful fractionation, however, certain conditions have to be met. Before starting the experiment, the liquid phases should be mutually saturated. Furthermore, the concentration of the sample in each solvent should be so low that the partition coefficient is independent of concentration.

The countercurrent distribution method has been applied to a wide range of compounds, including lipophilic hydrocarbons [65], steroids [66], hydrophilic aminoacids [67], peptides [68], serum proteins, and synthetic polymers [69]. With the development of high-speed liquid chromatography (see Chapter 4, Section IV.D), however, microscale separation by countercurrent distribution has become less attractive. Nevertheless, coun-

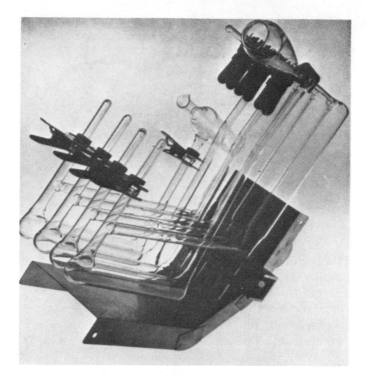

Fig. 5.25. Four-tube manual countercurrent distribution apparatus [63]. (Courtesy *Anal. Biochem.*)

tercurrent distribution is still an indispensable technique for the fractionation and purification of milligram amounts of material in very complicated mixtures.

REFERENCES

1. A. Weissberger (Ed.), *Technique of Organic Chemistry,* Vol. IV, 2nd ed. Wiley-Interscience, New York, 1965.
2. E. Krell, *Handbuch der Laboratoriumdistillation.* VEB Dtsch. Verlag, Berlin, 1958.
3. F. Paulik, L. Erdey, and S. Gal, *Z. Anal. Chem.,* **163,** 32 (1958).
4. E. L. Colichman, R. F. Fish, and G. O. Bjarke, *Anal. Chim. Acta,* **16,** 250 (1957).
5. A. R. Javes, C. Liddell, and W. H. Thomas, *Anal. Chem.,* **27,** 991 (1955).
5a. British Standards Institution, Methods for the determination of distillation characteristics, BS 4591:1971.
6. N. D. Cheronis, A. R. Ronzio, and T. S. Ma, *Micro and Semimicro Methods,* p. 62. Wiley-Interscience, New York, 1954.

7.   J. Nemec, *Chem. Listy,* **62,** 591 (1968).

8.   S. A. Schrader and J. E. Ritzer, *Ind. Eng. Chem., Anal. Ed.,* **12,** 54 (1939).

9.   A. Dobrowsky, *Mitt. Chem. Forsch. Inst. Österreich,* **8,** 126 (1954).

10. J. Pinkava, *Chem. Listy,* **48,** 455 (1954).

11. N. W. Jacobson and J. Miller, *Chem. Ind.,* **1957,** 1621.

11a. A. E. Sherwood, *Lab. Pract.* **21,** 353 (1972).

12. K. C. D. Hickman, *J. Phys. Chem.,* **34,** 637 (1930).

13. G. S. Riley, *Pharm. J.,* **194,** 320 (1965).

14. S. Sevick, *Chem. Listy,* **62,** 35 (1968).

15. H. D. Martin, *G. I. T.,* **12,** 1293 (1968).

16. R. Lee and D. K. Lynch, *Chem. Ind.,* **1969,** 693.

17. H. L. Chin, *Chem. Ind.,* **1968,** 1313.

18. R. W. Henderson and H. A. Kamphausen, *J. Chem. Educa.,* **41,** 572 (1964).

19. A. W. Schrecker, *Anal. Chem.,* **29,** 1113 (1957).

19a. Aldrich Kugelrohr, Aldrich Chemical Co., Milwaukee, Wis.

20. C. R. Engell, B. Kimland, and A. Rosengreen, *Acta. Chem. Scand.,* **24,** 1462 (1970).

21. Ref. 6, p. 190.

22. R. Roper and T. S. Ma., *Microchem. J.,* **1,** 248 (1957).

23. A. A. Morton and J. F. Mahoney, *Ind. Eng. Chem., Anal. Ed.,* **13,** 494 (1941).

24. M. J. Babcock, *Anal. Chem.* **21,** 632 (1949).

25. J. Nemec, *Chem. Listy,* **62,** 596 (1968); British Patent, 1,156,115 (1968).

26. F. Walls, *Microchem. J.,* **16,** 684 (1971).

26a. M. G. Larrabee, *Anal. Biochem.,* **1,** 151 (1960).

27. H. Stage, *CZ Chemie-Tech,* **1,** 263 (1972).

28. S. Huwyler, *Experimentia,* **27,** 1376 (1971).

29. C. W. Gould, Jr., G. Holzman, and C. Nieman, *Anal. Chem.,* **20,** 361 (1948).

30. H. Abbeg, *Chima,* **2,** 133 (1948).

31. Podbielniak Inc., Bensenville, Ill. see Ref. 6, p. 68.

32. V. Brezina, *G.I.T., Fachz. Lab.,* **14,** 493 (1970).

33. G. Juchheim, *G.I.T. Fachz. Lab.,* **10,** 914 (1966).

34. N. J. Levin, *Zh. Prikl. Khim.,* **31,** 1655 (1958).

35. C. Craig, *Ind. Eng. Chem., Anal. Ed.,* **9,** 441 (1937).

36. M. I. Rozengart, *Usp. Khim.,* **17,** 204 (1948).

37. W. G. Fischer, *Fette Seifen Anstrichm.,* **72,** 444 (1970).

38. W. G. Fischer, *G.I.T. Fachz. Lab.,* **13,** 535 (1969); **14,** 23 (1970).

39. W. G. Fischer, *Chemikerztg.—Chem. Appar.,* **94,** 157 (1970).

40. Rinco Instrument Co. Inc., Greenville, Ill.

41. L. J. Williamson, *J. Appl. Chem.,* **1,** 33 (1951).

42. A. G. Nernheim, *Anal. Chem.,* **29,** 1546 (1957).

43. A. L. Irlin, *Zh. Anal. Khim.,* **5,** 44 (1950).

44. Kontes Glass Co., Vineland, N. J.

45. M. Boivin, *Chim. Anal.*, **35**, 182 (1953).
46. W. F. Pease, A. H. Gilbert, and A. Cahn, *Anal. Chem.*, **32**, 894 (1960).
47. R. E. Jentoft, W. R. Doty, and T. H. Gouw, *Anal. Chem.*, **41**, 223 (1969).
48. R. G. Nester and R. M. Nester, in Pittsburgh Conference on Analytical Chemistry and Applied Spectroscopy, 1966.
49. Nester-Faust Glass Products Products Division, Newark, Del.
50. R. Watt, *Chem. Ind.*, **1961**, 680.
51. K. C. D. Hickmann, *Chem. Rev.*, **34**, 51 (1944).
52. R. F. Pascke, J. R. Kerns, and D. H. Wheeler, *J. Am. Oil Chem. Soc.*, **31**, 5 (1954).
53. D. F. Rushman and M. G. Simpson, *J. Oil Col. Chem. Assoc.*, **37**, 319 (1954).
54. G. R. Sanders and H. L. Helwig, *Anal. Chem.*, **31**, 484 (1959).
55. B. Loev, K. M. Snader, and M. F. Kormendy, *J. Chem. Educ.*, **40**, 426 (1963).
56. W. G. Fischer, *G.I.T. Fachz, Lab.*, **15**, 753 (1971).
57. K. A. McCully and W. P. McKinley, *J. Assoc. Off. Anal. Chem.*, **47**, 652 (1964).
58. O. W. Grussendorf, A. J. McGinnis, and J. Solomon, *J. Assoc. Off. Anal. Chem.* **53**, 1048 (1970).
59. G. Gorbach, *Mikrochim, Acta,* **1962**, 1035.
60. J. P. Friedrich, *Anal. Chem.*, **33**, 974 (1961).
61. J. Shapiro, *Science,* **133**, 2063 (1961).
62. L. C. Craig and D. Craig, in A. Weissberger (Ed.), *Technique of Organic Chemistry,* Vol. III. Interscience, New York, 1956.
63. M. T. Bush and O. W. Post, *Anal. Biochem.*, **32**, 145 (1969).
64. G. P. Wall, A.E.R.E., CE/R 1730 (1955).
65. P. Plattner, E. Heilbronner, and S. Weber, *Helv. Chim. Act,* **32**, 574 (1949).
66. L. L. Engel and W. R. Slaunwhite, *J. Biol. Chem.*, **191**, 621 (1951).
67. L. C. Craig, W. Hausman, E. H. Ahrens, and E. F. Harfenist, *Anal. Chem.,* **23**, 1236 (1951).
68. D. W. Wooley, *J .Biol. Chem.*, **179**, 593 (1949).
69. V. P. V. Tavel, *Chimia,* **23**, 57 (1969).

SEPARATION OF ONE SPECIFIC PORTION FROM A MULTICOMPONENT MIXTURE

I. GENERAL REMARKS

When a complex mixture is submitted for analysis, it often suffices to separate only a specific portion of the working material. Usually the particular portion obtained after separation from the bulk is still a mixture. This may be due to (1) the separation device (e.g., a membrane for dialysis), (2) the technique (e.g., steam distillation), or (3) the nature of the intended separation (e.g., unsaponifiable matter from butter oil [1], fatty acids from hydrolysates [2], ether-soluble food preservatives [2a], organic mercury compounds in polluted water [3]). Such an experiment in general represents the first step in the analysis of a complicated multicomponent system.

In the present chapter, we discuss the methods and techniques that are suitable for microscale manipulations with the aim of isolating a specific portion from a multicomponent mixture. The physical methods for separating such a fraction from a solution (Section II) and from solid materials (Section III) are described first, followed by the separation of vapors from solid or liquid phase (Section IV). The chemical methods, which are applicable to both solid and liquid samples, are discussed last.

The separation of a specific portion from micro amounts of a complex mixture merits special attention for several reasons. First, as in other microscale manipulations, the capacity of the vessel and volume of reagent used are naturally reduced with the sample size, but there is a limit to how far this can be extended. (For instance, the smallest separatory funnel with a stopcock has a capacity of about 2 ml.) Secondly, since the microsample is a complex mixture, the difficulty of recovering all the desired material from the vessel is increased owing to the adhesion of liquid and adsorption of ions and molecules on the walls. Thirdly, whereas the unwanted portion of the complex mixture should be completely removed, the material that is to be retained should not be lost. Two examples may be cited: (1) Organics that interfere with the determination of formaldehyde can be removed by porous polymer adsorbents [4], but all the formaldehyde present in the original sample may not elute from the column. (2) A number of metal ions can be extracted quantitatively from iodide-containing aryl sulfonic acid solutions by 2-ethyl-1-butanol; some metals are partially extracted, others not at all [5]. Thus, a method for group separation of metals presents itself. It should be ascertained, however, that the presence of certain metal ions does not affect the extractability of the other ions.

It should be noted that some methods described in the preceding chapters are adaptable to the separation of one particular portion from a complex mixture. For instance, sublimation (Chapter 3, Section III) can be employed to isolate all volatile components from solid material [6]; freeze-drying separates liquids from solids; ion exchange (Chapter 4, Section VIII) removes ionic species from nonionic compounds. These techniques are not treated again in the present chapter. On the other hand, certain devices (e.g., molecular sieve, centrifuge) that are mentioned briefly in previous chapters are discussed further for reasons stated in the respective sections.

It is evident that a good deal of the equipment described in Chapters 2 to 5 may be utilized in the methods discussed in the present chapter. Besides, there are apparatus which have been designed to perform several techniques of separation. Thus, Mohanty and co-workers [7] have constructed an all-glass assembly for liquid-liquid extraction, azeotropic distillation, and steam distillation. Kramer [8] has fabricated a "Combitrockner," which can be set up for Soxhlet extraction, liquid-liquid extraction, and the continuous drying of solvents by means of molecular sieves.

It may be mentioned that the separation of a multicomponent mixture into portions is sometimes performed for the purpose of analysis without the objective of separating the material for the sake of recovering a particular portion of the original sample. For example, a standard method for classifying gasoline [9] involves distilling the sample at atmospheric pressure by reading the vapor temperature (with a thermometer for the range 2 to 300°C) simultaneously with the collection of each 10-ml portion. Hummel and Crummett [10] have utilized extraction to determine the purity of widely different organic compounds by extraction-solubility experiments.

From the viewpoint of experimental organic chemists, the present chapter is concerned with separation methods that are generally less selective than those discussed in Chapters 3, 4, and 5. Therefore, application of the techniques described below results only in obtaining a portion of the original working material characterized by certain common properties (e.g., ether solubility, steam volatility, the presence of a carbonyl function). This particular portion still constitutes a mixture, which can be further fractionated by appropriate specific methods (e.g., gas chromatography, recrystallization) described in the previous chapters.

II. SEPARATION OF ONE PORTION OF MATERIAL FROM A SOLUTION

A. LIQUID-LIQUID EXTRACTION

1. Principle

In liquid-liquid extraction processes [11, 12], the partition coefficient is a measure of the distribution of molecules between the two liquid phases. At low concentrations and at constant temperature, the partition coefficient (usually referred to as the distribution coefficient K_D) is a constant:

$$K_{D(A)} = \frac{[A]_M}{[A]_N}$$

where $K_{D(A)}$ is the partition coefficient of compound A distributed between solvents M and N.

When the solution contains only one solute (i.e., compound A), the situation is very simple: a large value of K_D indicates that solvent M is an efficient means of separating compound A. When there are two or more compounds in the solution, the K_D values of the all components should be considered. The simplest case for the separation of two compounds, A and B, occurs if $K_{D(A)}$ is very large and $K_{D(B)}$ very small ($K_{D(A)} > 1$, $K_{D(B)} < 1$); then single-step extraction will be a satisfactory technique to achieve separation after two or three repeated equilibrations with fresh solvents. In another case, compound A has a very small partition coefficient (M is the extracting solvent), while the coefficient for compound B is close to 1 ($K_{D(A)} < 1$, $K_{D(B)} \approx 1$). If solvent N is water, then compound A is retained in the aqueous phase while compound B is extracted into solvent M. However, since $K_{D(B)}$ is close to 1, the separation of compound B is incomplete with single-step extractions, and the repetitive or continuous liquid-liquid extraction process is recommended. In still another case, both compounds have K_D values close to 1; then neither the single-step nor the continuous technique can achieve separation. In a multicomponent mixture, compounds with various K_D values are generally present; therefore the extract usually contains more than one solute.

In practice, the complex mixture is usually in an aqueous solution to be extracted with organic solvents which are immiscible with water. It should be recognized that the characteristics of the water–organic-liquid extraction system are dependent on the properties of the organic solvent and its interaction with the substrate molecules. The organic liquid can be polar (e.g., n-butyl alcohol, benzyl alcohol, phenol, nitrobenzene, ethyl acetate, nitromethane), nonpolar (e.g., hexane, cyclohexane, benzene, carbon tetrachloride), or of intermediate polarity (e.g., diethyl ether, methylene chloride, anisole, chloroform); it may be a hydrogen donor (e.g., alcohols, phenols, chloroform) or a hydrogen acceptor (e.g., ethers, esters, cyclohexanone, collidine). The K_D values change from one solvent pair to another as result of the differences in solvent-substrate interactions. Furthermore, water can be partly or entirely replaced by polar organic solvents such as methanol, formamide, dimethylformamide, dimethylsulfoxide [13], and sulfolane, and the resulting solution can be extracted by a nonpolar liquid such as petroleum ether.

There are some empirical rules for the selection of the extraction solvent system. Since ionic species, either organic or inorganic, have a strong affinity for water, the separation of nonionic compounds from ionic species usually takes place readily. This principle can be further extended by using a buffer, strong base, or strong acid in the system. The pH of these aqueous solutions provides conditions for either dissociation of weak acids or protonation of weak bases. In both cases, ionic species are produced from neutral molecules, and their affinity toward water is increased. Thus, Gupta and co-workers [14] have studied the effect of pH on the extraction of thiamine dye salt and derived an equation to express the relationship.

For neutral molecules (organic compounds), the affinity of the extraction system for solvents depends primarily on the presence of polar groups and the molecular weight. The presence of polar groups, particularly those with labile hydrogen available for hydrogen bonding (e.g., —OH, —NH$_2$, —SH) and those with basic functions carrying unshared electron pairs (e.g., $\ddot{\text{O}}$:, $\equiv$N:, C=$\ddot{\text{O}}$:), enhances the affinity for water, while an increase in the number of carbon atoms in the molecule operates in the opposite direction. Consequently, the differences in K_D for a water-organic liquid system can be approximated by the $C_m/$(hetero atom O, N, S)$_n$ ratio. A ratio of about 1 is characteristic of a compound with highly hydrophilic properties (e.g., carbohydrates, amino acids, peptides). In contrast, large numerical values of this ratio indicate compounds of low hydrophilicity.

The ionic character of metal cations can be partially or completely obliterated by chelation; hence chelated ions [14a] are extractable from aqueous solutions into organic liquids. This technique is utilized frequently in the analytical and preparative separation of many metallic elements. Similarly, the properties of organic ligands can be changed by chelation with an inorganic partner, and this principle has been employed for the selective separation of certain organic compounds. For example,

1,2-diols can be chelated with boric acid to form strongly hydrophilic species [14b]:

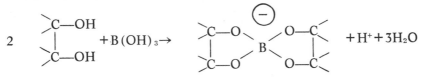

Sometimes inorganic salts in solid form are added to the aqueous solution to effect extraction into the organic phase (salting-out technique). For instance, Coward and Smith [15] have described methods of preparing urinary extracts using acetonitrile, ethyl acetate, or diethyl ether by adding Na_2SO_4 or $(NH_4)_2SO_4$ crystals.

Owing to its simplicity in equipment and operation as well as quantitative aspects, solvent extraction is a popular separation and isolation technique. It is frequently used in microscale manipulations, especially in those fields where clean and quantitative work is required, such as the separation and purification of isotope-labeled materials [16]. Reinhardt and Rydberg [16a] have described a rapid and continuous system for measuring the distribution ratios in solvent extraction.

2. Single-Step Liquid-Liquid Extraction

In the single-step liquid-liquid extraction process, equilibrium is established in one operation for all components of the system. After the solution containing the working material is placed in the same vessel with the extracting solvent, agitation is usually necessary to facilitate the establishment of equilibrium. If the system has a tendency to form emulsions, rotary swirling is recommended instead of shaking the vessel. If an emulsion has already formed, it may be broken by churning the material in the vessel with a glass rod in slow strokes. Hautke and Folk [17] have found that stable emulsions formed during extraction of beer and wort with hydrocarbons can be broken by centrifugation or by adding anhydrous sodium sulfate after the bulk of the aqueous phase has been removed. Sometimes clear liquid can be obtained from an emulsion by filtration to remove the "emulsifiers."

The separatory funnel is the common apparatus for single-step extraction. A PTFE (Teflon) stopcock is preferred, since it obviates the use of stopcock grease, which may dissolve in the organic liquid, causing leakage and contamination. The polyethylene separatory funnel shown in Chapter 2, Fig. 2.7 is easily fabricated. Karamian [18] has recently described an improved separatory funnel (Fig. 6.1), which facilitates draining the upper layer and is operated as follows: Following an extraction, the sub-

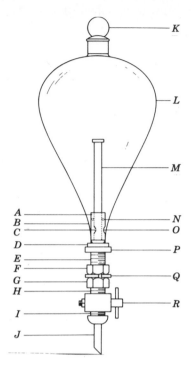

Fig. 6.1. The Karamian separatory funnel [18]. (Courtesy *Anal. Chem.*)

stance being sought is transferred from the aqueous layer to the organic layer (lighter than water). During this procedure, the drain holes C and O are disaligned. The height of the conduit M is adjusted to the lower level of the organic layer, which is then removed by means of the stopcock R. At this stage, if desired, a second extraction can be performed by adding another portion of the organic liquid, and the procedure is repeated. The aqueous lower layer is removed from the separatory funnel by aligning the drain holes C and O and draining via the stopcock R. This alignment is accomplished by manipulation of the flare nuts F and G and the lock washer Q.

Two examples of microscale separatory funnels are shown in Chapter 2, Fig. 2.23. The liquids are transferred into these funnels by suction. The simplest device for liquid-liquid extraction is the test tube. Agitation can be achieved manually or mechanically (see Chapter 2, Fig. 2.54). After equilibrium has been established, separation of two layers can be facilitated by spinning the test tube in the centrifuge. Either of the two liquids can be recovered by means of a pipet (see Chapter 2, Fig. 2.22). The lower layer is removed more conveniently from a conical test tube. The upper layer is pipetted off more readily by utilizing the capillary elevation of the liquid at the walls. Another procedure for separating the aqueous from the organic layer is to use phase-separating filter paper [19], which fits into a filter funnel (Chapter 2, Fig. 2.34). Since the paper is water repellent, organic liquids heavier or lighter than water both go through, while the aqueous phase remains in the funnel. Foams also can be broken in this manner. Rajama and Mäkalä [20] have employed a test tube to extract preservatives (benzoic and sorbic acids) from fruit juices; the generation of emulsions is reduced by using a large proportion of diethyl ether (20 ml) in comparison with the aqueous phase (2 ml of 1:4 diluted juice saturated with sodium chloride). The extraction is carried out at pH 6 in the presence of cyclamates or saccharin.

The use of an extraction vessel, shown in Fig. 6.2, prevents losses upon shaking the mixture [21]. The pipet is used to transfer the liquids. Thornett [22] has described an apparatus for extraction with solvent volume 1 ml. It consists of a capillary U tube connecting two vessels, in which liquids may be mixed and separated by operation of a plunger; it is particularly suited for use with liquids that are dangerous to handle. An extractor fabricated by Klc [23] is a centrifuge tube cut into two parts, which are rejoined by a piece of capillary tubing (32 mm × 0.33-mm bore); the combined volume of the lower conical section and the capillary is 2.5 ml. Laessig [24] has performed solvent extraction of blood alcohol using a Unopette (see Chapter 2, Fig. 2.41).

Marchat [25] has described two microextractors (Fig. 6.3) made from polyethylene tubing. For volumes over 1 ml, the tube is closed with a cap at one end and a capillary fitted into a cap (carrying a filter paper) at the other end. After being shaken, the device is held vertically, with the capil-

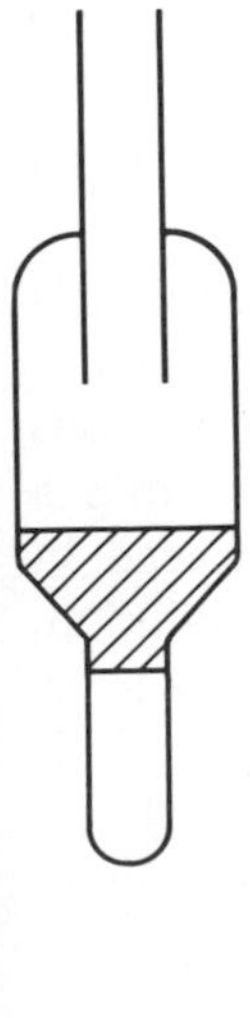

Fig. 6.2. Extraction vessel of Holeysovsky [21].

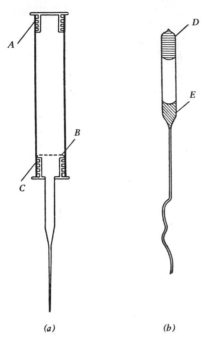

(a)   (b)

Fig. 6.3. Polyethylene microextractors. (*a*) For volume over 1 ml: *A*, polyethylene plug; *B*, siliconized filter paper; *C*, polyethylene plug joined to glass tubing. (*b*) For volume under 0.1 ml: *D*, organic layer; *E*, aqueous layer.

lary downwards, and the aqueous phase passes through the filter paper. When the organic phase is denser than the aqueous, a silicone-treated filter paper is used. For volumes less than 1 ml, the polyethylene tube is sealed at one end and drawn out to form a capillary at the other end. After squeezing the tubing, the sample solution and extracting liquid are sucked into the device, which is then agitated and put in a horizontal position. Owing to surface-tension effects, the aqueous phase is collected at the closed end and the organic phase at the capillary end. The organic liquid can be squeezed out and replenished with fresh solvent. Recovery of the aqueous phase is achieved by cutting the polyethylene tube.

Morgan [26] has described a method for extraction in a beaker that is divided into two compartments, making it possible to float an aqueous solution on an organic liquid and so extract a solute from one aqueous solution into the organic phase and then from the organic phase into a second aqueous solution. It has been used to transfer 20 μg of lead in 25 ml of alkaline solution through dithizone-$CHCl_3$ into a HNO_3 solution, with 100% recovery after agitation for 150 min by a magnetic stirrer. Niedermaier [27] has constructed a double separatory funnel that it is claimed prevents loss of solution during extraction. Ingamells [28] has fabricated a separatory flask by attaching the bottom half of a separatory funnel to a conical flask. A similar device has been used by Hubbard and Green [29] for multiple solvent extractions, as illustrated in Fig. 6.4.

Bailey and co-workers [30] have described a semiautomatic extraction procedure using a rotary extractor. Milwidsky [31] has designed an apparatus for the automated extraction, with subsequent washing, of unsaponified or unsulfonated compounds from detergents. Automatic detection and phase separation have been achieved by the use of sensing electrodes or centrifugation, as reported by Trowell [32] and Sutton and Vallis [33]. Reinhardt and Rydberg [34] have designed a system for continuous unit operations, including mixing, separation, and on-line analysis.

3. Continuous Liquid-Liquid Extraction

Continuous liquid-liquid extraction involves running a steady stream of fresh extracting liquid through the solution containing the working material, and combining the extracts. In most devices, the fresh solvent is obtained by boiling the extract; the vapors are condensed in a reflux condenser, and the liquid flows into the extraction chamber, whence it carries more of the extracted compounds into a distilling flask holding the extract. It is apparent that partition equilibrium is seldom established during the process. Nevertheless, given sufficient time, practically all extractable materials will leave the original solution.

Usually the working material is in aqueous solutions to be extracted continuously by low-boiling organic liquids such as diethyl ether, petro-

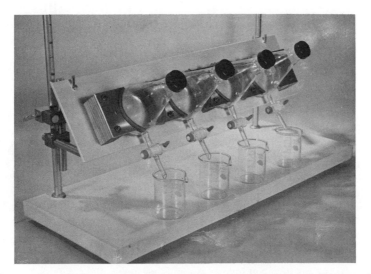

Fig. 6.4.   Extraction flasks [29]. (Courtesy G. L. Hubbard and T. E. Green.)

leum ether, chloroform, benzene, and ethyl acetate. Since the extracting liquid flows by gravity, the arrangement of the extraction device is dependent on whether the organic liquid is lighter or heavier than water. Some simple apparatus are shown in Chapter 2, Figs. 2.25–2.28. Various styles of commercial extractors for use with organic liquids lighter than water are available, three samples [35] being given in Fig. 6.5. The extraction is facilitated by increasing the contact between the liquids in a spiral upward path or by dispersing the extracting liquid into minute droplets through the fritted disc. Figure 6.6 shows an assembly in which the aqueous solution is placed in a test tube surrounded by the vapors of the extracting liquid; hence it cannot be used with extracting liquids that boil at temperatures above 100°C. The apparatus can be modified to handle 50 ml or less of the sample solution. They are employed for the extraction of trace constituents (in the range of ppm) in order to obtain sufficient material for subsequent microscale manipulations [35a].

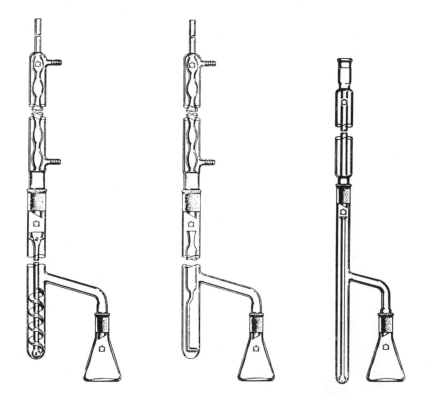

Fig. 6.5. Apparatus for continuous liquid-liquid extraction with solvent lighter than water [35]. (Courtesy Kontes Glass Co.)

Fig. 6.6. Continuous liquid-liquid extraction at boiling temperature of extracting liquid. (Courtesy Kontes Glass Co.)

Hollander and Zanten [36] have described an extraction device that consists of two calibrated vessels interconnected at overflow level through a PTFE stopcock; a reflux condenser is mounted in one vessel through a ball joint, and a ball-jointed tube leads from the upper end of the other vessel into the condenser. Other extractors have been constructed by Martin [37], Riley [38], and Piorr [39]. Pan [40] has designed a novel continuous liquid-liquid extractor that consists of a column packed with strands of a rayon-cotton yarn, which confines one liquid by capillary action. The second liquid flows countercurrently outside the strands. The strands behave as wall-less tubing, permitting solute flow in the radial direction. Because gravity is the only driving force involved, this extractor can run unattended for prolonged periods.

Werner and Waldichuk [41] has constructed a device for extracting small amounts of material from large volumes of solution. Goldberg and DeLong [42] have described two apparatus (Fig. 6.7) that can extract and simultaneously concentrate organic solutes from water; a concentration factor of up to 10^5 is obtained with this technique.

Figure 6.8 shows two commercial apparatus for extracting aqueous

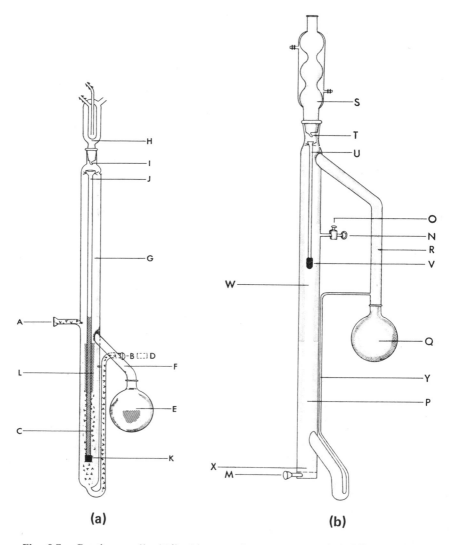

(a)   (b)

Fig. 6.7. Continuous liquid-liquid extraction apparatus of Goldberg and DeLong [42]: (a) For solvent lighter than water. (b) For solvent heavier than water. (Courtesy *Anal. Chem.*)

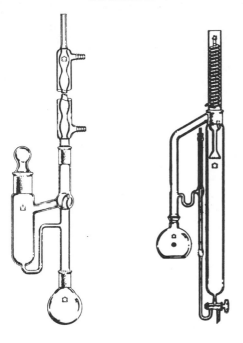

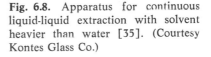

Fig. 6.8. Apparatus for continuous liquid-liquid extraction with solvent heavier than water [35]. (Courtesy Kontes Glass Co.)

solutions with organic liquids heavier than water. Palecek and Skala [43] have designed an apparatus with interchangeable parts, thus permitting continuous extraction with light or heavy solvents in volumes from 20 ml upward, at reduced pressure if required, as illustrated in Fig. 6.9. Popov and co-workers [44] have described a method for the continuous extraction of copper and iron from vegetable oils with an azeotropic mixture of HCl and H_2O (21% HCl).

4. Modifications

Fuginaga and co-workers [45] have proposed a method that involves liquid-liquid extraction at elevated temperature, followed by solid-liquid separation at room temperature. For example, copper in aqueous solution is extracted at 90°C into 1% 8-hydroxyquinoline in biphenyl (melting point 70°C); as the mixture cools, the biphenyl layer solidifies and the aqueous phase is removed by decantation. This technique can be utilized to remove interfering major components in trace analysis.

Difficulties are sometimes encountered in the extraction of concentrated solutions or those containing insoluble residues. The former may form a second organic phase or emulsions upon mixing; the latter may lead to the presence of solid particles at the interface. Ko [46] has described a

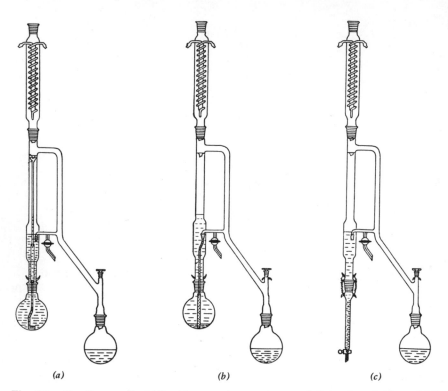

(a)   (b)   (c)

Fig. 6.9. Continuous liquid-liquid extraction apparatus of Palecek and Skata [43]: (a) for extraction with solvent lighter than water, (b) for extraction with solvent heavier than water, (c) for azotropic separation of water (Courtesy *Chem. Listy.*)

static extraction technique to circumvent these situations. The extraction is performed in polyethylene bottles by adding 10 ml of the aqueous solution and 2 ml of the organic liquid. The bottle is capped and set aside overnight. The organic layer is then removed by tilting the bottle; a light shining through the bottle clearly shows the separation of the phases. Fresh organic liquid is added and the process is repeated.

Roedel and co-workers [47] have extracted volatile substances (e.g., esters, aldehydes, ketones) from dilute aqueous solutions by passing diethyl ether–pentane (2:1) upwards through the sample in the form of fine bubbles, thus avoiding the use of a large volume of inflammable solvent; it is claimed that the extraction time is thus reduced from several hours to about 20 min. Bowen reported that flexible polyurethane foams can be used as selective absorbents for a number of substances from dilute aqueous solutions [48]. These substances include those that are normally extracted by diethyl ether (e.g., iodine, benzene, chloroform, phenol); Hg(II) and Au(III) from 0.2 M HCl; and U(VI) from saturated Al(NO$_3$)$_3$ solution. The capacities of different foams for absorption range from 0.5 to 1.5 mmol per gram of foam. Desorption can be achieved without difficulty. Kienberger [48a] has constructed an automated extraction apparatus for the recovery of uranium in solutions.

B. STEAM DISTILLATION OF LIQUID SAMPLES

1. Principle

By steam distillation it is usually understood that a current of live steam is conducted through the sample, whereupon the water vapor leaves the solution together with some of the volatile substances, which are subsequently recovered by condensation. When a complex mixture is subjected to steam distillation, the volatile portion may contain four categories of compounds:

1. Water-soluble compounds that are easily vaporized (e.g., NH$_3$, methanol, acetic acid).
2. Water-insoluble compounds that are easily vaporized (e.g., hexane, benzene). Distillation for this category should be away from open flames because of the fire hazard.
3. Water-insoluble compounds that form azeotropes with water (e.g., bromobenzene).
4. Water-insoluble compounds that exhibit appreciable vapor pressure at 100°C (e.g., essential oils) without forming an azeotrope.

An azeotrope is a mixture of two or more liquids characterized by constant composition at a constant pressure and hence by a constant boiling temperature. In most cases of steam distillation, water contributes more than 80% of the total vapor pressure at the boiling point. Nevertheless, owing to the low molecular weight of water, steam distillation is still an efficient method of separating the particular substance from the working material. For example, the boiling point of water-bromobenzene azeotrope is 95°C at 760 torr. At this temperature, the vapor pressure of water is 640 torr and that of bromobenzene 120 torr. Therefore the composition of the azeotrope is 85:15 for the molar ratio, and 38:62 for the weight ratio (molecular weights: H$_2$O = 18, C$_6$H$_5$Br = 157).

Many organic compounds of category 4 are separated by steam distillation, instead of by direct distillation, primarily to prevent their possible decomposition at high temperatures. The working material is exposed to an environment of only about 100°C during steam distillation, although a long period of distillation is required to separate compounds of low volatility.

The volatility of a compound with steam is structure dependent. For instance, o-nitrophenol and p-nitrophenol differ considerably in volatility, since the former can engage in intramolecular hydrogen bonding, while the latter undergoes only intermolecular hydrogen bonding, resulting in large molecular associates. The steam distillation of certain mixtures may be controlled by adjusting the pH of the solution. By use of the fact that charged particles are strongly solvated in water and hence nonvolatile, organic bases and acids can be selectively separated from neutral steam-volatile compounds. Thus, by making the solution strongly acidic, the basic substances are retained by forming cations while the neutral compounds and organic acids are steam-distilled. Similarly, phenols and steam-volatile acids can be separated from neutral compounds by making the solution alkaline before steam distillation.

2. Apparatus, Operation, and Applications

Two simple steam-distillation apparatus for microscale manipulations are given in Chapter 2, Figs. 2.29 and 2.30. An assembly [49] for working with 10 ml or more of the liquid sample is illustrated in Fig. 6.10; it consists of the steam generator A; the steam trap B, which can be closed

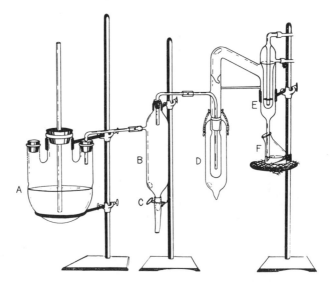

Fig. 6.10. Assembly for steam distillation [49]. (Courtesy *Microchem. J.*)

by means of the pinch clamp C; the distillation vessel D, which has a vacuum jacket; and the condenser E. A commercial [35] apparatus is shown in Fig. 6.11. It may be mentioned that the micro Kjeldahl distillation apparatus [50] is a steam-distillation device, but the distillation vessel cannot be detached for cleaning. An efficient condenser is recommended, since the steam usually passes through rapidly, and its condensation involves the exchange of a large amount of heat content because of the large heat of vaporization for water. Other devices for steam distillation have been proposed, such as the apparatus described by Bohm [51], Seehofer and Borowski [52], and Antonacopoulos [53]. Herlihy and Hayes [54] have constructed an apparatus that performs six distillations simultaneously.

The operation of steam distillation is simple. If the source of steam comes from the pipe line, it should be purified. Attention should be given to the increasing volume of the liquid in the distillation vessel due to steam condensation, foaming, and splashing of the contents. All volatile materials from the sample should be condensed in the condenser.

Steam distillation is used frequently in preparative organic chemistry as a technique for purification. It is a common method of separation in the investigation of natural products. It is also employed in quantitative organic functional-group analysis, such as the determination of acetyl [55] group, and the micro Kjeldahl procedure for nitrogen determination. Rougieux and Cesaire [56] have described a method for the isolation of volatile acids, ethanol, ketones, or ammonia from 5 ml of test solution within 3 min. Ridle [57] has separated 0.5 to 100 μg of organic acids in small samples by steam distillation. Herbert [57a] has reported a collaborative study of the distillation procedure for the determination of formaldehyde in maple syrup. Halvarson [57b] has described a method for analyzing volatile carbonyls in fats by steam distillation in a stream of H_2O-saturated nitrogen.

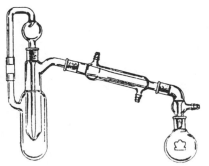

Fig. 6.11. Steam-distillation apparatus [35]. (Courtesy Kontes Glass Co.)

3. Continuous Steam Distillation

Williams [58] has designed an apparatus for continuous steam distillation by recycling. It enables the steam distillate to be continuously extracted before returning to the distillation vessel for further steaming of the working material. In this way, the steam distillation may be carried on as long as desired with a limited quantity of water, all the steam-volatile substances eventually passing into the relatively small amount of extracting solvent. The apparatus is shown in Fig. 6.12. The working material is mixed with 100 ml of water, placed in the flask A, and brought to the boil. The steam-volatile substances are condensed by the condenser C and allowed to run via D and E into the receiver F. When the liquid level in F reaches that of the side arm G, it flows back into the flask A. The level of the side arm G is such that it is approximately 0.5 cm lower than that of the other side arm H; thus water runs back into the flask A, rather than to the extractor flask I. As this steam cycle is taking place, the

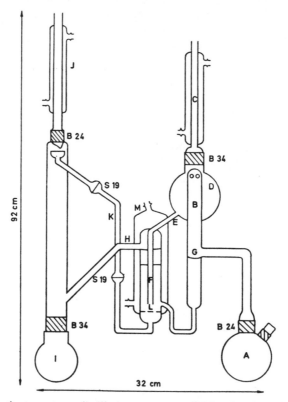

Fig. 6.12. Continuous steam-distillation apparatus [58]. (Courtesy *Chem. Ind.*)

water in the receiver F is continuously solvent extracted. The extracting solvent is circulated from the flask I (100-ml capacity) via the condenser J and takeoff arm K, is dispersed by the sinter L, bubbles through the steam distillate, forming an upper layer, and finally returns via the side arm H to the flask. The condenser M is incorporated to cool the hot water as it enters the receiver, thus preventing the boiling of the extracting solvent.

C. DIALYSIS

1. Principle

Dialysis is a separation process in which one portion of the complex mixture diffuses through a permeable membrane by the principle of osmosis while the other portion is retained behind the membrane. The separation is dependent on molecular-weight differences and is controlled by the pore size of the permeable membrane. The sample solution is dialyzed against the dialyzate, which is the liquid medium that receives the substances leaving the membrane. On the basis of pores diameter, membranes are classified by the molecular-weight cutoff values, which usually range between 2000 and 30,000. Therefore these membranes are all permeable to inorganic ions and monomeric organic molecules, while polymers (e.g., proteins) are retained.

A variation of dialysis, known as electrodialysis [59], involves placing the dialyzing chamber between two electrodes, with pure water in compartments on either side. Under the influence of a direct current, the charged ions migrate from the sample solution to the oppositely charged electrodes. Thus, the transport of particles through the permeable membrane is accelerated in a homogeneous electric field.

2. Apparatus and Operation

Cellulose and its derivatives, such as acetates and nitrates, are the common materials for making dialysis membranes. Commercial cellophane casings and membranes are generally used, although special dialysis bags can be made in the laboratory [59a]. The most common form of dialysis vessel is a cellophane tube closed at one end. After the sample is introduced, the cellophane tube is placed in a suitable container that holds the dialyzate, usually deionized water. The dialyzate is conveniently stirred by means of a magnetic stirrer. Alternatively, the cellophane tube can be agitated in a rocking machine; for this purpose the tube should be sealed at both ends and a glass bead placed inside to facilitate mixing. The dialyzate is frequently or continuously replaced with fresh solvent.

The apparatus for dialysis can be easily fabricated. An example, described by Wood and co-workers [60], is illustrated in Fig. 6.13. It consists of a circular plate with six grooved, tapered tubes (a) onto which lengths of wet Visking dialysis tubing (size: ¼ in.) are slipped and secured with O rings (b) that fit into the grooves. The lower ends of the tubes are tied off, and the tubes are filled with the sample (2 to 5 ml) to be dialyzed. The plate rests on the top of a beaker whose contents are agitated with a magnetic stirrer.

Toms [61] has described a dialysis assembly in which the cellophane tubing is bent to form a U-shaped vessel. Other workers [62, 63] have prepared dialysis bags by fastening the membrane to glass tubes. A method that involves putting the membrane between two plastic plates has been advocated [64, 65]; the commercial model [66] is shown in Fig. 6.14. The volume of working material required is reduced by this arrangement. Various designs for handling microsamples have also been proposed. For instance, Wilkinson [67] has described a device for processing 0.2 to 1.0 ml of body fluids; Reitz and Riley [68] have fabricated an apparatus which accommodates cells with a capacity of 0.025 to 0.5 ml. Hesketh and Dore [69] have used disposable polyethylene disalysis cells to hold 50 μl of liquids. Heatley [70] has described a simple device for closing

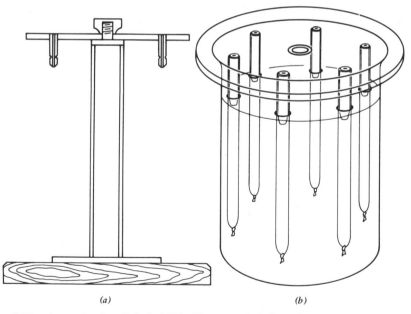

(a)   *(b)*

Fig. 6.13.   Apparatus for dialysis [60]. (Courtesy *Lab. Pract.*)

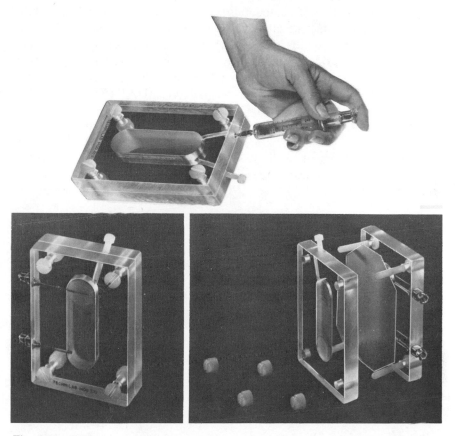

Fig. 6.14. Dialysis cell between plastic plates [66]. (Courtesy Cole-Parmer Instrument Co.)

small-bore cellophane tubing without knotting and also an apparatus for the dialysis of 5 to 50 μl of sample through a 7-mm membrane disc.

Neuhoff and Kiehl [71] have constructed, from plastic, three types of equipment suitable for the dialysis, electrodialysis, and equilibrium dialysis of 10 to 500 μl of working material. Sauermann [72] has used a V-shaped chamber to electrodialyze 10 μl of solution. Rüst [73] has described a dialyzer in which up to 98 samples can be processed at the same time across a single sheet of membrane.

3. Modifications

Craig and co-workers [74, 75] have found that standard cellophane dialysis casing can be modified so that it has the range of pore sizes optimal for the separation of compounds with molecular weights from 100 to 135,000. Pore sizes can be altered in a useful way by mechanical stretching while the membrane is wet. They can be made much larger by treatment with zinc chloride or made much smaller by acetylation. The diffusion rates of the species that can pass the membrane increase with temperature and generally vary with molecular weight (e.g., $H_2O > CH_3OH > HCl > KCl > NaCl > CH_3COOK > glycine > glucose > EDTA$). This technique is known as thin-film dialysis. An improved apparatus described by Harrington and O'Conner [76] is illustrated in Fig. 6.15.

A recent development in dialysis technique involves the use of hollow fibers, which tremendously increase the area for permeation. The assembly [77] is shown in Fig. 6.16. It consists of a beaker and bundles of

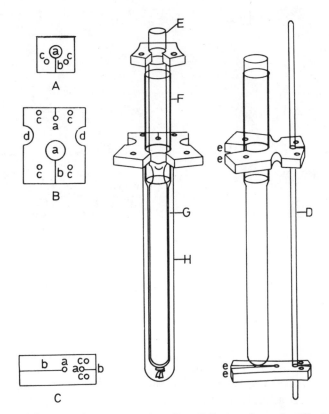

Fig. 6.15. Modification of Craig-type thin film dialysis cell to simplify its manipulation. The new components consist of small rubber collar A, large rubber collar B, rubber clamp C, and glass rod D. These components are shown attached to dialysis cell, which consists of spreader tube E, glass collar F, cellulose sac G, and outside cuvet H [76]. (Courtesy *Anal. Biochem.*)

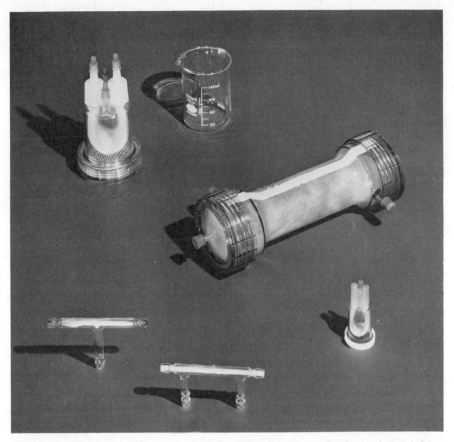

Fig. 6.16. Bio-Fiber devices: (*a*) Minitube; (*b*) T tube; (*c*) beaker; (*d*) Miniplant; (*e*) Minibeaker.

cellulose capillaries (Hollow-Fiber 50, i.d. 180 to 200 μm, wall thickness 25 μm, molecular-weight cutoff 5000), which are connected with a flow system. The working material is placed outside in the beaker, while the capillaries are flushed with the dialyzate (solvent) inside, or vice versa. The increase in volume of the sample solution resulting from the osmosis of the solvent (water) from the dilute solution to the concentrated solution can be minimized by applying slight air pressure against the osmotic pressure. The commercially available devices [77] have capacities (inside the hollow fibers) of 0.38, 5.2, and 150 ml accompanied by containers of 10, 100, and 500 ml, respectively. The rate of dialysis using the hollow-fiber device is demonstrated by the 90% desalting of 100 ml of serum in

15 minutes, and it can be further speeded up by a recycling process [77]. Besides cellulose, new materials have been utilized to prepare the hollow fibers [78]. Whereas cellulose fibers are nondirectional in flow, slow in solvent flux, and subject to degradation, the capillaries made of synthetic plastics (e.g., Diafiber by Aminco) are unidirectional and provide very fast flow rates at low operating pressure (50 psi).

4. Applications

Dialysis is used mainly for the separation of small molecules and ions from macromolecules and colloids. The most common application of dialysis is for desalting biological materials. This is carried out frequently in biochemical investigations and routine analytical experiments. For instance, dialysis provides a simple method for the concentration of urine [79], or the isolation of steroids from 1 ml of blood plasma [80]. Equilibrium dialysis [81] is a versatile technique for studying the interactions between high-molecular-weight and low-molecular-weight species, such as the binding of metal or small molecular species to proteins. The thin-film device [74] is particularly well suited for this purpose, as reported by Bush and Alvin [82]; this technique is also applicable to certain bioassays [83].

Dialysis can be applied to the preparative separation of organic compounds. For example, Klasek [84] has separated the alkaloids from cinchona bark as follows: The plant extract is acidified to convert the alkaloids into the corresponding salts, and the mixture is dialyzed against water. At 24-hr intervals the diffusate is changed for fresh water. After a fivefold dialysis, practically the entire alkaloid content has passed into the diffusate.

III. SEPARATION OF ONE PORTION OF MATERIAL FROM A COMPLEX SOLID MIXTURE
A. SEPARATION BY SOLVENT EXTRACTION

1. Principle

The isolation of materials from solids by means of solvent extraction is based on the differences in solubility of individual components in the particular solvent. Since solubility is generally temperature dependent, the procedures for solvent extraction may be classified into four types:

1. Single-step extraction at laboratory temperature.
2. Single-step extraction in the boiling solvent.

3. Exhaustive extraction at ambient temperature.

4. Exhaustive extraction at the boiling temperature of the solvent.

The rate and effectiveness of the extraction process from solid samples are not only a function of solubility, but are also related to the dissolution process, which depends particularly on the ease of penetration of the solvent into the matrix. Obviously, fine particles are extracted more easily than large particles, since the former provide more surface area. Sometimes it is worth while to regrind the sample after extraction for a period. Grinding is more efficient if the sample is frozen or mixed with Dry Ice. Difficulty may arise when a wet sample is extracted with solvents that are immiscible with water. Therefore preliminary drying is usually recommended. It should be noted, however, that dehydration at elevated temperature and *in vacuo* may cause losses of volatile materials. For instance, Chiba and Morley [85] have found that drying causes significant loss of pesticides in residue analysis. The presence of foreign matter may influence the extraction of the desired components. For example, the extraction of alkaloids from drugs is affected by surfactants in the solution. [86].

Some insoluble compounds in the original complex mixture may be rendered extractable by changing the pH of the medium. For instance, the alkaloids in plants often exist as salts of organic acids and therefore are not extracted by nonpolar liquids. The free alkaloids, however, can be liberated by the addition of ammonia and pass into solution. On the other hand, the basicity of alkaloids can be utilized to separate them from neutral compounds in a two-step extraction procedure as follows: The complex mixture is first treated with an acidic substance (e.g., aluminum sulfate solution). Since the alkaloids now become insoluble in nonpolar solvents, neutral compounds (e.g., fats and waxes) can be selectively extracted. Subsequently, the alkaloids can be regenerated and recovered by extraction with an organic liquid.

2. Single-Step Solvent Extraction

a. Selection of Solvent

A general discussion of the choice of solvent is given in Section II.A.1, which is also applicable to the extraction of materials from solids. The common solvents (arranged according to decreasing polarity) are water, ethanol, chloroform, diethyl ether, benzene, and petroleum ether. If the working material is extracted stepwise with different solvents, usually the extraction starts with the least polar solvent. Various liquids and mixed solvents can be employed, however, some being specified for particular extractions. For example, Dorer and Lubej [87] recommended the extraction of alkaloids from belladonna leaves by means of aqueous acetic acid (2:3), while Vargese and co-workers [88] extracted metal ions from tissues with an acetic-trichloroacetic acid mixture. Carbon disulfide has has been used to extract anesthetics such as ether and chloroform from animal tissues [89]. Propylene carbonate (boiling point 241°C) has been proposed as an extractant for organochlorine and organophosphorus pesticides in a wide variety of solid materials [90].

b. Extraction at Laboratory Temperature

Concerning the techniques for extraction at laboratory temperature, attention should be called to the fact that in dealing with multicomponent materials, a large portion of the sample may be insoluble. Therefore, after the solvent and microsample are mixed in the test tube or flask, the vessel should be swirled or shaken for a period of time.

Some materials require preliminary treatment to make the entire sample accessible to the solvent. Thus, Puech and co-workers [91] have extracted alkaloids from galenical powders by triturating with sand before treatment with the solvent. Gaind and co-workers [92] have performed small-scale ethanolic extraction of drugs in a Waring Blender by operating it for 10 min, setting it aside for 10 min, and repeating the process. Schnorbus and Phillips [90] have described a system for extracting pesticide residues by macerating the finely chopped sample for 1 min with 2 ml of propylene carbonate per gram of sample.

The procedures for extraction from thin-layer plates and paper chromatograms are given in Chapter 7, Section III.A. Ultrasonic waves are very effective in extracting complex solid mixtures. In a study by DeMaggio and Lott [93], it was shown that (1) there is a clear distinction between ultrasonic maceration and ultrasonic extraction, and (2) ultrasonic energy definitely has utility in the extraction of alkaloid-containing plants, provided that the ultrasound is of sufficient intensity and is accurately applied. Ultrasonic vibrators with probes that can fit into a beaker or test tube are commercially available [94]. According to Sawicki [95], this is a convenient device for extracting materials that are imbedded tightly in fibers such as those of filter paper. Skauen and co-workers have demonstrated the efficacy of the ultrasonic extraction of alkaloids [96] and aglycones [97] from plant roots, bark, or leaves. The ultrasonic extraction assembly is shown in Fig. 6.17.

After extraction, the solution is separated from the insoluble portion of the original working material by filtration at ambient temperature. For air- and moisture-sensitive samples, the operations of extraction and filtra-

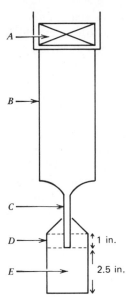

Fig. 6.17. Assembly for ultrasonic extraction [96]: *A,* transducer; *B,* sonic converter; *C,* step horn; *D,* polyethylene bottle; *E,* extraction mixture. (Courtesy *J. Pharm. Sci.*)

tion should be performed in a "dry box" (see Chapter 9, Section III.B). Gibbins [98] has described an apparatus that permits protracted manipulation of organic solvents during the extraction and filtration of air- and water-sensitive solids in all-glass system free of stopcocks; the solution and solids are transferred from one compartment to another by gravity.

c. Extraction at Elevated Temperature

Extraction at elevated temperature is usually carried out by heating the mixture in a test tube or flask under a reflux condenser. Since organic solvents such as diethyl ether, benzene, petroleum ether, and carbon disulfide are very flammable, heating the vessel over an incandescent bulb or in warm water is recommended. In certain analytical extraction procedures, the temperatures are specified. For example, in the determination of active components in detergents [99], the organic components are separated from inorganic salts by extraction with ethanol under reflux. In the assay of cascara, Fairbairn [100] has reported that the use of boiling water, instead of cold water as specified in the standard method, gives a higher yield of glycosides.

The technique of hot filtration (Chapter 3, Section II.B.3) is applicable to the separation of insoluble material from the extract. Rapid operation is necessary in the case of very volatile solvents such as diethyl ether and carbon disulfide. No free flame can be tolerated nearby. The extraction should be repeated at least three times, even though the compounds to be extracted are very soluble in the solvent used, in order to achieve high recovery of the extractable portion from the complex mixture.

3. Exhaustive Extraction

a. Apparatus and Operation

The dynamic principle for the continuous extraction of solid materials with a solvent is similar to that used for continuous liquid-liquid extraction (Section II.A.3). The sample is exposed to the solvent, which flows down from the reflux condenser. The solution thus formed then runs down into the boiling flask, which serves as a reservoir and from which the solvent is recycled. Thus the sample is continuously washed with fresh solvent until all extractable matter of the original material has been transferred into the flask. Hence this procedure is known as exhaustive extraction.

If the vapors of the solvent bypass the extraction chamber, the extraction occurs at about laboratory temperature. This is the operation of the popular Soxhlet apparatus, an example of which is shown in Fig. 6.18. The solid sample is placed inside a thimble (not shown), which receives the condensate from the condenser. When the liquid level in the extraction compartment reaches the top of the siphon tube, it overflows into the flask. The capacity of the smallest thimble is about 5 ml; of the flask, 10 ml.

If the extraction chamber comes in direct contact with the vapors of the solvent, the extraction occurs at a temperature slightly below the boiling point of the solvent. Fig. 6.19 shows a commercial apparatus [35] for which either a glass thimble or a fiber thimble can be used. The extraction of microsamples can be achieved by means of the universal apparatus shown in Chapter 2, Fig. 2.1d. Another method is to hang the thimble from a metal coil, which serves as the condenser, as shown in Fig. 6.20. A similar technique has been described in Chapter 3, Fig. 3.12. Two more devices suitable for extraction of solid samples are shown in Fig. 6.21; the extraction tubes illustrated may be used in the liquid-liquid extraction apparatus shown in Fig. 6.6.

b. Modifications

Prosky and O'Dell [101] have described a modified Soxhlet thimble holder for multiple fat extractions; it consists of a glass cyclinder with a stainless steel wire platform on which can stand eight porous plastic thimbles, each accommodating about 100 mg of solid sample. Wix and Hopton [102] have constructed a semimicro apparatus in which 24 samples

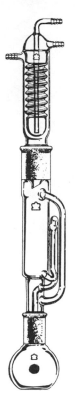

Fig. 6.18. Soxhlet apparatus. (Courtesy Kontes Glass Co.)

(a)

(b)

Fig. 6.19. (*a*) Apparatus for extraction at boiling temperature of the solvent. (*b*) Glass thimble with fritted glass bottom. (Courtesy Kontes Glass Co.)

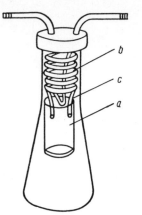

Fig. 6.20. Extraction in hanging thimble: *a*, glass tube with fritted bottom; *b*, copper spiral condenser; *c*, spring holder.

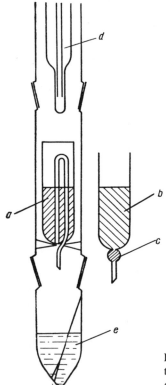

Fig. 6.21. Microextraction device: *a*, Soxhlet-type thimble; *b*, filtration-tube-type thimble; *c*, cotton; *d*, condenser; *e*, extraction liquid.

of fatty material can be assayed at one time. Franks [103] have described an extraction apparatus for the determination of fatty matter in cotton fibers. Raymond and co-workers [104] have improved the technique for the extraction of coal with pyridine; the swelling and poor drainage are avoided by mixing the sample with potassium chloride and using a glass rod to raise the thimble out of the area of bad drainage. Grussendorf [105] has designed an apparatus for the extraction of grain or soil; solvent from a reservoir flows, under gravity, through the base of a tube containing the sample; from the top of the tube it is siphoned into a distillation vessel, the vapor being condensed and returned to the reservoir. Streitwieser and co-workers [105a] have described a side-arm Soxhlet extractor (Fig. 6.22a) used for the purification of actinide and lanthanide complexes.

Zimmermann [106] has described a simple device to extract micro amounts of material from plant tissues for paper chromatography. As illustrated in Fig. 6.22b, the condensing solvent causes a continuous flow from the condenser through the side arm that contains the tissue pieces and back to the boiling solvent in a flask that has capacity of 5 ml. The capacity of the side arm can be adapted to the size of tissue by the choice of position as indicated. After extraction, the solution can be transferred to the filter paper and evaporated on the spot, using the heating block with suction channel as described by Gorbach and Gollob-Hausmann [107]. Another simple microextraction device for handling dried fruits and fresh or powdered leaves has been proposed by Bhatnagar and co-workers [108]. Grdinic and Gertner [109] have designed an apparatus for the continuous extraction of microsamples from circular zones on filter paper, such as those obtained by the ring-oven method (Chapter 3, Section IV); the extraction vessel, which consists of a tube fixed vertically to the inside bottom of a cup (with communication between the tube and cup), is held inside a tube connected to heated vessel for solvent below, and to a reflux condenser above. Voss and Blass [109a] have described a modified Soxhlet apparatus in which solvent is distilled and reused, thus minimizing both solvent losses and the time taken for solvent removal from the extract.

4. Comparison of Extraction Methods

Evaluation of the efficiency of various solvent extraction methods has been undertaken by a number of workers. These studies are of particular importance in the estimation of drugs in biological materials and pharmaceutical preparations. For example, Solomon and Crane [110] have compared extraction procedures in the determination of atropine and hyoscine and found that the extraction procedure is a larger source of variation

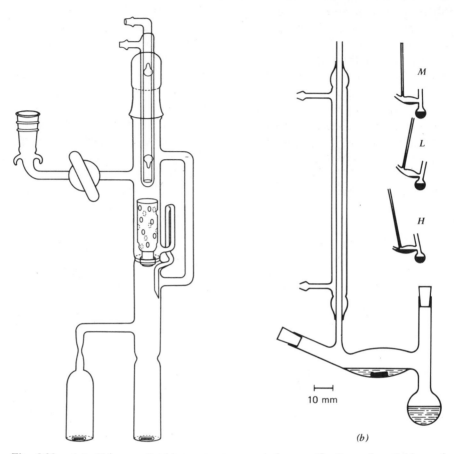

10 mm

(b)

Fig. 6.22. (a) Side-arm Soxhlet extractor used for purification of actinide and lanthanide complexes [105a]. (Courtesy *J. Am. Chem. Soc.*) (b) Apparatus for extraction for plant tissues [106]. Positions for medium, low, and high capacity of side arm: *M, L,* and *H,* respectively. (Courtesy *Science.*)

than the GLC determination, and the single-step extraction method is less variable than the percolation and continuous-extraction procedures. Fricke [111] has described various extraction techniques for the analysis of drugs in formulations and stated that the method to be used depends on the type of sample.

In a comparative study between Soxhlet extraction and ultrasonic extraction (see Fig. 6.17) of alkaloids, Ovadia and Skauen [96] have made some interesting observations: (1) The rate of extraction was more rapid with ultrasound than with Soxhlet extraction, in all cases. (2) The amount of alkaloids obtained from the ipecac root in 30 sec was greater than that produced by Soxhlet extraction in 5 hr. (3) Alkaloidal extraction reached a maximum in 5 min for ipecac root and 30 sec for jaborandi leaf with ultrasound; for both materials, Soxhlet extraction gave a maximum (but lower) yield in 5 hr. (4) Jaborandi leaf alkaloids showed degradation after 30 sec of ultrasonic extraction and after 5 hr of Soxhlet extraction. Prost and Wrebiakowski [112] have made a statistical analysis of the determination of fat in meat by simple extraction with chloroform-methanol and by Soxhlet extraction, with the conclusion that the former method is more accurate.

The possible complications in the extraction of drugs from biologic tissues can be examplified by the study of Feldman and Gibaldi [113]. It was found that recovery of phenol red from homogenates of rat intestinal tissue, using aqueous extraction procedures, was near quantitative. However, significantly poorer recovery was observed after incubation of phenol red with intact intestinal sacs. It appears that phenol red binds to tissue when in intimate contact, and the bound material is relatively resistant to aqueous extraction.

The reliability of extraction procedures in the residue analysis of pesticides has drawn much attention. Watts [114] has studied three procedures for extracting organophosphorus pesticides from bean leaves: (1) blending for 5 min with ethyl acetate, (2) with acetonitrile, (3) 24-hr Soxhlet extraction with $CH_3OH-CHCl_3$; all yield recoveries of more than 90%. Evaluating the extraction and cleanup methods for analysis of DDT and DDE in alfalfa using $CH_3OH-CHCl_3$, Ware and co-workers [115, 116] have found that allowing the sample to stand in the solvent about 16 hr after mincing is the most thorough extraction method, while exhaustive Soxhlet extraction after extraction yields only an additional 4.6% residue. Comparing the method for extracting dieldrin from soil, Saha [117] has reported recoveries of (1) more than 85% by shaking air-dried samples for 1 hr with hexane-acetone, (2) 97% by extracting with hexane-isopropyl alcohol after adding water (20%) to the sample, and (3) 100% by extracting the air-dried sample in a Soxhlet apparatus with $CH_3OH-CHCl_3$ for 8 hr. Dieldrin undergoes biodegradation in moist soil, and the degradation products are difficult to extract. Chiba and Morley [118] and Saha and co-workers [119] have investigated the factors influencing the extraction of aldrin and dieldrin residues from different soil types. The most promising and practical technique appears to be overnight contact between soil and solvent, followed by mixing in a high-speed blender. The fortification of air-dried soils with dieldrin should not be used for measuring true recovery rates from field samples extracted in the air-dried state [120].

B. SEPARATION BY STEAM DISTILLATION

1. Apparatus and Operation

The principle for separating one portion of material from a solid sample by steam distillation is similar to that discussed in Section II.B.1 for the steam distillation of a liquid mixture. However, these two processes differ in apparatus used and experimental operations. In general, the solid sample is mixed with water in the same distillation vessel, which is heated, whereupon water vapor and steam-volatile matter leave the vessel together and condense in a condenser. The distillate is then separated.

Franklin and Kayzer [121] have designed two apparatus for the estimation of the essential oil content of plant samples; in operation, the water is returned to the distillation vessel for further use. As illustrated in Fig. 6.23, the working material (e.g., plant leaves) is placed in the round flask (250 ml) containing sufficient water. The contents of the flask are boiled. The condensate enters C, and its momentum forces the oil (i.e., water-insoluble distillate) into D, where the oil separates as the top layer while

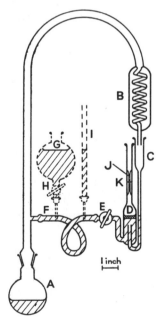

Fig. 6.23. Steam distillation of solid material to collect 50 μl of oil [121]: A, still pot; B, condenser; C, condensate inlet; D, receiving chamber; E, stopcock; F, rubber tube; G, separatory funnel; H, stopcock; I, graduated pipet; J, measuring capillary, 1-mm i.d.; K, calibration mark. (Courtesy *Anal. Chem.*)

the water returns through the rubber tube F to the flask. The volume of oil collected can be as little as 50 μl. Fig. 6.24 shows the apparatus for collecting oil in the range of 2 to 50 μl. The plant tissue is placed into an appropriate flask. Water (about 3 ml per gram of working material) is added, and the flask is attached to Part 1. The mixture is refluxed for 1.25 to 1.5 hr to open the oil sacs in the plant tissue, at a rate of reflux such that condensation occurs below the vapor inlet E. The reflux rate is then increased so that condensation occurs within D and F. After 15 min, the oil has collected at the bottom of F. The distillation flask is removed and replaced by part 2. Saturated salt solution, at 55°C, is injected into the assembly, using a 25- to 50-ml syringe J attached to H. When sufficient solution has been injected to fill the apparatus to a level near the bottom of D, the assembly is maintained at 55°C in a water bath. A 10- or 50-μl syringe is then attached to H and used to effect measurement of the volume of oil by difference.

Krishnamurthy and co-workers [122] have described an assembly for the steam distillation of essential oils in which steam generated in a round-bottomed flask is passed through small holes into a vertical glass column packed with the coarsely powdered sample (e.g., spices). The mixture of steam and extracted oils passes through an almost horizontal condenser and into a separatory funnel. Höltzel [123] has construtced a device in which vapor from water boiling in a flask passes up through the sample (e.g., leaves, flower heads, roots), which is wrapped in muslin, and the

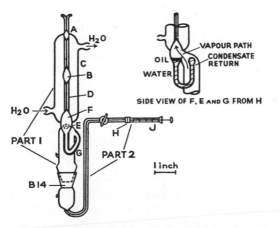

Fig. 6.24. Apparatus for steam distillation of solid material to collect 2–50 μl of oil [121]: A, B, bubble beakers, approximately ⅜-in. o.d.; C, condenser jacket; D, capillary measuring and condensing tube, 1-mm. i.d.; E, vapor inlet, ³⁄₁₆-in. o.d.; F, condensing chamber; G, condensate return; H, Luer Lock syringe needle permanently attached to part 2 (with Araldite); J, syringe. (Courtesy *Anal. Chem.*)

steam-distillate passes through a condenser sloping down to a microburet. The condensate separates into two layers in the buret, and the lower aqueous layer is returned to the flask and re-boiled. The upper end of the buret is open to the atmosphere through a condenser, and the lower end is fitted with a three-way stopcock through which the oily layer of distillate can be recovered.

2. Applications and Modifications

The main use of the steam distillation of solid materials is to separate essential oils and volatile matter from plant tissues. Sometimes better results are obtained by boiling the material with dilute acids. For instance, Piquet and co-workers [124]) have recommended 0.5 M HCl for the distillation of volatile oil from camomile flowers, to reduce foaming. Steam-volatile phenols are distilled with dilute phosphoric acid [125]. Schmidt [126] has studied the possible errors in the azeotropic distillation of essential oils from leaves and flowers.

Separation by steam distillation has been utilized to isolate nicotine from tobacco [127] and N-nitrosamines from foodstuffs [128, 129]. Gyenes [130] has employed codistillation with benzene to separate amines, such as ephedrine, from complex mixtures. Storherr and Watts [131] have described a codistillation cleanup method for the separation of organophosphate pesticides.

Hughes [132] has described a modified receiver for heavier-than-water essential oils. The apparatus (Fig. 6.12) designed by Williams [58] for continuous steam distillation can be used with solid samples. Maraval-has [133] has constructed an apparatus for the steam distillation of essential oils in which the condensing oil does not come into contact with hot vapor during distillation, so avoiding the possibility of decomposition. The apparatus of Wasicky and Akisue [134] (Chapter 2, Fig. 2.31) is fabricated for the same purpose.

C. SEPARATION BY PHYSICAL DIMENSIONS

1. Sieving and Centrifuging

When the components of a complex solid mixture exhibit discernible differences in particle size, a portion of the working material can be separated by means of a sieve or centrifuge. The techniques of separation by sifting and by centrifugation have been discussed in Chapter 3, Sections VI.A and VI.C, respectively. Sieving is applicable to dry materials, while centrifuging is for suspended solid particles in a liquid. General speaking, in a single-stage sifting or centrifuging operation, the separated portions still contain more than one species from the original sample.

Bricky and co-workers [135] have reviewed the various procedures, including sifting and filtering, for the microanalysis of foods for extraneous material. Wilcocks and Oliver [136] have described a sifting method to separate solid biological materials (e.g., grass, roots, pest larvae) from soil samples. Withey and Van Dyk [137] have constructed an inexpensive sieve pump as illustrated in Fig. 6.25: Two acoustic loudspeakers, connected in parallel, but out of phase, to a low-voltage 50- to 60-Hz source, are tightly mounted on opposite ends of a cylinder to cause a vertical pumping action across a sieve mesh fixed within the cylinder. Using this device, the time required to sift 400 mg of a drug through a 62-μm mesh sieve is reduced from 1 week with conventional mechanical shaking to about 1 min.

Spencer [138] has fabricated a micro basket-centrifuge apparatus in order to obtain a greater recovery when separating finely homogenized leaf tissue from the extractant. An extraction thimble is supported (on the perforated base of a polypropylene tube) inside a 50-ml metal bucket with a 10-mm hole drilled in its base. This bucket is supported in a 100-ml metal bucket; the mixture is poured into the extraction thimble and centrifuged as usual, the piece of plastic tubing preventing collapse of the thimble through the hole in the 50-ml bucket. Casciato [139] has reviewed the analytical and preparative uses of the ultracentrifuge in biomedical research. Bowers and Haschemeyer [140] have constructed a versatile small-volume ultrafiltration cell, which has a capacity of 0.25 to 3 ml.

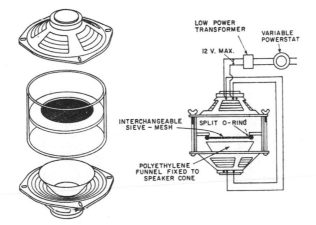

Fig. 6.25. Construction of an inexpensive sieve pump [137]. (Courtesy *Lab. Pract.*)

Hudson and Ison [141] have described a simple technique for centrifuging less than 100 μl of material by using 1-mm-bore polyethylene tubing as follows: The sample is drawn up into a 4-in. length of 1-mm tubing by means of a micro cap filler, and the tube is effectively sealed by folding it into two equal lengths. The 2-in. U tube so formed is inserted into a cylindrical wooden block (see Fig. 6.26). The wooden blocks, balanced with one another, have been drilled with six $\frac{1}{8}$-in.-diameter holes to a depth of $1\frac{3}{4}$ in. About $\frac{1}{4}$ in. of the open ends of the U tube projects from the block to facilitate subsequent removal. The blocks fit snugly into the 50-ml polyethylene tubes used in the 8×50-ml angle head of the MSE-25 centrifuge; thus 48 samples may be centrifuged at one time.

2. Clathrates and Molecular Sieves

The separation of particles by physical dimensions at the molecular level can be achieved by means of molecular sieves [142] and clathration [143–145a]. It should be noted that these two separation techniques are not restricted to solid samples.

Molecular sieves are prepared by the dehydration of synthetic or naturally occurring metal aluminosilicates known as zeolites. The dehydrated zeolite crystals are honeycombed with regularly spaced cavities interlaced

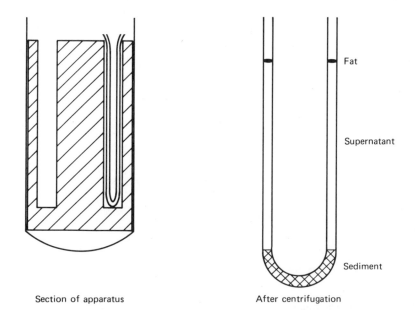

Section of apparatus   After centrifugation

Fig. 6.26. Centrifugation of 100 μl of biological homogenate in polythene tubing [141]. (Courtesy *Lab. Pract.*)

by channels of molecular dimensions, which offer a very large capacity for the absorption of foreign molecules. By virtue of the uniform pore size of a given molecular-sieve crystal type, molecules having a minimum projected cross section larger than the effective diameter of the zeolite pore are excluded from the internal labyrinth. Molecules having a cross section smaller than the effective pore diameter are absorbed internally. For example, molecular sieve type 5A has an effective pore size such that straight-chain hydrocarbons, which can enter the pores, are absorbed. Branched-chain and cyclic hydrocarbons are excluded and hence separated from the straight-chain compounds. Molecular sieves with pores of various dimensions are commercially available [146].

Clathration involves the formation of molecular complexes, known as clathrates or inclusion compounds, in which "guest" molecules occupy the empty space in the crystalline structure built by the "host" substance. The shape of the molecule is the primary factor in determining the formation of a clathrate. For example, the crystalline urea clathrates are formed from urea as host and unbranched paraffins or their derivatives as guests. In the hexagonal urea crystal, there is a channel 5.3 Å across, which fits the molecular dimensions of these compounds (straight-chain saturated hydrocarbons, halides, alcohols, aldehydes, ketones, carboxylic acids). The presence of one double or triple bond still permits clathration with urea. However, spatial requirements caused by two double bonds (e.g., in linolenic acid), branching, or ring structure obstruct the urea clathrate formation. This phenomenon is utilized for the separation of one portion of material from a mixture containing various types of organic compounds. A saturated solution of urea (in methanol or water) is commonly employed as the reagent. The formation of clathrates is spontaneous after agitating the sample in the milieu. The crystals formed are addition compounds in definite molecular ratios, but these are not necessary integral. The molar ratio of urea to the substrate is determined by the length of the hydrocarbon chain, not by any functional group present in the guest molecule. The urea clathrates are usually thermolabile and cannot be recrystallized; they are irreversibly decomposed in boiling water and sometimes even by grinding.

Thiourea clathrates are similar to the above molecular complexes but with different characteristics. Owing to the larger dimensions of the thiourea crystal channels, only branched paraffins, cycloparaffins, and their derivatives participate in the clathrate formation. Contrastingly, the unbranched molecules are too small and hence cannot be held rigidly.

Other representatives of lattice inclusion compounds are the choleic acids, which are built out of deoxycholic acid with fatty acids or other lipoic substances. Hydroquinone and some phenols form open crystal lattices, which can accommodate smaller gas and solvent molecules. Sybilska and co-workers [147] have separated o-, m-, and p-xylenes, ethylbenzene, o- and p-diethylbenzenes, and benzene and thiophene on a column filled with clathrates of the type $Ni(4\text{-picoline})_4(SCN)_2 \cdot 0.6quinol$ on a Celite support.

IV. SEPARATION OF VAPORS FROM SOLIDS OR LIQUIDS

A. HEAD-SPACE METHOD

Separation of vapors from solids or liquids by the head-space method involves the removal of the vapor phase from the vessel containing the working material. It can be carried out with simple apparatus and techniques. A syringe may be used to withdraw the head-space gas from ampuls, bottles, cans, or plastic bags containing the samples. Fig. 6.27 illustrates an assembly described by Drawert and co-workers [148] in which the head-space vapors are condensed in a coil cooled by liquid air.

The head-space method has been employed for the separation of aroma compounds of processed and raw foods, such as vanilla extracts [149], strawberry [150], apple [151], pear [152], vegetables [153], and canned fish [154]. It is a convenient separation technique suitable for use in the investigation of the fragrance of flowers, perfumes, volatile matter of medicinal plants, odor of polluted water, and so on.

Attention is called to the fact that the walls of the container may act as an adsorption surface for some of the vapors. For instance, Maier [155] has found that measurable amounts of acetone, butylamine, and ethyl acetate are adsorbed on the rubber stopper of an 80-ml head-space vessel. Therefore, if sufficient sample is available, the first batch of vapors should be discarded and fresh sample added into the vessel. In this manner, the walls of the vessel will be saturated with the adsorbed compounds, and

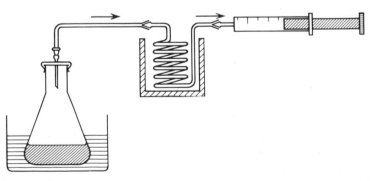

Fig. 6.27. Assembly for separation of head-space vapors [148]. (Courtesy *Chromatographia*.)

thus the vapors obtained subsequently by separation from the working material represent the total yield from the original sample.

B. OTHER METHODS

Several methods are available for the separation of gases from liquids or solids, depending on the characteristics of the sample and the nature of the vapor phase to be separated. If the vapors are easily condensable, a current of inert gas can be swept through the sample to conduct the vapors into the condenser. Another method uses the vacuum manifold system and condenses the vapor with liquid nitrogen [155a]. If the vapors can be adsorbed on a suitable surface (e.g., active charcoal, silica gel), they may be separated from the working material by adsorption and recovered by subsequent desorption. It should be recognized that the adsorption-desorption process may cause chemical alteration of the compounds originally present in the vapor phase [156, 157].

Another technique involves the diffusion of the vapors into another area in the same enclosure, where they are absorbed physically or chemically. For example, water vapor can be removed from plant material by means of silica gel, and amines from fish by means of phosphoric acid; subsequently, the vapors can be recovered by regeneration. Conway dishes (Fig. 6.28) are generally used for the diffusion of microsamples [158]. Cooney and co-workers [159] have described a procedure in which disposable plastic centrifuge tubes are employed to diffuse ammonia, carbon dioxide, and water vapor from 20 μl of an enzyme reaction mixture; the results are about as accurate as with Conway dishes, though the rate of diffusion is slower.

Oxygen, hydrogen, nitrogen, and other inert gases (e.g., hydrocarbons) that are trapped in a liquid can be separated from it by passing a stream of carbon dioxide through the sample (see Chapter 8, Fig. 8.9). The

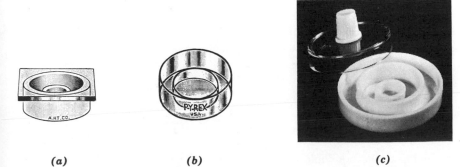

(a)   *(b)*   *(c)*

Fig. 6.28. Conway diffusion dishes [158]: (*a*) porcelain; (*b*) glass; (*c*) plastic. (Courtesy A. H. Thomas Co.)

gases are recovered after the removal of carbon dioxide by its absorption in potassium hydroxide solution. Gases trapped in metals are separated by the vacuum-fusion technique. An apparatus described by Lilburne [160] is capable of separating 2 μl of gases from 0.05 to 0.15 g of metal. The sample is introduced through a lock into a graphite crucible contained in a cooled, evacuated silica furnace-head, and is heated to 2400°C by a coil powered by a 5-kHz generator. The gases liberated are conducted to the pumping line, which is connected to a gas chromatograph.

V. SEPARATION BY CHEMICAL METHODS

While the ideal procedure for separating one portion of a complex mixture is by physical methods, so that the original species can be recovered without alteration, this is not always possible. For instance, micro amounts of a mixture containing several crystalline inorganic compounds may be separated under the microscope according to color or morphology (e.g., in mineralogy or air-pollution studies), but there is no physical method of separating them as individual compounds if these substances are in solution. The common technique for separating metals and organic compounds into groups by chemical reagents involves the formation of complexes, from which the original species may be regenerated. Another technique for organic compounds consists of the preparation of derivatives. As is noted in the following sections, some of the chemical transformations can be carried out advantageously on polymeric solid supports.

A. SEPARATION BASED ON THE FORMATION OF COMPLEXES

1. Chelates For Metals

Chelation is currently the most frequently recommended technique for the separation of metals in microscale manipulations. It involves the formation of heterocyclic organic structures that contain the specific metal cation chelated with ligand groups [161]. For instance, a bivalent metal cation forms a chelate with β-diketone as depicted below:

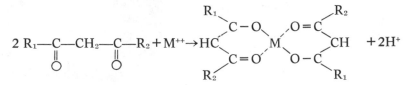

Owing to the non-ionic character of the chelate, the metal becomes extractable into nonpolar organic solvents and thus can be separated from the original mixture containing other ionic species. Since the example shown above is a general reaction for divalent metal cations, all metallic elements that can exist in the oxidation state of $+2$ may be extracted into the organic liquid by using a suitable chelating agent and suitable experimental conditions [162, 163]. Thus, numerous organic reagents and schemes have been proposed [164–166a] for the separation of metals in multicomponent samples.

It should be recognized that the metal chelates of the same structure differ in stability [167]; e.g., the β-liketone chelates generally follows the order

$$Hg^{++} > Cu^{++} > Ni^{++} > Co^{++} > Zn^{++} > Pb^{++} > Mn^{++} > Cd^{++} > Mg^{++} > Ca^{++} > Sr^{++} > Ba^{++}$$

Obviously, if the microsample contains only one metallic element that undergoes chelation with the given organic reagent, it provides a simple method of separating this metal from the mixture. For example, procecedures have been described for the extraction of copper [168] or chromium [169] in blood samples, and cesium [170] in environmental materials. When the mixture contains two metals that belong to the same valence group but form complexes of vastly different stability, separation of the individual metal is still practicable (e.g., cadium from mercury [171]). On the other hand, separation by complexation of a mixture containing several metals that yield chelates of nearly equal stability always produces a portion that consists of more than one species from the original sample. The separation of such chelates requires efficient techniques (e.g., chromatography).

2. Ion Pairs and Related Products For Organic Substances

The organic complexes known as clathrates have been discussed in Section III.C.2; they are inclusion complexes the formation of which is controlled by physical size and shape. The complexes described below are formed on different principles.

Aromatic compounds (homo- and heterocyclic) and some of their derivatives especially those with electron-rich groups (e.g., —OR) can form complexes of the charge-transfer type [172, 173]. Crystalline products are usually obtained in reactions between these compounds (called donors) and 1,3,5-trinitrobenzene, picric acid, chloranil, or 2,4,7-trinitrofluorenone (called acceptors) as 1:1 complexes [174, 175]. These complexes can be recrystallized, preferably in the presence of excess amounts of the acceptor. In order to regenerate the donor compound, the most efficient way is to use a short column of alkaline alumina; the donor species will be eluted with nonpolar solvent while the acceptor is retained.

Alkyl and aryl sulfides form insoluble complexes with mercuric chloride. According to Večera and co-workers [176], the molar ratios of the products are dependent on the conditions of their preparation. These complexes cannot be purified by recrystallization. Treatment with H_2S regenerates the organic sulfides.

Ion pairs may be considered as complexes formed by the union of organic cations and anions. Modin and co-workers [177] have investigated the separation method using ion-pairing agents, such as the extraction of organic cation–carboxylate ion and organic anion–quaternary ammonium ion into organic solvents, aminophenols and amino alcohols by bis(2-ethylhexyl)phosphate, pencillins with tetrabutylammonium ion, and choline with picrate. Biswas and Mandel [178] have studied the extraction of dye-base anions by hexadecyltrimethylammonium ion into chloroform and the effect of inorganic anions. The extent of interference decreases in the order ClO_4^-, I^-, Br^-, NO_3^-, Cl^-, which is also the order of extraction of these anions into chloroform by the quaternary ammonium ion. Other anions, such as F^-, SO_4^{2-}, and acetate, do not interfere.

Rothwell and Whitehead [179] have described the extraction of polycyclic aromatic hydrocarbons and carbazole from cyclohexane into 90% aqueous formic acid containing 15% caffeine; in the absence of caffeine, only basic polycyclic N-heterocyclic compounds are extracted. Gautier and co-workers [180] have recommended the separation of organic bases in the form of thiocyanates that are insoluble in water. Hedin and co-workers [181] have utilized trichloracetyl isocyanate as reagent to separate terpene alcohols from essential oils. The esters which form are extracted from the oil into 2% KOH in CH_3OH-H_2O (4:1), with concomitant hydrolysis to carbamates. The alcohols are then regenerated by refluxing with 10% KOH in the same solvent mixture.

B. SEPARATION OF ORGANIC SUBSTANCES BASED ON THE PREPARATION OF DERIVATIVES

1. Uses of Derivatives

The separation of organic substances through the formation of derivatives is dependent on the reactions of functional groups [55]; hence all compounds which possess the particular functional group respond to the selected reagent and are simultaneously separated from the working material. There are various reasons for the preparation of derivatives in microscale manipulations:

1. If the microsample is a liquid, it is advantageous to convert the working material into solid derivatives. Manipulation with milligram amounts of solid is much easier than with microliter amounts of liquid.

Solids can be recrystallized, and a few crystals may be used for identification by melting-point or spectral analysis.

2. Certain classes of compounds, which are easily oxidized, may be converted to more stable derivatives. For example, phenols and amines may be stabilized as the corresponding acetates or benzoates.

3. Derivatization is often utilized to modify the polarity of the molecule. This may be performed to achieve two different objectives. In the first case, a derivative is prepared in order to eliminate the strongly polar group (e.g., —OH, —NH$_2$, —SH, —COOH) in the molecule. Such modification is important because the polar groups cause the molecules to interact, via hydrogen bonding, with one another as well as with the solvent, or the adsorbent (e.g., in liquid and thin-layer chromatography), or the stationary liquid (e.g., in gas chromatography) and produce tailing. The derivatives are less polar, and the absence of active hydrogen results in a normal chromatographic process, forming symmetrical zones without tailing. The groups that are introduced should have low polarity and small molecular weight (e.g., methyl, acetyl, trimethylsilyl). With these groups in the derivatives, smaller differences in the molecular structure will be recognized in chromatographic separation processes than in the original compounds.

In the second case, polar and hydrophilic groups are introduced into lipophilic molecules in order to effect the separation of one type of compounds from the other types in the multicomponent mixture. For example, the lipophilic steroidal ketones [182] may be converted to water-soluble derivatives by using the Girard reagents, and selectively separated from the lipophilic alcohols, hydrocarbons, and so on. It should be noted that the introduction of large and highly polar groups into small molecules may wipe out the small structural differences within a homologous series, this being the case of 2,4-dinitrophenylhydrazones derived from the lower members of aliphatic aldehydes and ketones.

4. The group introduced to form the derivatives may serve as tracer for monitoring. These groups may be chromophores in the visible and ultraviolet regions, or carry a radioactive atom. For example, azobenzene derivatives absorb visible light and therefore are useful in liquid or thin-layer chromatographic separation; fluorescein derivatives are well-known fluorescent labels; C^{14}-labeled compounds can be qualitatively and quantitatively monitored. Electron-spin-resonance labels have been utilized in biochemical investigations [182a].

2. Selection of Derivatives

Table 6.1 lists the common derivatives of the most common types of organic compounds. The reader is referred to the manuals [183, 184] and

TABLE 6.1. Derivatives of Organic Compounds

Type of compound	Derivative	Remarks	References
Olefins	Vic. chlorodinitrophenylsulfides	2 isomers	186
	Vic. chlorodinitrophenylsulfones	2 isomers	186
	Vic. methoxymercuriacetates	2 isomers	187
1,3-Dienes	N-phenylmaleimide adduct		188, 189
	N-(p-phenylazo)-phenylmaleimide adduct	Colored	190
Acetylenes	Carbonyl compound by hydration	2 isomers	191
	Mercury acetylides	R·C:C·H only	192
	Olefinic chlorodinitrophenyl-sulfides	2 isomers	186
Aromatic hydrocarbons	Nitro derivatives	More isomers	183
	Sulfonamido-derivatives	More isomers	193, 194
Alkyl halides	S-alkylthiuronium picrates	Ter. halides sluggish	195
	S-alkylthiuronium 3,5-dinitro-benzoates		196
	4-Nitrophthalimides	Hydrolyzed to imidoacids	197, 197a
Aryl halides	Anilides via Grignard reag.		198
	Nitro derivatives	See aromatic hydrocarbons	
Alcohols	Phenylurethanes	Ter. dehydrated	199
	a-Naphthylurethanes	Crystallize well	200
	Acetyl derivatives	Good for GC	201
	3,5-Dinitrobenzoates	Crystallize well	183
	4-Phenylazobenzoates	Colored	202
	Trimethylsilyl deriv.	For GC	203
	Carbonyl comp.	Prim. and Sec. only	183
	N-Phthalimidomethyl ethers		204
Phenols	See derivatives for alcohols		
	2,4 Dinitrophenylethers	Crystallize well	205
	Bromoderivatives	Polysubstitution	183
	Aryoxyacetic acids		206
Alkyl ethers	Alkyl 3,5-dinitrobenzoates	Cleavage of the ether linkage	207
	S-alkyl thiuronium 3,5-dinitrobenzoates	Cleavage of the ether linkage	208
Aryl ethers	Nitro derivatives	See aromatic hydrocarbons	
	Bromo derivatives	See phenols	

Table 6.1 (cont.)

Type of compound	Derivative	Remarks	References
Aldehydes and ketones	2,4-Dinitrophenylhydrazones	Colored	183
		Modifications of the test	209-213
		Carbonyl comp. in presence of carbohydrates	214
	p-Nitrophenylhydrazones	Separation by TLC	215
	Semicarbazones		183
	Oximes		183
	Girard reagents	P,T,D types, water-soluble	216
		T-type prep.	183
		Regeneration of aldehydes from deriv.	217
	Methone (dimethone) deriv.	Aldehydes only	183
	Bisulfite deriv., S benzyl thiuronium salts	Aldehydes, methyl-ketones, some cyclic ketones	218
Carboxylic acids	S-benzyl thiuronium salts		219
	S-α-naphthylmethylthiuronium salt	Good even for low-molecular acids	220
	Amides and anilides	Via acylchlorides	183
	Phenacylesters		183
	4'-Nitrophenylazo-4-phenacyl esters	Colored	221
	Methylesters	For GC	222, 223
	Trimethylsilyl deriv.	For GC	224
	2-Alkylbenzimidazoles		225
	Hydroxamic acids	Easy detection with Fe^{3+}	226-228
Amines primary and secondary	Benzene- or toluenesulfonamides	Primary and secondary distinguished	229
	p-Bromobenzenesulfonamides	Crystallize well	230
	m-Nitrobenzenesulfonamides	Crystallize well	231
	p-Benzeneazosulfonamides	Colored	232
	Phenylthioureas		233
	α-Naphthylthioureas		234
Amines tertiary	Picrates		183
	Tetraphenylboronates		183
	Chloroplatinates		183

307

references cited for the detailed experimental procedures, which are usually described for preparations on or above the milligram scale. Some techniques for the preparation of derivatives in the microgram region are available in the literature [185].

The choice of the derivative is dependent on other factors besides the objectives mentioned in the previous section. If the original compound has to be recovered, a derivative with a functional group that can be cleaved easily is preferred. Hence the preparation of such a derivative cannot involve a structural change of the original molecule that renders it difficult to reconstitute. For example, a methyl ketone may be converted to a carboxylic acid, bisulfite adduct, or benzilidene derivative:

$$R\text{—}CO\text{—}CH_3 \rightleftharpoons \begin{array}{l} R\text{—}COONa \\ R\text{—}C(CH_3)(OH)SO_3Na \\ R\text{—}CO\text{—}CH=CH\text{—}C_6H_5 \end{array}$$

It is apparent that the second derivative should be selected, since it can be easily reverted to the original compound. For the same reason, complexation is in general preferred to derivatization, if applicable. On the other hand, if regeneration of the original sample is not required after separation, stable derivatives will serve the purpose.

Reactions that give unambiguous results and one specific product are better than those that give several products. In practice, however, the derivatization reaction often produces a mixture of products from one compound, although one of the products may predominate overwhelmingly. For instance, the carbonyl group produces two 2,4-dinitrophenylhydrazones, but the thermodynamically less stable stereoisomer is formed only in small quantities when the sample is protected from radiation.

Derivatives of low stability may cause difficulties in the subsequent operations. For example, some trimethylsilyl derivatives are hydrolyzed in the presence of traces of moisture. Hence, if these derivatives are used in the gas chromatograph, they are preferably prepared *in situ* (i.e., directly in the column).

The importance of employing pure reagents in the preparation of derivatives in microscale manipulations cannot be overemphasized. Thus, as reported by Gadbois and co-workers [235], when Girard *T* reagent is used to separate carbonyl compounds for subsequent separation by gas chromatography, large extraneous peaks interfere with the interpretation of the chromatograms. Such peaks are prevented by purifying the reagent used. Liu [235a] has reported extraneous peaks, due to some common impurities, in the infrared and nuclear-magnetic-resonance spectra of numerous organic materials.

C. CHEMICAL REACTIONS ON POLYMERIC SOLID SUPPORTS

It has been noted in the ion-exchange separation methods (Chapter 4, Section VIII), acid-base reactions occur in the resin beads. Thus, all acidic substances in a complex mixture can be separated from the sample by means of ion exchange. By using the appropriate resins and conditions, other chemical reactions have been carried out. Various classes of compounds thus can be selectively separated. This method has found applications in gas chromatography; it is called the "subtractive technique" when the unwanted components are removed from complex samples in a precolumn [236]. Jeltes and Veldink [237] have described a procedure for the selective retention of polar compounds such as alcohols and ketones. Fritz and Chang [238] have reported resins that strongly retain H_2S and SO_2 at room temperature and release the gases at 100°C. Walton and Murgia [239] have used metal-loaded cation-exchange resins to perform ligand exchange chromatography of alkaloids and drugs. Mescaline and atropine are the most strongly retained, followed by methadone; morphine and its derivatives are retained relatively weakly. Rabel [240] has developed a pellicular packing using polyamide [241, 242], which is an excellent adsorbent for amino and hydroxy compounds that are capable of hydrogen bonding.

Lehmann and Hahn [243] have used polyamide powder to adsorb food dyes from aqueous or alcoholic solutions and desorb them with methanolic NaOH. Berret and co-workers [244] have selectively extracted synthetic coloring matter from pharmaceuticals by filtering the fluid through discs of aminoethylcellulose. Okita [245] has demonstrated that glass-fiber filters impregnated with $Hg(CN)_2$ or $HgCl_2$ solution are suitable for the retention of thiols, dimethyl sulfide, and methylamine. Myasoedova and co-workers [246] have described a method of introducing complex-forming groups into the cyanurated polymers to give adsorbents that are polymeric analogs of the monomeric chelating agents. Polymeric supports carrying reactive groups have been used extensively by Merrefield and co-workers [247] in peptide synthesis. These reactions are typical examples of the formation of derivatives that can be cleaved easily and quantitatively. Certain qualitative tests have been improved by adsorbing either the reagent or the test material on ion-exchange resin beads [248, 249].

Solid-phase extraction can be carried out by means of analytically pure, physically rigid, surface-modified polymers, which are commercially available [250]. The desired components in a complex mixture are rapidly extracted by a few milligrams of the polymer. The polymer is separated from the matrix by decantation, filtration, or centrifugation. The components of interest are then desorbed by using milliliter quantities of solvents.

REFERENCES

1. D. P. Schwartz, L. H. Burgwald, and C. R. Brewington, *J. Am. Oil Chem. Soc.*, **43**, 472 (1966).
2. C. H. Doering and H. Torver, *Anal. Biochem.*, **9**, 498 (1964).
2a. J. Rajama and P. Makela, *J. Chromatogr.*, **76**, 199, (1973).
3. S. Nishi and Y. Horimoto, *Jap. Anal.*, **17**, 1247 (1968).
4. L. S. Frankel, P. R. Madsen, R. R. Seibert, and K. L. Wallisch, *Anal. Chem.* **44**, 2401 (1972).
5. E. Gagliardi and Tümmler, *Talanta*, **17**, 93 (1970).
6. E. Stahl, *J. Chromatogr.*, **37**, 99 (1968); *Analyst*, **94**, 723 (1967).
7. B. C. Mohanty, S. C. Basa, and S. N. Mohopatra, *Chem. Ind.*, **1972**, 467.
8. M. T. Kramer, *G.I.T. Fachz. Lab.*, **14**, 1265 (1970).
9. British Standards Institution, BS 4717 (1972).
10. R. A. Hummel and W. B. Crummett, *Talanta*, **19**, 353 (1972).
11. L. C. Craig and D. Craig, in A Weissberger (Ed.), *Technique of Organic Chemistry*, Vol. III, Part 1. p. 149. Interscience, New York, 1956.
12. H. Schreiner, *Chem. Ztg—Chem. Appar.*, **91**, 667 (1967); **93**, 971 (1969).
13. R. L. Stedman, R. L. Miller, L. Lakritz, and W. J. Chamberlain, *Chem. Ind.*, 1968, 394.
14. V. D. Gupta, D.E. Cadwallader, H. B. Herman, and I. L. Honigberg, *J. Pharm. Sci.*, **57**, 1199 (1968).
14a. J. A. Zolotov, *Extraction of Chelate Compounds*, Israel Program for Scientific Translations, Keter, 1970.
14b. R. Köster, in E. Müller (Ed.), *Houben-Weil Methoden der organischen Chemie*, Vol. 6/2 p. 221. Thieme, Stuttgart, 1963.
15. R. F. Coward and P. Smith, *J. Chromatogr.*, **39**, 496 (1969).
16. N. D. Cheronis, A. R. Ronzio, and T. S. Ma, *Micro and Semimicro Methods*, p. 378. Interscience, New York, 1954.
16a. H. Reinhardt and J. Rydberg, *Chem. Ind.*, **1970**, 488.
17. P. Hautke and F. Folk, *Mschr. Brau.*, **22**, 22, (1969).
18. N. A. Karamian, *Anal. Chem.* **45**, 2154 (1973).
19. Whatman No. 1 PS Phase Separating Filter Paper.
20. J. Rajama and P. Makala, *Lab. Pract.*, **18**, 149 (1969).
21. V. Holeysovky, *Chem. Listy*, **53**, 858 (1959).
22. W. H. Tornett, U.K.A.E.A. Rep. RCC-M, 173 (1964).
23. J. Klc, *Collect. Czech. Chem. Commun.*, **31**, 1395 (1966).
24. R. H. Laessig, *Anal. Chem.*, **40**, 2205 (1968).
25. H. Marchart, *Mikrochim. Acta*, **1962**, 913.
26. E. Morgan, *Anal. Chem.*, **38**, 1093 (1966).
27. T. Niedermaier, *Z. Anal. Chem.* **207**, 186 (1965).

28. C. O. Ingamells, *Chem.-Anal.,* **53,** 55 (1964).
29. G. L. Hubbard and T. E. Green, *Chem.-Anal.,* **53,** 119 (1964).
30. R. E. Bailey, J. H. Beck, and H. P. Pieters, *Nature,* **204,** 588 (1964).
31. B. M. Milwidsky, *Chem. Ind.,* **1969,** 411.
32. F. Trowell, *Lab. Pract.,* **18,** 44 (1969).
33. D. W. Sutton and D. G. Vallis, *J. Radioanal. Chem.,* **2,** 377 (1969).
34. H. Reinhardt and J. Rydberg, *Chem. Ind.,* **1970,** 488, International Symposium on Solvent Extraction, Antwerp, May, 1972.
35. Kontes Glass Co., Vineland, N. J.
35a. F. Drawart and A. Rapp, *Chromatographia,* **2,** 446 (1969).
36. W. den Hollander and B. van Zanten, *Chem. Ind.,* **1970,** 742.
37. D. Martin, *G.I.T. Fachz. Lab.,* **14,** 1404 (1970); **15,** 33 (1971).
38. G. S. Riley, *Pharm. J.,* **1965,** 320.
39. W. Piorr, *Z. Lebuns Forsch.,* **132,** 140 (1966).
40. S. C. Pan, reported in ACS 166th National Meeting, Chicago, September, 1973.
41. A. E. Werner and M. Waldichuk, *Anal. Chem.,* **34,** 1674 (1962).
42. M. C. Goldberg and L. DeLong, *Anal. Chem.,* **45,** 89 (1973).
43. J. Palecek and V. Skala, *Chem. Listy,* **58,** 1095 (1964).
44. A. D. Popov, B. S. Ivanova, and N. V. Yanishlieva, *Nahrung,* **13,** 39 (1969).
45. T. Fujinaga, T. Kuwamoto, and E. Nakayama, *Talanta,* **16,** 1225 (1969).
46. R. Ko, *Anal. Chem.,* **39,** 1903 (1967).
47. W. Roedel, D. Zoell, and G. Woelm, *Nahrung,* **15,** 425 (1971).
48. H. J. Bowen, *J. Chem. Soc., (A),* **1970,** 1082.
48a. C. A. Kienberger, *Anal. Chem.,* **29,** 1721 (1957).
49. T. S. Ma and R. Breyer, *Microchem. J.,* **4,** 484 (1960).
50. T. S. Ma., in F. J. Welcher (Ed.), *Standard Methods of Chemical Analysis,* 6th ed., Vol. II, p. 385. Nostrand, Princeton, 1963.
51. E. Bohm, *Dtsch. Lebensm. Rundsch,* **59,** 132 (1963).
52. F. Seehofer and H. Borowski, *Tabakforschung,* **2,** 37 (1963).
53. N. Antonacopoulos, *Z. Lebensm. Unters.,***113,** 113 (1960).
54. M. Herlihy and P. Hayes, *Lab. Pract.,* **19,** 190 (1970).
55. N. D. Cheronis and T. S. Ma, *Organic Functional Group Analysis,* Wiley, New York, 1964.
56. R. Rougieux and G. Cesaire, *C. R. Hebd. Seances Acad. Agric. Fr.,* **53,** 736 (1967).
57. V. M. Riddle, *Anal. Chem.,* **35,** 853 (1963).
57a. C. Herbert, *J. Assoc. Off. Anal. Chem.,* **56,** 132 (1973).
57b. H. Halvarson, *J. Chromatogr.,* **76,** 125 (1973).
58. A. Williams, *Chem. Ind.,* **1969,** 1510.
59. R. E. Stauffer, in A. Weissberger (Ed.), *Technique of Organic Chem.* Vol. III, Part 1, p. 65. Interscience, New York, 1956.
59a. H. Elford, *Trans. Faraday Soc.,* **33,** 1094 (1937).
60. E. J. Wood, J. Simler, and B. B. Bonello, *Lab. Pract.,* **17,** 57 (1968).
61. G. C. Toms, *Lab. Pract.,* **22,** 291 (1973).

62. A. C. Arcus and B. E. W. Browne, *Lab. Pract,.* **17**, 1239 (1968).
63. D. A. Burns, J. N. Budna, and J. M. Chamberlain, *Science,* **138**, 138 (1962).
64. S. A. Katz and H. A. Weusberger, *Experientia,* **25**, 672 (1969).
65. K. Koci, *Chem. Listy,* **55**, 1229 (1961).
66. Cole-Parmer Instrument Co., Chicago, Ill.
67. R. H. Wilkinson, *J. Clin. Pathol.,* **13**, 268 (1960).
68. R. H. Reitz and W. H. Riley, *Anal. Biochem.,* **36**, 535 (1970).
69. T. R. Heskath and C. F. Dore, *Anal. Biochem.,* **49**, 298 (1972).
70. N. G. Heatley, *Analyst,* **94**, 1021 (1969).
71. V. Neufodd and F. Kiehl, *Arzneim.-Forsch.,* **19**, 1898 (1969).
72. G. Sauermann, *Anal. Biochem.,* **48**, 491 (1972).
73. P. Rüst, *Anal. Biochem.,* **17**, 316 (1966).
74. L. C. Craig, "Thin-Film Analytical Dialysis," in C. N. Reilly (Ed.), *Advances in Analytical Chemistry and Instrumentation,* Vol. 4, p. 37. Wiley, New York, 1965.
75. K. K. Stewart and L. C. Craig, *Anal. Chem.,* **42**, 1257 (1970).
76. M. G. Harrington and P. E. O'Connor, *Anal. Biochem.,* **27**, 330 (1969).
77. Bio-Rad Laboratories, Richmond, Calif., Tech. Bull. 1009 (1972).
78. W. F. Blatt, *Am. Lab.,* p. 78. October, 1972.
79. H. McFarlane, *Clin. Chim. Acta,* **9**, 376 (1964).
80. H. Kalent, *Biochem. J.,* **69**, 99 (1958).
81. S. A. Katz, C. Parfitt, and R. Purdy, *J. Chem. Educ.,* **47**, 721 (1970).
82. M. T. Bush and J. D. Alvin, reported at the Conference on Drug Protein Binding, New York Academy of Sciences, New York, January, 1973.
83. L. C. Craig and H. C. Chen, *Anal. Chem.,* **41**, 590 (1969).
84. A. Klasek, *Sep. Sci.,* **3**, 319 (1968).
85. M. Chiba and H. V. Morley, *J. Assoc. Off. Anal. Chem.* **51**, 55, (1968).
86. B. Durek-Kluczykowska and K. Krowczyniski, *Diasnes Pharm. Warsz.,* **20**, 221 (1968).
87. M. Dorer and M. Lubej, Arch. Pharm. Berl. **305**, 273 (1972).
88. F. T. N. Vargese, A. Lipton, and G. J. Huxham, *Lab. Pract.,* **18**, 419 (1969).
89. D. N. Robertson and D. S. Erley, *Anal. Bichem.,* **2**, 45 (1961).
90. R. R. Schnorbus ad W. F. Phillips, *J. Agric. Food Chem.,* **15**, 661 (1967).
91. A. Puesch, M. Jacob, and J. J. Serrans, *Ann. Pharm. Fr.* **27**, 201 (1967).
92. K. N. Gaind, H. C. Mital, and H. K. Dhir, *Indian J. Pharm.,* **24**, 252 (1962).
93. A. E. DeMaggio and J. A. Lott, *J. Pharm. Sci.,* **53**, 945 (1964).
94. Branson Sonic Power Co., Danbury, Conn.; Tekmar Co., Cincinnati, Ohio.
95. E. Sawicki, personal communication, June, 1973.
96. N. E. Ovadia and D. M. Skauen, *J. Pharm. Sci.,* **54**, 1013 (1965).
97. I. C. Patel and D. M. Skauen, *J. Pharm. Sci.,* **58**, 1135 (1969).
98. S. G. Gibbins, *Anal. Chim. Acta,* **56**, 486 (1971); **60**, 242 (1972; *Anal. Chem.,* **43**, 1349 (1971); *Sep. Purif. Methods,* **1**, 237 (1972).

99. M. Nigrin, *Prum. Potravin.*, **19**, 578 (1968).
100. J. W. Fairbairn, *J. Pharm. Pharmacol.*, **22**, 778 (1970).
101. L. Prosky and R. G. O'Dell, *J. Assoc. Off. Anal. Chem.*, **56**, 226 (1973).
102. P. Wix and J. W. Hopton, *Chem. Ind.*, **1957**, 805.
103. F. Franks, *J. Text. Inst.*, **47**, 369 (1956).
104. R. Raymond, I. Wender, and L. Reggel, *Fuel*, **43**, 299 (1964).
105. O. W. Grussendorf, *Chem. Ind.*, **1966**, 52.
105a. A. Streitwieser, Jr., U. Muller-Westerhoff, G. Sonnichson, F. Mares, D. G. Morrell, and C. A. Harmon, *J. Am. Chem. Soc.*, **95**, 8645 (1973).
106. M. H. Zimmermann, Science, **122**, 766 (1955).
107. G. Gorbach and G. Gallob-Hausmann, *Mikrochim. Acta*, **1962**, 102.
108. J. K. Bhatnager, K. C. Laroia, and C. K. Atal, *Indian J. Pharm.*, **27**, 10 (1965).
109. V. Grdinic and A. Gertner, *Acta Pharm. Jugosl.*, **20**, 199 (1970).
109a. G. Voss and W. Blass, *Analyst*, **98**, 811 (1973).
110. M. J. Solomon and F. A. Crane, *J. Pharm. Sci.*, **59**, 1680 (1970).
111. F. L. Fricke, *J. Assoc. Off. Anal. Chem.*, **55**, 1162 (1972).
112. E. Prost and H. Wrebiakowski, *Z. Lebensm. Forsch.*, **149**, 193 (1972).
113. S. Feldman and M. Gibaldi, *J. Pharm. Sci.*, **57**, 1234 (1968).
114. R. R. Watts, *J. Assoc. Off. Anal. Chem.* **54**, 953 (1971).
115. G. W. Ware and M. K. Dee, *Bull. Environ, Contam. Toxicol.*, **3**, 375 (1968).
116. F. M. Whiting, J. W. Stull, W. H. Brown, M. Milbrath, and G. W. Ware, *J. Dairy Sci.*, **51**, 1039 (1968).
117. J. G. Saha, *Bull. Environ. Contam. Toxicol.*, **3**, 26 (1968).
118. M. Chiba and H. V. Morley, *J. Agric. Food Chem.*, **16**, 916 (1968).
119. J. G. Saha, B. Bhavaraju, Y. W. Lee, and R. L. Randell, *J. Agric. Food Chem.*, **17**, 877 (1969).
120. J. G. Saha, B. Bharvaraju, and Y. W. Lee, *J. Agr. Food Chem.*, **17**, 874 (1969).
121. W. J. Franklin and H. Keyzer, *Anal. Chem.*, **34**, 1650 (1962).
122. N. Krishnamurthy, E. S. Nambudiri, and Y. S. Lewis, *Lab. Pract.*, **17**, 912 (1968).
123. C. Höltzel, *Dtsch. Apoth-Ztg.*, **103**, 1207 (1963).
124. J. S. Piquet, C. van Wijlick, and J. H. Zwaving, *Pharm. Weekbl. Ned.*, **105**, 329 (1970).
125. W. Kleber and N. Hume, *Brauwissenschaft*, **25**, 98 (1972).
126. F. Schmidt, *Dtsch. Apoth-Ztg.*, **109**, 137 (1969).
127. F. Seehofer and H. Borowswi, *Beitr. Tabaforsch.*, **2**, 37 (1963).
128. T. G. Alliston, G. B. Cox, and R. S. Kirk, *Analyst*, **97**, 915 (1972).
129. N. T. Crosby, J. K. Foreman, J. F. Palframan, and R. Sawyer, *Nature*, **238**, 342 (1972).
130. I. Gyenes, *Acta. Chim. Hung*, **70**, 189 (1971).
131. R. W. Storherr and R. R. Watts, *J. Assoc. Off. Anal. Chem.*, **48**, (1965).
132. A. Hughes, *Chem. Ind.*, **1970**, 1536.
133. N. Maravalhas, *Chem.-Anal.*, **53**, 23 (1964).

134. R. Wasicky and G. Akisue, *Rev. Fac. Farm. Bioquim. Univ. São Paulo,* **7**, 399 (1969).

135. P. M. Brickey, Jr., J. S. Gecan, J. J. Thrasher, and W. V. Eisenberg, *J. Assoc. Off. Anal. Chem.,* **51**, 872 (1968).

136. C. R. Wilcocks and E. H. A. Oliver, *N. Z. J. Agric. Res.,* **14**, 725 (1971).

137. R. J. Withey and M. Van Dyk, *Lab. Pract.,* **19**, 614 (1970).

138. R. Spencer, *Lab. Pract.,* **17**, 1356 (1968).

139. R. J. Casciato, *Anal. Chem.,* **41**, (Nov.), 99 A (1969).

140. W. F. Bowers and R. H. Haschemeyer, *Anal. Biochem.,* **25**, 549 (1968).

141. G. J. Hudson and R. J. Ison, *Lab. Pract.,* **18**, 1188 (1969).

142. C. K. Hersch, *Molecular Sieves.* Reinhold, New York, 1961.

143. M. Hagan, *Clathrate Inclusion Compounds.* Reinhold, New York, 1962.

144. E. C. Makin, in E. C. Perry (Ed.), *Separation and Purification Methods,* Vol. 1, p. 371. Dekker, New York, 1973. *Sep. Purif. Methods,* **1**, 371 (1972).

145. W. C. Child, Jr., *Quart. Rev.,* **18**, 321 (1964).

145a. C. R. Landolt and G. T. Kerr, *Sep. Purif. Methods,* **2**, 283 (1973).

146. Linde Company, Buffalo, N. Y.

147. D. Sybilska, K. Malinowski, M. Siekierska, and J. Bylina, *Chem. Anal.,* **17**, 1031 (1972); **18**, 157 (1973).

148. F. Drawert, W. Heimann, R. Emberger, and R. Tressl, *Chromatographia,* **2**, 77 (1969).

149. W. H. Stahl, W. A. Voelker, and J. H. Sullivan, *Food Technol.,* **14**, 14 (1960).

150. D. K. K. Graham, personal communication, June, 1972.

151. D. S. Brown, J. R. Buchanan, and J. R. Hicks, *Proc. Am. Soc. Horti. Sci.,* **88**, 98 (1966).

152. R. J. Romani and L. Ku, *J. Food Sci.,* **31**, 558 (1966).

153. R. G. Buttery and R. Teranishi, *Anal. Chem.,* **33**, 1439 (1961).

154. R. E. Hurst, *Chem. Ind.,* **1970**, 90.

155. D. F. Maier, *J. Sci. Food Agric.,* **8**, 313 (1957).

155a. See Ref. 16, p. 376.

156. D. E. Heinz, M. R. Sevenants, and W. G. Jennings, *J. Food Sci.,* **31**, 63 (1966).

157. S. R. Palemand, K. S. Markl, and W. A. Hardwick, *Proc. Am. Soc. Brew. Chem.,* **1968**, 75.

158. E. J. Conway, *Microdiffusion Analysis.* Crosby and Lockwood, London, 1947.

159. D. A. Cooney, H. A. Milman, and R. Truitt, *Anal. Biochem,* **41**, 583 (1971).

160. M. T. Lilburne, *Talanta,* **14**, 1029 (1967).

161. E. P. Dwyer and R. P. Mellor, *Chelating Agents and Metal Chelates.* Academic, New York, 1964.

162. Y. A. Zolotov, *Extraction of Chelate Compounds* (translated from Russian). Halsted, New York, 1971.

163. G. H. Morrison and H. Freiser, *Solvent Extraction in Analytical Chemistry*. Wiley, New York, 1957.

164. J. S. Fritz and G. I. Latwesen, *Talanta*, **17**, 81 (1970).

164a. J. D. Kinrade and J. C. Van Loon, *Anal. Chem.*, **46**, 1894 (1974).

165. E. Gagliardi and P. Tümmler, *Talanta*, **17**, 93 (1970).

166. S. Shibata, M. Furukawa, E. Kameta, and K. Goto, *Anal. Chim. Acta*, **50**, 439 (1970).

166a. K. Burger, *Organic Reagents in Metal Analysis*. Pergamon, Oxford, 1973.

167. J. Bjerrum, G. Schwartzenback, and L. G. Sillen, *Stability Constants of Metal-ion Complexes*. Chemical Society, London, 1958.

168. J. M. Carter and G. Nickless, *Analyst*, **95**, 148 (1970).

169. L. C. Hansen, W. G. Scribner, T. W. Gilbert, and R. E. Sievers, *Anal. Chem.*, **43**, 34 (1971).

170. W. W. Flynn, *Anal. Chim. Acta*, **50**, 365 (1970).

171. C. W. McDonald and F. L. Moore, *Anal. Chem.*, **45**, 983 (1973).

172. L. J. Andrews and R. M. Keefer, *Molecular Complexes in Organic Chemistry*. Holden-Day, San Francisco, 1964.

173. C.N.R. Rao, S. N. Bhat, and P. C. Dwivedi, *Appl. Spectrose. Rev.*, **5.**, 1 (1971).

174. O. L. Baril and E. S. Hauber, *J. Am. Chem. Soc.*, **53**, 1087 (1931).

175. D. E. Laskowski, D. G. Grabar, and W. C. McCrone, *Anal. Chem.*, **25**, 1400 (1953); **26**, 1497 (1954).

176. M. Večera, J. Gaspari, D. Snolbi, and M. Jurecek, *Collect. Czech. Chem. Commun.*, **21**, 1284 (1956); **24**, 640 (1959).

177. R. Modin, M. Johansson, M. Schroeder-Nielsen, and S. Bock, *Acta. Pharm. Suec.*, **8**, 509, 561, 573, 585 (1971).

178. H. K. Biswas and B. M. Mandel, *Anal. Chem.*, **44**, 1636 (1972).

179. K. Rothwell and J. K. Whitehead, *Chem. Ind.*, **1969**, 1628.

180. J. A. Gautier, M. Micoque, C. Combet-Farnoux, and J. F. Girardeau, *Ann. Pharm. Fr.*, **30**, 715 (1972).

181. P. A. Hedin, R. C. Gueldner, and A. C. Thompson, *Anal. Chem.*, **42**, 403 (1970).

182. J. A. B. Darling and R. A. Harkness, *Acta Endocr. Copenh.*, **72**, 391 (1973).

182a. I. C. P. Smith, *Biological Applications of Electron Spin Resonance Spectroscopy*. Wiley-Interscience, New York, 1972.

183. D. J. Pasto and C. R. Johnson, *Organic Structure Determination*. Prentice-Hall, Englewood, Cliffs, N. J. 1969.

184. N. D. Cheronis, J. B. Entrikin, and E. M. Hodnet, *Semimicro Qualitaive Organic Analysis*, 3rd ed., Wiley-Interscience, New York, 1965.

185. T. S. Ma, B. J. Marcus, and J. Polesuk, *Mikrochim. Acta*, **1968**, 436; **1969**, 352, 815; **1970**, 677; **1971**, 267; **1972**, 313.

186. N. Kharasch and C. M. Buess, *J. Am. Chem. Soc.*, **71**, 2724 (1949).

187. W. Huber, *Mikrochim. Acta.*, **1960**, 44.

188. M. C. Kwitzel, *Org. React.,* **4**, 1 (1948).
189. H. L. Holmes, *Org. React.,* **4**, 60 (1948).
190. P. Nayler and M. C. Whiting, *J. Chem. Soc.,* **1955**, 2970.
191. J. G. Sharefkin and E. M. Beghosian, *Anal. Chem.,* **33**, 640 (1961).
192. J. R. Johnson and W. L. McEwen, *J. Am. Chem. Soc.,* **48**, 469 (1926).
193. E. H. Huntress and J. S. Autenvieh, *J. Am. Chem. Soc.,* **63**, 3446 (1941).
194. A. Newtown, *J. Am. Chen. Soc.,* **65**, 2439 (1943).
195. E. L. Brown and N. Campbell, *J. Chem. Soc.,* **1937**, 1699.
196. M. Jurecek and M. Vecera, *Collect. Czech. Chem. Commun.,* **19**, 77 1954).
197. P. P. T. Sah and T. S. Ma, *Berchite,* **65**, 1630 (1932).
197a. J. H. Billman and R. V. Cash, *J. Am. Chem. Soc.,* **75**, 2499 (1953).
198. H. W. Underwood, Jr., and J. C. Gale, *J. Am. Chem. Soc.,* **56**, 2117 (1934)
199. B. T. Dewey and N. F. Witt, *Ind. Eng. Chem., Anal. Ed.,* **12**, 459 (1940); **14**, 648 (1942).
200. V. T. Bickel and H. E. French, *J. Am. Chem. Soc.,* **48**, 747, 1736 (1926).
201. J. W. Berry, in J. C. Giddings and R. A. Keller (Eds.), *Advances in Chromatography,* Vol. II, p. 271. Dekker, New York, 1966.
202. E. O. Woolfolk, F. E. Beach, and S. P. McPherson, *J. Org. Chem.,* **20**, 391 (1955).
203. A. E. Pierce, *Silylation of Organic Compounds,* p. 72. Pierce Chemical Co., Rockford, 1968.
204. H. H. Hopkins, *J. Am. Chem. Soc.,* **45**, 541 (1923).
205. H. Zahn and A. Würz, *Z. Anal. Chem.,* **134**, 183 (1951).
206. K. Schlögel and A. Siegel, *Mikrochim, Acta,* **40**, 202 (1953).
207. H. W. Underwood, Jr., O. L. Barfil, and G. C. Toone, *J. Am. Chem. Soc.,* **52**, 4089 (1930).
208. M. Vecera and M. Gasparic, *Chem. Listy,* **48**, 1360. (1954).
209. H. J. Shine, *J. Org. Chem.,* **24**, 252 (1959).
210. G. D. Johnson, *J. Am. Chem. Soc.,* **73**, 5888 (1951).
211. C. Neuberg, A. Grauer, and B. Pisha, *Anal. Chim. Acta,* **7**, 238 (1952).
212. K. Hayes and C. O'Keefe, *J. Org. Chem.,* **19**, 1894 (1954).
213. H. O. House, *J. Am. Chem. Soc.,* **77**, 5083 (1955).
214. E. B. Sanders and J. Schulbert, *Anal. Chem.,* **43**, 59 (1971).
215. E. D. Barber and E. Sawicki, *Anal. Chem.,* **40**, 984 (1968),
216. A. Girard and G. Sandulesco, *Helv. Chim. Acta,* **19**, 1095 (1936).
217. D. F. Gadbois, P. G. Shauer, and F. J. King, *Anal. Chem.,* **40**, 1362 (1968).
218. A. von Wacek and K. Kratzl, *Chem. Ber.,* **76**, 1209 (1943).
219. J. J. Donleavy, *J. Am. Chem. Soc.,* **58**, 1004 (1936).
220. V. Horak, *Chem. Listy,* **48**, 1416 (1954).
221. El Sayd Amin and E. Hecker, *Chem. Ber.,* **89**, 496 (1956).
222. L. D. Metcalf and A. A. Schmitz, *Anal. Chem.,* **33**, 363 (1961).
223. L. F. Fieser and M. Fieser, *Reagents for Organic Synthesis,* Vol 1, p. 192. Wiley-Interscience, New York, 1967.

224. Ref. 203, p. 160.
225. W. O. Pool, H. J. Harwood, and A. W. Ralston, *J. Am. Chem. Soc.*, **59**, 178 (1937).
226. R. F. Goddu, N. F. LeBlanc, and C. M. Wright, *Anal. Chem.*, **27**, 1251 (1955).
227. K. J. Morgan, *Anal. Chim. Acta*, **19**, 27 (1958).
228. Y. Inoue and M. Noda, *J. Agric. Chem. Soc. Jap.*, **24**, 291 (1951).
229. F. Bell, *J. Chem. Soc.*, **1929**, 2787; **1930**, 1071.
230. C. S. Marvel and F. E. Smith, *J. Am. Chem. Soc.*, **45**, 2696 (1923).
231. C. S. Marvel, F. L. Kingsbury, and F. E. Smith, *J. Am. Chem. Soc.*, **47**, 166 (1925).
232. E. O. Woolfolk, W. E. Reynolds, and J. L. Mason, *J. Org. Chem.*, **24**, 1445 (1959).
233. T. Ottenbacher and F. C. Whitmore, *J. Am. Chem. Soc.*, **51**, 1909 (1929).
234. C. M. Suter and E. W. Moffett, *J. Am. Chem. Soc.*, **55**, 2497 (1933).
235. D. F. Gadbois, J. M. Mendelsohn, and J. Ronsivalli, *Anal. Chem.*, **37**, 1776 (1965).
235a. K. T. Liu, *Chemistry*, (Taipei), **1972**, 8.
236. I. H. Williams, *Anal. Chem.*, **37**, 1723 (1965).
237. R. Jeltes and R. Veldink, *J. Chromatogr.*, **32**, 413 (1968).
238. J. S. Fritz and R. C. Chang, reported at the ACS National Meeting, Chicago, Aug., 1973.
239. H. F. Walton and E. Murgia, reported at the ACS National Meeting, Chicago, Aug. 1973.
240. F. M. Rabel, *Anal. Chem.*, **45**, 957 (1973).
241. Y. T. Lin, K. T. Wang, and T. I. Yang, *J. Chromatogr.*, **20**, 610 (1965).
242. K. T. Wang, P. H. Wu, and T. B. Shih, *J. Chromatogr.*, **44**, 635 (1969).
243. G. Lehmann and H. G. Hahn, *Z. Anal. Chem.*, **238**, 445 (1968).
244. R. Berret, A. Gavanden, and J. Hirtz, *Ann. Pharm. Fr.*, **25**, 365 (1967).
245. T. Okita, *Atmos. Environ.*, **4**, 93 (1970).
246. G. V. Myasoedova, S. B. Savin, and N. I. Uryanskaya, *Zh. Anal. Khim.*, **26**, 1820 (1971).
247. J. M. Stewart, *Solid Phase Peptide Synthesis*. Freeman, San Francisco, 1969.
248. A. Tsuji and H. Kakihana, *Mikrochim. Acta*, **1962**, 475.
249. M. Fujimoto, H. Nakayama, M. Ito, H. Yanai, and T. Suga, *Mikrochim. Acta*, **1974**, 151.
250. Selectron Laboratories, Wesport, Conn.

COLLECTION OF THE ANALYTICAL SAMPLE

I. GENERAL REMARKS

A. IMPORTANCE OF PROPER COLLECTION TECHNIQUES

In this chapter are gathered together the techniques and apparatus that are generally employed to collect the analytical samples. As can be seen from the discussions in the preceding chapters, most materials are subjected to separation operations prior to the final process of identification (qualitative analysis) or determination (quantitative analysis). It is obvious that only a certain quantity of the analytical species need be collected. However, after separation, the amount of the substance obtained may be still insufficient for the subsequent experimental operation (e.g., material from thin-layer chromatographic spots for nuclear-magnetic-resonance spectrometry). Contrastingly, some separation procedures usually provide too much sample (e.g., vacuum distillation condensate for carbon and hydrogen analysis). In the former case, the separation is repeated and the collected samples are accumulated; in the latter case, a suitable receiver should be chosen so that an aliquot of the collected sample can be conveniently taken for analysis. While certain devices have been mentioned in previous chapters, it is worth while to discuss such devices in detail and gather them together here for ready reference. The present chapter also includes methods for collecting the analytical sample when no separation operation is involved.

Owing to increasingly sophisticated instrumentation in the finishing mode of chemical or physical measurement, and the requirement of very accurate and extremely sensitive analytical methods (e.g., the lowering of tolerance limits in environmental controls), it is prudent to use the correct technique to collect the analytical sample. Otherwise serious errors may be committed and remain undetected until it is too late to remedy them. For example, an automated ultramicro biomedical system is available that has the capability of simultaneously determining twelve constituents from 10 μl of blood [1]. If the twelve components are not properly collected for their respective spectrophotometric measurements, the analytical data print-out will be erroneous, leading to false diagnosis.

B. DIFFICULTIES IN COLLECTING ANALYTICAL SAMPLES

In chemical analysis related to environmental and pollution controls, the threshold limiting value (TLV) usually falls in the region of parts per million (e.g., the TLV for formaldehyde in a chemical-plant atmosphere [2] is 3 ppm). For such low concentrations, improper technique in collecting the analytical samples will undoubtedly lead to discrepancies and generate argument. While some standard procedures [3, 4] have been published, it is not unusual for a special situation to arise that invalidates the collecting method. Sometimes the discrepancy is unexpected. For instance, whereas infrared spectroscopy is capable of detecting atmospheric pollutants at the parts-per-billion level [5], faulty sample-collecting technique may easily vitiate the data. Thus Gruenfield and Ginell [6] have observed that chlorinated hydrocarbons collected in infrared gas cells did not give reproducible results and found that it was due to sorption on the plastic gaskets. Furthermore, the carbon dioxide absorption band was always present in the spectra, although this gas was not intentionally introduced. It appears that carbon dioxide can diffuse into the gas cell through the plastic gaskets. Chiou and Smith [7] have studied the adsorption of acidic and neutral organic compounds onto filter papers; the amounts adsorbed after filtration of 3 ml of solution ranged from zero to almost 100%, depending on the compound and paper used. Hence the effects of adsorption and impurities from filters on the analytical results cannot be ignored.

A few examples are cited below to illustrate the problems which one might encounter in collecting analytical samples from complex mixtures. It has been established that the tumor-initiating activity of cigarette-smoke condensate is in the neutral fraction, which contains trace amounts of polynuclear aromatic hydrocarbons. However, the amounts found by chemical analysis are too small to account for the observed carcinogenic activity. [8]. This discrepancy may be due to synergistic interaction of the hydrocarbons or to the biological contribution of unidentified neutral compounds in the working material. A major obstacle in differentiating between these effects is the difficulty of collecting the polynuclear aromatic hydrocarbons from the bulk material. A method has been proposed by Stedman and co-workers [9] (see Section VII) that ameliorates this situation. In the determination of total nitrogen in plant materials, Hamlyn and Gasser [10] have found that grinding to minimize errors due to inhomogeneous samples is effective in certain specimens but ineffective in others. Thus, for 100-mg samples analyzed by the Dumas method, grinding in a hammer mill to pass through a 1.0-mm screen was satisfactory for clover, beet tops, and grass. Grinding in a ball mill to pass through a

0.25-mm screen was satisfactory for spruce, oats, barley, and kale. Contrastingly, neither method of grinding have good precision for 50-mg samples analyzed by the Kjeldahl method, although grinding to pass through finer screens gave improved results for spruce, oats, grass, hay, and beet tops, but not for clover, barley, kale, or maize. In residue analysis, the reliability of sample collection is crucial. Bergstrom-Nielsen [11] has put forward a mathematical model to study the relationship between recovery and concentration, using the determination of sulfamerazine residues in trouts as the test case. For the analysis of content uniformity in pharmaceutical tablets, Lodge [12] has proposed a method that utilizes an extractant that contains an internal standard different from the substance to be determined. Thus, for the estimation of norethisterone in individual tablets, the procedure consists of extracting the hormone into acetone containing pregnenolone acetate, followed by gas-chromatographic analysis of the resulting solution. In this way, the extent of the collection of the analytical species can be ascertained.

II. COLLECTION OF PRECIPITATES AND SUSPENDED PARTICULATES

A. FILTRATION METHODS FOR COLLECTING PRECIPITATES

The techniques of crystallization and recrystallization have been discussed in Chapter 3, Section II. A number of simple filtration devices are given in Chapter 2, Section VI (see Fig. 2.33 to Fig. 2.34). For analytical purposes, if the precipitate is to be subsequently processed, it is desirable to keep the product in crystalline form. This is not always possible, however, and the analyst sometimes is confronted with the problem of handling amorphous or gelatinous precipitates.

1. Handling Crystalline Precipitates

For the purpose of qualitative analysis, one may employ any device shown in Chapter 2, Section VI, or Chapter 3, Section II, to collect the analytical sample. Needless to say, the collector should be commensurate with the amount of the precipitate. The filter funnel illustrated in Fig. 2.33 has the advantage of being able to collect between a few milligrams and 500 mg of precipitate and recover the crystals after drying in the same apparatus, with insignificant losses. Erdy and Buzas [13] have described an analytical glass filter that consists of a woven glass ribbon wound around a glass rod; filters of various porosities can thus be prepared.

The quantitative collection of milligram amounts of precipitate is usually performed by means of a filter tube with a sealed-in sintered glass

plate [14, 15]. In order to eliminate the troublesome adherence of particles of precipitate to the walls of the pressure (delivery) tube, Corliss [16]
has recommended a device shown in Fig. 7.1. A 2-ml syringe C is inserted
by means of a 22-gauge needle into the rubber stopper connecting the
siphon B and filter tube D. A variable air space is thereby introduced
above the solution in the pressure tube A. Rapid plunger action in the
syringe produces agitation in the solution in the pressure tube, thus free-
ing adhering particles and also hastening attainment of solubility equi-
libria so that wash volumes can be readily controlled.

If the filtrate is to be processed after filtration, the device employing
a bell jar placed on a rubber plate (see Fig. 2.36) is recommended. Alternatively, a cylindrical glass tube with side-arm and open ends (Fig. 2.35
can be used.

Potterat and Eschmann [17] have designed a filter flask for sugar determinations. It consists of a shortened sintered glass filter tube fused into
the side of a round 100-ml flask to make an angle of 120 degrees with the
neck of the flask. In the boiling position, the neck of the flask, which
carries a reflux condenser, is inclined at 45 degrees; in the filtering position, the inside of the flask can easily be washed by a jet of water. Boswall
and Mackay [18] have described a vacuum-filtration assembly for soil
samples; it has been used for the determination of such soil properties as
exchange capacity, exchangeable-cation content, water-equilibrated potassium status, and nitrate-production capacity after incubation. For performing the simultaneous vacuum filtration of many samples, a commercial

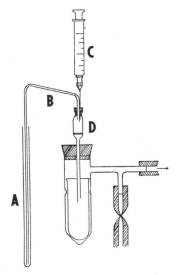

Fig. 7.1. Device for quantitative collection of
precipitate [16]. (Courtesy *Chem.-Anal.*)

apparatus [19] is available that can handle 30 fluid samples simultaneously through Millipore filters; it is applicable to gravimetric analysis, radiological assays, and the like.

Owing to the recent development of organometallic chemistry, analytical operations in the absence of air are frequently required. The apparatus shown in Fig. 7.2 has been used by Spencer [20] in the investigation of organolithium compounds. It is used for the quantitative analysis of lithiated silicas by titration with methanolic HCl, involving the following series of operations: Physisorbed water is completely removed by drying at 80°C under vacuum. Dried and degassed n-hexane is vacuum transferred into the flask. Excess n-butyllithium in n-hexane is added. The suspension is stirred magnetically under argon for the reaction period. The solvent is filtered, and all excess organolithium reagent is removed by repeated washing with anhydrous degassed solvent. Filtration is accomplished by rotating the flask so that the filtering arm is in the vertical posi-

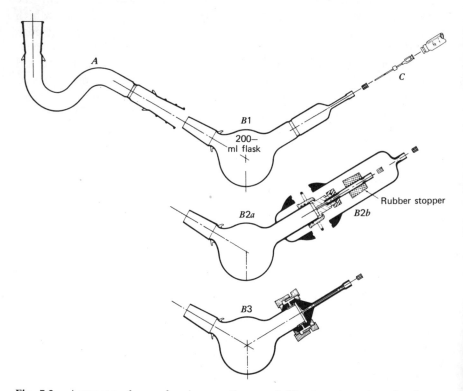

Fig. 7.2. Apparatus for preforming reactions and filtration operations in absence of air [20]. (Courtesy *Anal. Chem.*)

tion. The lithiated silica is then vacuum dried, taking care to continue excluding air.

2. Handling Gelatinous Precipitates and Protein Solutions

The collection of gelatinous precipitates for analytical purposes requires patience and ingenuity. Pietsch [21] has described a technique for filtering gelatinous hydroxides, which involves mechanical stirring of the liquid in the filtering vessel. Fletcher [22] has constructed an apparatus using an Alundum thimble for the suction filtration of such materials as oxidized products from lubricating oils and hydrolyzed cellulose. Gallinger [23] has employed Pellicon membranes for the ultrafiltration of protein solutions. Samuels [24] has described a method of collecting insoluble protein on tared funnels by the use of a specially designed 5-ml centrifuge flask.

B. COLLECTION OF DUST, POWDER, AEROSOLS, AND FLY ASH

The collection of dust, fly ash, and atmospheric aerosols is essential in air-pollution control. Industrial wastes, trace metals, pollens, and soot containing polynuclear aromatic hydrocarbons are the major substances to be collected and analyzed by microscopic [25], chemical, or physical methods. The collection is usually carried out by the filtration technique [26]. The filter material may be filter paper, glass fiber paper, Millipore filter discs, or cellulose ester membranes. For example, determinations of lead in airborne particulates have been performed by collection using a membrane [27] or on a filter inside a graphite cup [28], followed by atomic absorption spectrometry. McHugh and Stevens [29] have used Millipore 0.45-μm filters and air-filtering apparatus to collect samples of airborne particulates for ion-microprobe mass-spectrometric analysis. The filter is dissolved in acetone and centrifuged, and the particulate residue taken for analysis. Golden and Sawicki [30] have collected polynuclear aromatic hydrocarbons on filter paper and recovered them by ultrasonic extraction. Patterson [31] has used fired and organically bound glass-fiber filters to collect atmospheric aerosols for C-H-N analysis. A variety of filters for airborne particulates are commercially available; two examples are shown in Fig. 7.3. Sugimae has collected trace metals on silver membrane filters. After dissolving the sample filter in HNO_3 and precipitating AgCl, the metals are analyzed by emission spectrometry [31a].

Wyndham [32] has described a method for collecting micro amounts

Fig. 7.3. Filters for collection of airborne particulates. Portable sampler. (Courtesy Microchemical Specialties Co.)

of dust, powder, and fiber for microscopic examination. The apparatus consists of a tube supporting a lightly greased microscope cover slip and a small funnel; the specimen is collected on the cover slip, via the funnel, by applying suction. Solbacken [33] has described a device for collecting powder from a reaction vessel at high temperature (see Chapter 9, Fig. 9.11). Sherwood [34] has designed an apparatus for the collection of minute solid samples from otherwise rather inaccessable places. As illustrated in Fig. 7.4, this micro vacuum sample collector consists of a B 19 socket with a bulb into which are fused a reentrant jet and a hook-shaped delivery tube. A B 19 cone, into which is fused a glass tube terminating at the inner end with a sintered glass filter stick, fits into the socket. The outer end of the tube is connected to a vacuum pump, which draws air through the sinter and collecting jet. Any loose dust in the air stream is drawn through the jet into the body of the apparatus and held onto the sinter plate. This apparatus is used to collect corrosion deposits near thick grease areas on a large casting; 20 mg of clean corrosion salt can be collected without grease contamination. Makovsky [35] has described a technique for isolating small particles embedded in minerals for Raman

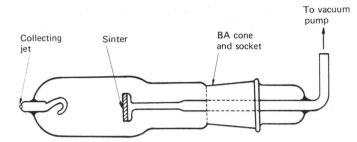

Fig. 7.4. Micro vacuum sample collector [34]. (Courtesy *Lab. Pract.*)

spectroscopy without mechanically destroying the heterogeneous matrix. Brachfeld [36] has proposed a technique for collecting composite metal powders for X-ray fluorescence analysis by compressing the powder into a pellet.

An automated apparatus for airborne dust collection and radioactive analysis has been patented [37]. In this apparatus a filter paper is mounted in an aperture in a data-processing card. The card is inserted in a holder in which a rubber O ring and a plastic ring hold the filter paper snugly. The holder is connected to a vacuum pump, and an air sample is drawn through the filter paper for a predetermined length of time. The card, with the collected particulate matter, is then transferred to a laboratory for processing through automatic analytical equipment, and the results are keypunched on the card.

C. COLLECTION OF SUSPENDED PARTICULATES IN LIQUID MEDIA AS RECEIVED

Unlike crystalline precipitates, suspended particulates in liquid media as received (e.g., polluted water) are not intentionally produced in the experiment, but have to be separated from the clear fluid in order to perform the analysis. One technique for collecting this kind of material is to pass the liquid through a fine filter. Membrane filters are commonly employed. Because of their surface retention ability, these filters not only accomplish a separation of particulate material from liquid media but also permit a nearly ideal presentation of the sample to an attenuated-total-reflection (ATR) crystal for infrared analysis. For instance, Hannah and Dwyer [38] have demonstrated the analysis of dispersed pigment in paints and of colloidal suspension in a cup of tea by collecting the analytical samples on Millipore filters of 0.2- to 0.8-μm pore size. For the ATR spectra, only microgram quantities of the suspected particulates need be collected.

If the clear liquid is the desired specimen, the in-line filter unit [39] shown in Fig. 7.5 can be used. It has a choice of four filter tubes with pore sizes from 0.3 to 8 μm, and is suggested for removing rust and sediment from water, or for clarifying liquid cosmetics, plating electrolytes, and so on. Another commercial micro suction ultrafiltration apparatus [40] uses a 15-Å pore membrane for the collection of proteins and macromolecules (molecular weight over 7000).

Centrifugation is a convenient technique for collecting suspended particulates from small volumes of liquid. Feldman and Ellenburg [41] have designed a two-piece centrifuge crucible that allows the suspended material to be further processed without changing containers; it consists of a centrifuge tube that can be drained and separated into two parts, the lower of which can be used for drying, ignition, or dissolution of the solid. Hudson and Ison [42] have described a simple technique for centrifuging a few microliters of material (e.g., the protein extract from a single fly) as follows: The sample is drawn into a polyethylene tube (100 mm long, 1-mm i.d.), which is then folded to form a U tube. Several such tubes may be inserted into drilled wooden blocks fitted into the centrifuge head, thus allowing a number of samples to be handled simultaneously. Riccobini and co-workers [43] have constructed a low-temperature laboratory centrifuge for use between $-20°$ and $-70°C$ with an accuracy of $\pm 1.5°C$. Casciato [44] has reviewed the ultracentrifuge, particularly with regard to its uses in biomedical research.

Fig. 7.5. In-line filter unit [39]. (Courtesy Metro Scientific, Inc.)

III. COLLECTION OF SOLID SAMPLES AFTER CHROMATOGRAPHIC SEPARATION

A. FROM A THIN-LAYER OR PAPER CHROMATOGRAM

The simplest way to collect solid samples from a thin-layer chromatogram is to remove the area containing the desired material by scraping with a razor blade or sharp spatula. In the case of an aluminum or plastic plate, it is possible to cut out the needed area. Another technique is to collect the adsorbent together with the analytical sample using one of the suction devices [45]. A simple device described by Clement [46] (see Chapter 9, Fig. 9.15) is made from a disposable Pasteur pipet, the tip of which has been turned in slightly and plugged with glass-fiber filter sheet [47]. There are various types of commercial TLC collectors. The spot collector [48] shown in Fig. 7.6 is equipped with a rounded glass tip to loosen the adsorbent thin layer. As the particles are loosened, they are immediately sucked into the collector and are deposited on the filter disc.

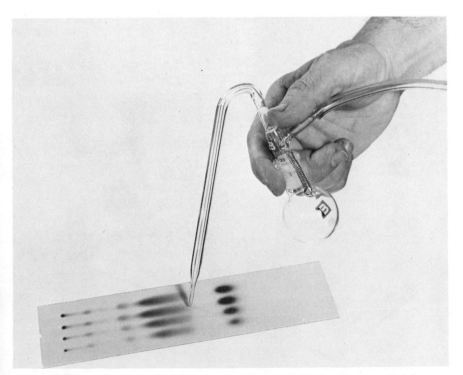

Fig. 7.6. Spot collector for thin-layer chromatogram [48]. (Courtesy Brinkmann Instruments, Inc.)

The analytical species is subsequently extracted by sucking an appropriate solvent into the tip of the collector and through the adsorbent, so that the resulting eluent is accumulated in the flask while the adsorbent remains in the filter disc. The zone collector [49] is illustrated in Fig. 7.7.

The analytical sample collected from a single spot of the thin-layer chromatogram is adequate for absorption spectrophotometry [50], ATR infrared spectrometry [51], or melting-point determination [52]. To obtain the mass spectrum, Rix and co-workers [53] transfer the separated zone to an elution column prepared from a shortened Pasteur pipet, drawn out near its lower end. The sample is eluted with methanol and collected in the lower part of the pipet. This is snapped off, and the eluate is evaporated on the tip of the solid-sample insertion probe of the mass spectrometer.

An arrangement for the simultaneous elution of one chromatographic plate with several solvents has been described by Sandroni and Schlitt [54]. The thin layer is divided into five bands, and polypropene barriers are

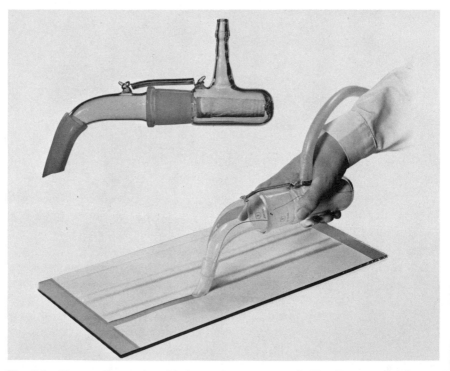

Fig. 7.7. Zone collector for thin-layer chromatogram [49]. (Courtesy Brinkmann Instruments, Inc.)

inserted between them. Thus five different solvents can be tested on the same plate. A further modification of the barriers and the solvent wicks permits the use of a continuous solvent gradient.

If the sample collected from one chromatospot is insufficient for the subsequent experiment, more material can be obtained by utilizing preparative-layer chromatography [55] (see Chapter 4, Section I.F). It should be noted that cellulose powder is difficult to remove from the plate. Therefore, when this adsorbent is employed, it is preferably mixed with silica gel [56].

The collection of steroids from thin-layer chromatograms has received much attention. Hunt [57] has found that finger marks could introduce contaminants, and has designed an apparatus for the elution of nanogram amounts of steroids that eliminates the risk of contamination from sebum during handling of the equipment. Masaracchia and Gawienowski [58] have described a procedure for the quantitative collection of 20 to 100 μg of steroids. Vandenheuwel [59] has devised a technique whereby 1 μg of adsorbed steroids can be recovered quantitatively. It is applicable to the serial elution of many fractions through the use of the equipment described. Thus it is particularly useful in clinical analysis and metabolic studies of urinary steroids. Quantitative collection of lecithins after argentation chromatography on thin-layers has been reported by Kyriakides and Balint [60]; the lecithins are freed from silver nitrate by means of ion exchange.

The techniques for collecting analytical samples from paper chromatograms are similar to those discussed above, bearing in mind that paper is pliable and can be cut apart easily. Lewis [61] has constructed a stainless-steel tank with accessory fittings for the collection in glass capillary tubes of 0.2-ml volumes of eluate from paper strips 7×1.5 cm. Reith [62] has described a method for the simultaneous collection of compounds of low solubility from many paper chromatograms. Laksminarayana and Sebastian [63] have put forward a procedure whereby 12 to 25 strips cut from paper chromatograms can be eluted simultaneously. It uses a semicircle of filter paper, shaped into a cone and placed in a glass funnel, to support the strips (see Fig. 7.8). This device is more stable than the apparatus described by Ballinger and Manes [64].

B. COLLECTION OF SOLID SAMPLES FROM A GAS CHROMATOGRAPH

Various techniques have been proposed for the collection of solids from a gas chromatograph. As summarized by Van Lier and Smith [65], the methods involve the use of (1) simple cooled traps; (2) traps with sol-

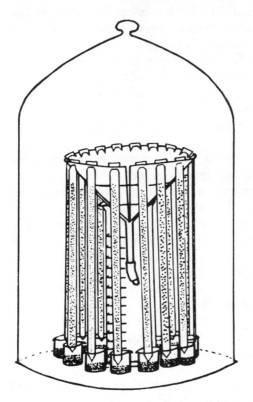

Fig. 7.8. Simultaneous elution of paper chromatograms [63]. (Courtesy *Lab. Pract.*)

vent, glass, or cotton wool; (3) capillary tubes; (4) electrostatic precipitators; (5) thermoelectric cooled devices; (6) centrifugal coolers; (7) traps with specific thermal gradients; (8) Millipore filters; (9) cellulose acetate filters. It should be noted, however, that each method has its limitations and may not be successful in collecting the particular analytical sample. DuPlessis [66] has described a micro cold-finger trap with capacity of 1 to 5 μl that can be cooled in liquid nitrogen or Dry Ice mixtures and also safely centrifuged at 750g (gravities). Crystalline sterols (0.1 to 1 mg) have been collected in capillary tubes [65].

Cooke [67] has investigated the collection of steroids in stainless-steel tubing (2-mm i.d.) packed with glass wool. It was found necessary to have the temperature of the arm of the fraction collector of not less than 202°C to prevent tailing of the eluted steroids. Silanizing the glass wool had little effect on the efficiency.

Hardy and Keay [68] have described a glass trap for collecting high-molecular-weight compounds that elute from the gas-chromatographic column as aerosols. The formation of mist or aerosol, however, usually complicates the collection of the analytical sample. Therefore Bloomer and Eder [69] have designed a collecting apparatus that utilizes electrostatic precipitation to prevent aerosol formation. As depicted in Fig. 7.9, the glass apparatus is connected to the outlet of the gas chromatograph by the 3-mm tube *A*, which is sealed to the U tube *B* of 5-mm o.d., the latter being placed in the cooling bath *I* contained in the beaker *J*. To provide maximum effectiveness in electrostatic precipitation, a Dewar seal *C* is made, in which a tube (3-mm o.d.) extends down almost to the level of the cooling bath and provides a very small gap between the well and the body of the tube. The electrode is a large-diameter nickel wire *D*, which fits snugly into the inner well. An ordinary vacuum leak detector (or Tesla coil) is used as a source of high-voltage alternating current. The outflow tube *E* is bent up in such a way that the sleeve *F* can pass over it easily.

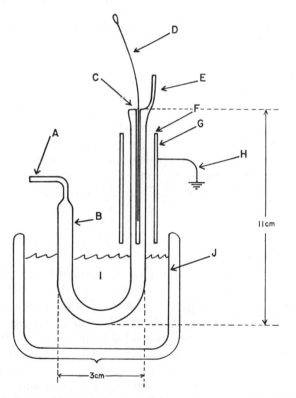

Fig. 7.9. Electrostatic collecting device for gas chromatography [69]. (Courtesy *J. Gas Chromatogr.*)

The ground electrode of the system is simply a layer of metal foil over the glass sleeve F; it is connected via copper wire to a suitable ground. This apparatus can collect quantitatively 1-mg samples of low volatility.

Curry and co-workers [70] have described a collection technique whereby the separated components that emerge from the column pass through a heated line via a stream splitter and condense independently on a glass surface. Willis [71] has applied the trapping technique for open tubular columns to samples that have been catalytically hydrogenated or subjected to preliminary separation.

IV. COLLECTION OF SUBLIMATES

As mentioned in Chapter 3, Section III, sublimation is a technique particularly suited for microscale fractionation or purification of organic materials. Collection of the sublimate for qualitative analysis presents no difficulty, since the compound that condenses on the receiver (cooled portion) of the glass apparatus can be plainly observed in milligram quantities. The receiver can be detached from the sublimation assembly during or after heating, and the crystals can be recovered from the glass surface by scraping or dissolution in an appropriate solvent. On the other hand, quantitative collection of the analytical species cannot be accomplished easily, because (1) the sublimate is spread thin over a relatively large area and (2) the loss of some vapor during heating is unavoidable. Parish [72] has described a method for quantitative collection of crystals from a cold-finger condenser by placing under the condenser a paper cone supported in a loop of wire; the pointed free end of the wire is pressed into the base of the stopper of the filter flask surrounding the cold finger.

The simultaneous collection of several fractions can be carried out in the gradient temperature-tube furnace [73, 74]. The apparatus depicted in Fig. 7.10 is employed by Shigetomi [75] for the separation of quinones and halogenated phenols. It consists of a ceramic tube (100 cm long, 5-cm i.d.) wound with 1-kW nichrome wire. The windings are closer together on the upper part than on the lower part of the tube. A Slidac-type transformer is attached to the nichrome wire for heating with electricity at an appropriate voltage. A copper spiral tube (0.5-cm i.d.) is placed inside the tube, and for cooling, air is passed into this tube at a uniform rate. The combined cooling and heating processes keep the temperature inside the ceramic tube at the desired level. The temperature in each section is indicated by the thermometer. The sublimation tube is a glass tube of 8- to 9-mm bore. The material to be sublimed is wrapped in an aluminum foil and placed in the upper part of the sublimation tube. After the pressure

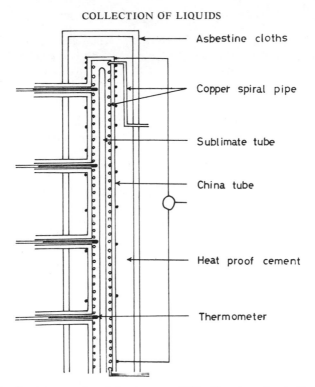

Asbestine cloths

Copper spiral pipe

Sublimate tube

China tube

Heat proof cement

Thermometer

Fig. 7.10. Gradient temperature-tube furnace [75]. (Courtesy *Anal. Chem.*)

inside the sublimation tube has been reduced to the desired level, the tube is then positioned in the gradient temperatur-tube furnace.

V. COLLECTION OF LIQUIDS

Probably the most difficult task in microscale manipulation is the collection of analytical samples that are in liquid form. The recent debate on "polywater" is a classic example: Much effort was wasted in making elaborate measurements while the purity and reproducibility of the microsample had not been validated. In the collection of liquid for microanalysis, it is practically impossible to eliminate the danger of contamination and losses due to volatility and during transfer. The quantitative collection of liquids is difficult because of the strong adherence of liquids to the walls of the containers.

The volume of liquid sample that is required to be collected depends on the analytical problem. Thus, the liquid may be a compound obtained by fractional distillation; in this case the microsample needed usually does not exceed a few milligrams. In contrast, when the liquid is a complex mixture (e.g., industrial waste water or an effluent from an elution column), the required collected volume is relatively large, since the liquid contains only minute amounts of the species to be identified or quantitated. Similarly, the analytical species in the gas chromatograph is mixed with such large proportions of the carrier gas that it takes ingenuity to collect a suitable amount of the desired component for subsequent processing.

A. COLLECTION FROM A DISTILLATION APPARATUS

The reader is referred to the discussion on distillation (Chapter 2, Section I.B; Chapter 5, Section II) and extraction (Chapter 6, Sections II and III), where various types of receivers are described. Needless to say, every device for collecting distillates has its limitations. For the purpose of collecting the analytical sample, it is prudent to select a suitable technique and appropriate design of the receiving vessel. A few additional collecting devices for specific applications are presented below.

Jacobson and Miller [76] have constructed a vacuum multifraction collector that is especially useful for collecting condensates that at normal temperatures are viscous liquids. As shown in Fig. 7.11, the assembly, which contains a simple brass rack, is made with ground-glass joints. The outer container is broad and firmly based, while the changing of collectors is carried out simply, without breaking the vacuum in any way, by rotating a control rod, flattened at the top and bottom and keyed into the brass rack.

Bitz [77] has described a fraction collector illustrated in Fig. 7.12, for handling air- or moisture-sensitive compounds. The tube A is sealed to the bottom of the collector and becomes an axle for the tube B. A Teflon washer (not shown) is placed between the end of tube A and the flat surface in the tube B at the position C, thereby reducing friction when the tube B is rotated. The tube D is for evacuating the apparatus, and the stopcock on the other side is for introducing the inert atmosphere. At the top of the apparatus a female ball joint is fused to a modified parallel, oblique-bore stopcock. A male joint is fused to one arm on the underside of the stopcock, and a through joint is fused to the other arm to provide a drip tip into the receivers. The forerun receiver E is fabricated from a female joint. A piece of 9-mm glass tubing F is fused to the end of the stopcock for a vacuum takeoff. The stopper portion of the stopcock has a groove cut in it to allow the continuous evacuation of vapor from the forerun receiver. It is essential that the volatile contents in the forerun receiver be kept isolated from the sample receivers to avoid contamination of the

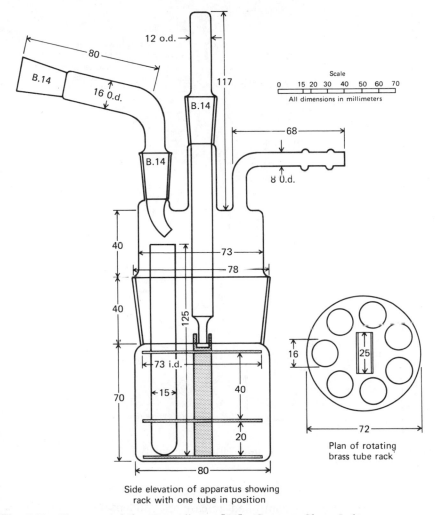

Side elevation of apparatus showing rack with one tube in position

Plan of rotating brass tube rack

Fig. 7.11. Vacuum multifraction collector [76]. (Courtesy *Chem. Ind.*)

fractions by vapor transfer. The fraction collector is connected to the distilling apparatus through the T 18/9 semi–ball joint. After the distillation is complete, the vacuum source is clamped off and an inert gas is introduced through the stopcock. The receivers are stoppered with the aid of a pair of tweezers through the 24/25 joint.

Vass [78] has constructed an all-walls-cooled cylindrical trap, shown in Fig. 7.13, in order to maximize the cooling surface for the condensate. Hyde and Redford [79] have described a resistance-wire-wound glass

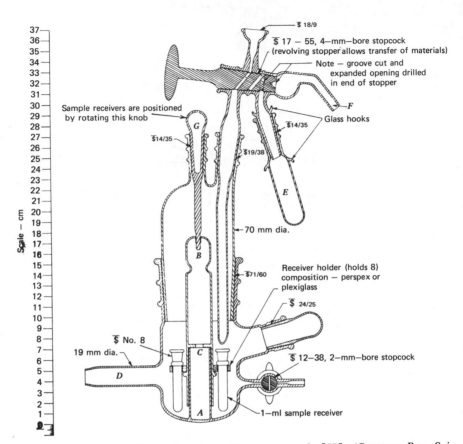

Fig. 7.12. Fraction collector for air-sensitive compounds [77]. (Courtesy *Rev. Sci. Instrum.*)

cold-finger trap for the controlled release of condensable gases, which eliminates the need for freezing mixtures such as Dry Ice–acetone and liquid-nitrogen-cooled ethanol and pentane, with their attendant fire hazards. The trap is capable of maintaining temperatures ranging from $-196°$ to $0°C$ with only a small temperature gradient. It has been used to condense carbon dioxide, sulfur dioxide, and water vapor and release the gases under controlled temperatures. Timsit and Kiang [79a] have fabricated an efficient continuously operated cold trap.

B. COLLECTION FROM ELUTION COLUMNS

The common practice in collecting analytical samples from elution columns (see Chapter 4, Section IV) is to use a fraction collector. There are

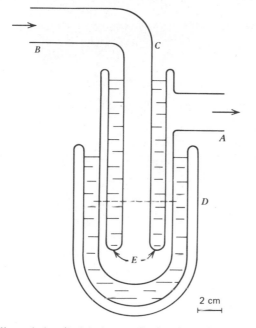

Fig. 7.13. All-walls-cooled cylindrical trap [78]: *A, B*, lateral connections; *C, E*, solder joints; *D*, line of final soldering. (Courtesy *Rev. Roum. Chim.*)

many commercial models of such collectors, most of them being designed for automatic operation. One manufacturer [80] has marketed a unit that holds 60 tubes (75- to 160-mm length, 15- to 18-mm diameter) and occupies minimal space (38×29×30 cm). The dispensing time can be adjusted between 1 and 48 min per fraction.

Večerek [81] has published an extensive review of automated fraction collectors. Many investigators prefer to construct their own fraction collector in order to suit specific purposes. Some can be easily assembled using inexpensive equipment. For instance, Popper and Nury [82] have described an apparatus that consists of collection units joined together by simple glass cross-connectors. The parts of one collection unit, shown in Fig. 7.14, comprise a test tube *D*, a two-hole rubber stopper *C* that carries the vertical air vent *B*, and the cross-connector *A*. The units, placed on a tilted rack, are connected by the tubing *E*, joining the arms of *A*. In use, the column effluent enters the first tube of the assembly through one arm of *A*, flows into *D*, and closes the air vent *B*. When *D* is filled, flow continues through the opposite arm of *A*, and so on. The vent on *A* prevents siphoning.

Soracco and Borris [83] have described a variable-volume siphon collector fabricated of borosilicate glass and a plastic disposable syringe. It

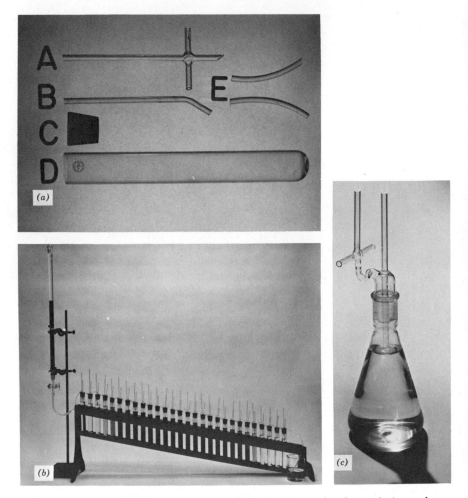

Fig. 7.14. Simple fraction collector to collect liquid samples from elution column [82]: (*a*) Parts of one unit; (*b*) the tilted rack; (*c*) the all-glass version. (Courtesy Dr. K. Popper.)

eliminates the need to use several constant-volume collectors and can be used to collect volumes from 1.5 to 5 ml with an accuracy comparable to that of the commercial instruments. Elsdon-Dew [84] has designed a versatile fraction collector consisting of a light wheel that holds the receiving tubes and is rotated by a clock spring. The rotation is controlled by a simple pin-and-gate solenoid-operated escapement mechanism actuated by an electric pulse from a drop counter or timer. Huber and co-workers [84a] have designed an automated fraction collector for the batchwise collection of fractions from high-pressure liquid chromatography.

With the increasing use of mass spectrometry for the characterization of microsamples, considerable attention has been drawn to the fact that the analytical species collected from the chromatographic apparatus can be fed directly into a mass spectrometer. While most of the investigations have been carried out with gas chromatographs (see Section V.C below), Lovins and co-workers [85] have described an interface that accepts the effluent from a liquid chromatographic column, isolates the solute from the solvent, and introduces the residue into the ion source of the mass spectrometer. The schematic diagram is shown in Fig. 7.15. The analytical

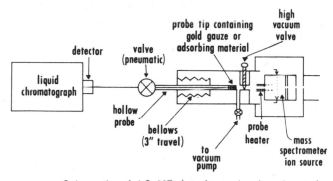

Schematic of LC–MS interface showing the major components, including the capillary line from LC to pneumatic valve located on the hollow probe, the motor driven hollow probe (i.d. = 0.032 $\times$ 12$^3/_8$ in. long) on 3-in. bellows, and the high vacuum inlet valve (motor driven)

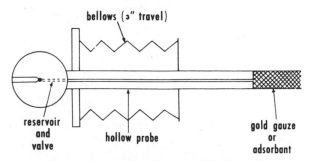

Schematic of probe assembly showing the major components: pneumatic valve and reservoir (5-ml capacity) coupled to the hollow probe (12$^3/_8$ $\times$ 0.16 in.; 0.032 in. i.d.) operating on the bellows (1.8-in. diameter) having a travel of 3 in. The probe tip (1.09 $\times$ 0.128 in. i.d.) is removable and can be packed either with gold gauze or absorbent

Fig. 7.15. Coupling of liquid chromatograph to mass spectrometer [85]. (Courtesy *Anal. Chem.*)

sample is collected in the reservoir and valve assembly, located at the end of the probe proper. The valve to the vacuum pump is opened, the probe is evacuated, and the pneumatic valve is opened to allow the sample to be drawn into the probe tip. For flash evaporation operation, the probe tip is heated to 70–125°C (depending on the eluting solvent). After the material is collected on the probe, the pneumatic valve is closed, the probe is evacuated to 10^{-5} torr, the source isolation valve is opened, and the probe is inserted into the ion source for mass-spectral analysis.

C. COLLECTION FROM A GAS CHROMATOGRAPH

Since the quantity of analytical species in the gas chromatograph is generally below the milligram region, it is extremely difficult to collect the specimen as pure liquid, because 1 mg of liquid occupies a volume of only about 1 μl. Hence, when the vapor of the compound is condensed by direct cooling, the collecting vessel is usually kept at a temperature below the freezing point of the compound, and the condensate is subsequently liquified by warming. Armitage [86] has studied some factors (the boiling point of the compound, the collection temperature, and the detector temperature) affecting the efficiency of collection of C_5–C_{16} normal paraffins at the detector exit in glass melting-point tubes. Lowering the detector temperature is shown to be very effective in condensing the effluent from the column.

A useful technique for collecting the microsample is to mix it with the vapor of another liquid with which it is miscible and condense both together. This method is known as "cocondensation with a solvent" [87, 88]. Beyermann and co-workers [89] have tested methane, water, carbon tetrachloride, and dichloromethane as the collector solvents, the last named being preferred because of its high volatility. Parliament [90] has proposed an apparatus shown in Fig. 7.16. It consists of a pear-shaped flask of 5- to 10-ml capacity, to which has been fused a short side arm

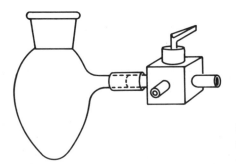

Fig. 7.16. Apparatus for collection by solvent cocondensation [90]. (Courtesy *Anal. Chem.*)

(3.5-mm i.d.). This is connected to a three-way valve by means of Teflon tubing. The distal end of the valve connects to the exit port of the gas chromatograph. A micro spiral condenser is placed above the flask to provide efficient reflux. The valve is heated with heating tape to the temperature of the gas-chromatograph manifold to prevent premature condensation. In operation, 0.5 to 1.0 ml of solvent is placed in the flask and brought to reflux. As undesired peaks pass through the exit port, they are vented to the atmosphere through the three-way valve. When the analytical sample elutes, the position of the valve is adjusted to pass the entire effluent into the flask containing the refluxing solvent. Freon, with boiling point of 47°C, is recommended as a solvent, since it is nonflammable.

Kubeczka [91] has proposed a method to collect the analytical sample leaving a gas chromatograph by adsorption on columns (10 cm×6 mm) of kieselguhr supporting 15% of SE-30 and maintained at ambient or subambient temperature. Dixon and co-workers [92] have described a technique of collecting trace components in bergamot oil by first removing the major components with high-capacity preparative columns. Papa and Varon [93] have determined cis- and trans-octafluorobut-2-enes in octafluorocyclobutane at the 0.5-ppm level by collecting, in a trap cooled by liquid nitrogen, the tail section of the major component and then warming the trap to release the condensate for analysis on a second column.

The collection of the components resolved by a gas chromatograph is frequently desired for subsequent spectral analysis. For instance, Moshonas and Shaw [94] have described a technique for collecting the essential oils of citrus fruits by means of a coiled hypodermic needle that is connected through an adapter to the exit port of the gas chromatograph. The needle trap is cooled in an 11-ml Dewar flask containing liquid nitrogen. The collected analytical sample is then delivered to the mass spectrometer by inserting the coiled needle into the septum of a gas-liquid sampling system of the spectrometer. The common practice nowadays, however, is to couple the gas chromatograph directly to the mass spectrometer. Various methods of performing the coupling have been reported [95–100]. The apparatus (Fig. 7.17) designed by Nota and co-workers [101] for coupling a capillary gas-chromatographic column to a mass spectrometer consists of three main parts: a joint, a stainless-steel capillary, and a splitter-junction, linked together through fittings and gaskets. The joint links the tared capillary to the ion source through a reentrant. Grob and Jaeggi [102] have described a technique that facilitates rapid column interchange by using platinum capillary and PTFE tubing for shrunk connections. Markey [103] has constructed an improved retractable glass-frit interface (Fig. 7.18), which permits a complete mass-spectral scan with 0.05 μg of decane or 0.5 μg of cholesterol. Bruner and co-workers [104]

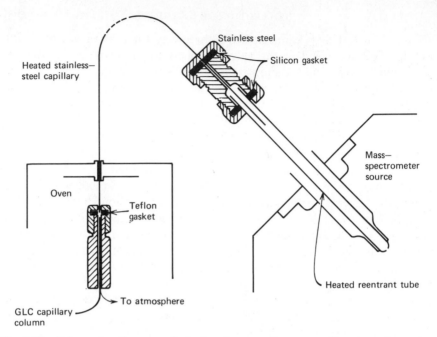

Fig. 7.17. Schematic diagram of device for coupling a capillary gas-chromatographic column to a mass spectrometer [101]. (Courtesy *Chem. Ind.*)

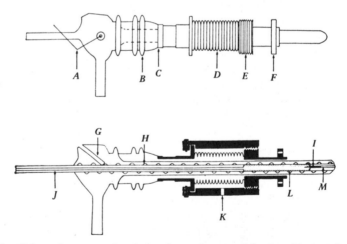

Fig. 7.18. Side and cross-sectional drawings of retractable glass-frit interface [103], overall length 12 in.: *A*, nichrome heating wire feedthrough; *B*, glass bellows; *C*, Kovar glass-to-metal seal; *D*, seamless, thin-wall flexible stainless-steel bellows; *E*, threaded portion of flange; *F*, ion-source housing flange; *G*, thermocouple well; *H*, nichrome heating wire; *I*, wire restriction in capillary *M*; *J*, glass capillary (2-mm i.d.) connected to chromatographic inlet; *K*, helium leak-testing port in steel driving ring; *L*, fritted-glass portion; *M*, pressure-restriction capillary. (Courtesy *Anal. Chem.*)

342

have described a double-detection interface for the analysis of complex mixtures. Dutton [105] has developed a method to collect odor constituents from 1 to 5 ml of heated oils and analyze them with a GC-MS-computer loop. Simmonds and co-workers [106] have suggested an automated pyrolysis-GC-MS system to perform organic analysis of the soil on Mars. Crosmer and co-workers [106a] have described a cold-trap fractionation technique for collecting the condensates from the gas chromatograph and then evaporate them into the mass spectrometer.

The collection of gas-chromatographic effluents for infrared analysis has also been studied [107, 108]. Thus, in the method described by Curry and co-workers [109], the effluent is split so that 90% of it is led, via a heated capillary tube and a needle that pierces a septum, to a cooled glass vessel. The condensate is dissolved in 25 μl of solvent and transferred by means of a syringe to 0.5 mg of potassium bromide powder to prepare the micropellet (0.5-mm diameter). Bus and Liefkins [110] have described a procedure for the collection of microgram amounts of the analytical sample in the following manner (see Fig. 7.19): The exit port of the gas chromatograph A is fitted with a hypodermic needle (7 mm long) with a cone B. The basic trap consists of the silver capillary tubing C (18 cm long, 0.8-mm i.d., 2.0-mm o.d.), one end of which is soldered to the fitting D and the other end to the fitting E, both without needle. Each fitting is closed with a silicone rubber septum. When the analytical sample starts to elute, the needle B is pushed into the rubber septum of D, and the trap C is immersed in the Dewar flask L containing liquid nitrogen. After elution, the trap is disconnected from the needle B and the tube C is taken out of the Dewar flask. At the same time, the drying tube F (packed with silica gel) is replaced by the tapered glass vessel G. Subsequently, the

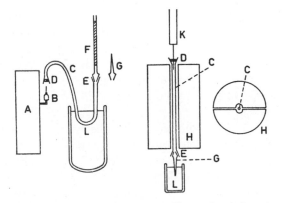

Fig. 7.19. Collection of gas-chromatographic effluent for infrared scanning [110]. (Courtesy *Z. Anal. Chem.*)

silver tube C is straightened and inserted in the heater H (kept at 230°C) which comprises two semicircular parts with a 7-mm hole in the center. The tapered vessel is now immersed in liquid nitrogen. After 3 min, 4 μl of carbon disulfide is injected into the tube C by means of a 10-μl syringe K. The carbon disulfide evaporates immediately in the upper part of the tube C, and within 1 or 2 min, the analytical sample cocondenses with carbon disulfide in the vessel G. The silver tube C and the vessel G are then removed from the heater and cooling bath, respectively. After the vessel G has warmed up to room temperature, its contents are transferred by means of a 10-μl syringe into an infrared microcavity cell (0.5-mm path length, 4-μl volume). This method is suitable for liquids with boiling points between 35 and 250°C. It is not satisfactory, however, for compounds that contain a polar group and boil above 200°C, probably because of interaction with the silver tube.

A simple stopped-flow gas-chromatic technique has been developed by Hammer and co-workers [110a]. When the elution of a single component monitered by the flame-ionization detector is complete, the flow of the carrier gas through the column is stopped. This is achieved by making the gas bypass the column. The particular fraction that is eluted is trapped in another column, which is heavily loaded with stationary phase (up to 40%) and cooled with ice. The sample is then transferred from the trap into the infrared cell or the inlet tube of the mass spectrometer by means of carrier gas after immersing it in a heated Wood's-metal bath. After the measurements are completed, the carried gas flow through the column is resumed, and the cold trap is put into operation.

D. COLLECTION OF FLUIDS AND POLLUTED WATER

The technique for the collection of fluids depends on the nature and consistency of the material. For example, heparinized microtubes are recommended for collecting blood samples for the analysis of electrolytes [111], since the anticoagulant prevents coagulation and subsequent hemolysis frequently observed in plastic vessels; however, the sample thus prepared cannot be used for electrophoresis. The recent micro sampling techniques [112, 113] have improved the sensitivity of atomic-absorption analysis by several orders of magnitude. Thus, Delves [114] has determined lead in whole blood by collecting 10-μl samples in microcrucibles made from nickel foil and, after partial oxidation with hydrogen peroxide, volatilizing the residue into a nickel absorption tube. Fitzgerald and co-workers [114a] have developed a cold-trap preconcentration procedure for the flameless atomic-absorption analysis of mercury in seawater.

A simple variable-volume collector that delivers a continuous range of volumes from 1.5 to 5.0 ml has been designed by Soracco and Borris [115]. Noll [116] has described an apparatus for collecting and analyzing fractions from a centrifuged sucrose gradient. The plastic centrifuge tube is pierced from below with a needle, and the contents are then pumped by a nonpulsing pump to the flowthrough cell of a spectrophotometer.

Standard methods for the collection of waste water for pollution-control analysis have been published [117]. Collection vessels for special analysis (e.g., mercury, arsenic, selenium [118]) are commercially available. Miget and co-workers [118a] have described a surface-film collector that permits the rapid, consistent, and efficient retrieval of petroleum spills in natural waters.

VI. COLLECTION OF GAS SAMPLES

In this section we discuss methods for collecting analytical samples from gaseous mixtures at ambient temperature (e.g., atmospheric hydrocarbons at an airport, not the "gas" in a gasoline tank). At present, these methods are most frequently employed for air-pollution analysis [3]. Various apparatus for the collection of air, known as "air samplers," are commercially available [119]. Generally speaking, the air-sampling devices may be classified into five categories:

1. Grab samplers
2. Filters
3. Cryogenic collectors
4. Impinging apparatus
5. Adsorption samplers

The use of filters for collecting atmospheric particulates and aerosols has been mentioned in Section II.B. Cryogenic collection consists of condensing the air pollutants at liquid-nitrogen temperature. Grab sampling involves simply collecting the air in an evacuated vessel (e.g., plastic bag); it is employed primarily for the analysis of pollutants that are present in high concentrations. The principles of impinging and adsorption sampling are discussed below. These methods are also applicable to other kinds of gaseous material submitted for analysis.

A. COLLECTION FROM AN OPEN SPACE OR GAS STREAM

1. Absorption Methods

The absorption methods are known as impinging methods for air-pollutant collection. The procedure involves passing the gaseous sample through a solvent or a solution containing an appropriate chemical reagent whereby the substance to be analyzed is retained in the liquid. Nitrogen dioxide, sulfur oxides, aldehydes, and some particulates in the atmosphere are collected in this manner. Wartburg and co-workers [120] have described a bubbler of glass and PTFE that has interchangeable parts and is easy to use in the field. It is less subject to damage than all-glass designs, but equally efficient in pollutant collection.

Obviously, the amount of air (or any gas mixture) that is conducted through the liquid medium should be measured accurately for quantitative evaluation of the analytical sample. Several techniques have been proposed [121–123]. A system using a differential flow controller is shown in Fig. 7.20. This unit requires a constant upstream pressure. With a flow controller, a 46-torr change in the upstream-side (bubbler) pressure drop produces a change of 8.9% in a flow rate of 150 ml/min. In actual practice, the bubbler pressure drop is not truly constant, because of pressure differences that can result from evaporation of the bubbler solution and differences in the ceramic frit used for dispersion. Consequently, the upstream pressure does not vary more than 20 torr; this represents a 3.2% error, which is quite acceptable [123].

2. Adsorption Methods

Solid reagents with or without coated chemicals are utilized in the adsorption methods. The compounds to be analyzed are thus retained on the solid surface. This technique can be used for the analysis of organic vapors, pesticides, polychlorinated biphenyls, sulfur dioxide, fluoride, and so on. For instance, Williams and Umstead [124] have collected hydrocarbons and halogenated hydrocarbons of low molecular weight on a column of porous polyaromatic-polymer beads at room temperature and subsequently eluted the components by temperature programming. Biasotti and Bradford [125] have collected ethanol vapor on anhydrous magnesium perchlorate and determined it by gas chromatography.

The adsorption technique has been applied indirectly to the determination of formaldehyde in a chemical-plant atmosphere. Frankel and co-workers [126] have found that the porous polymer Porapak Q can be utilized to retain virtually all interfering species (methylal, 2-propanol, etc.) in the determination of formaldehyde by the chromotropic acid reaction. As little as 3 ppm of formaldehyde is quantitatively coeluted with water when the air sample passes through a column packed with 500 mg of the polymer for 5 min at the rate of 1.0 liter/min, while the interfering compounds are retained.

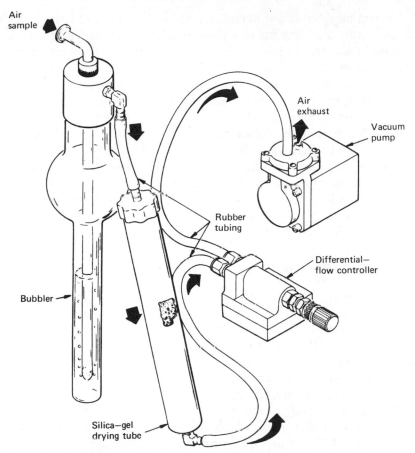

Air sample

Air exhaust

Vacuum pump

Rubber tubing

Differential— flow controller

Bubbler

Silica—gel drying tube

Fig. 7.20. Atmospheric sampling system using differential-flow controller [123]. (Courtesy *Anal. Chem.*)

3. Collection from a Gas Stream

Weber and Spanier [127] have constructed an apparatus that permits automatic gas collection at fixed intervals. It consists of a piston driven through a tube that acts as a gas manifold, the piston directing the gas stream through successive ports in the tube to a collection and analytical arrangement. Dudden [128] has described a device in which the plug of the bottom tap of the sampling bulb has one or more hemispherical depressions. As the plug is rotated, the depression contacts the gas stream, fills with gas, and is carried around until it contacts mercury in the sampling bulb. Sampling can be controlled automatically by various means.

A patent has been issued to Granger and co-workers [129] for a device for the continuous sampling of a gaseous mixture circulating at high temperature in a system of pipes (e.g., the effluent from a cracking furnace). One end of the sampling tube is introduced into the system, and the other end is connected to a circuit placed in a constant-temperature chamber fitted with a valve for isolating periodically a sample of given volume, which is carried from the chamber by a stream of carrier gas to a chromatograph for analysis. Kalab and Pinkava [130] have designed an apparatus that collects and delivers a gas mixture of constant composition even when there are pressure changes in the system. Seeler and Cahill [131] have described a quantitative sampling technique for the gas-chromatographic analysis of trace hydrocarbons (acetylene, ethylene, etc.) in methane. A valve equipped with a pressure transducer and related electronic devices for direct, instantaneous pressure readout is used.

B. COLLECTION FROM A CONFINED VOLUME

The techniques for collecting head-space gases have been reported by a number of workers in connection with the study of volatile materials in biological solutions [132], oxygen [132a] in pharmaceutical ampuls [133], volatile sulfur and aroma compounds in hops [134, 135] and the like. Hurst [136] has described a syringe and cooling jacket for the collection of volatiles from canned fish. The jacket is made from two aluminum blocks brought together and bored in such a manner that they fit firmly around the body of the syringe. The bottom block is screwed to a reservoir, also of aluminum, which is filled with liquid nitrogen. According to Maier [137], adsorption on rubber or silicone stoppers causes error in head-space analysis. Luberoff [137a] has suggested putting a small amount of mercury in the capped vessel, which is then inverted. Another remedy has been proposed by Martin [137b].

Zvonow and co-workers [138] have described a device to collect gases at high temperature and high pressure from internal-combustion engines for air-pollution studies. Kraus [139] has constructed a collector for small volumes of gas based on the principle of the Mariotte bottle. Radley [140] has described a technique for trapping gas in a glass capillary between two mercury globules. Tindle [141] has developed a variable-proportion gas-metering system that produces precise gas mixtures.

Allen [142] has described a quantitative gas-chromatographic sampler for static gaseous reaction systems (e.g., in the study of thermal or photochemical reactions). The apparatus is shown in Fig. 7.21. The operation of the sampling stopcock is performed as follows. With the stopcock in position 1, the bore is evacuated to about 10^{-2} torr. The stopcock is then

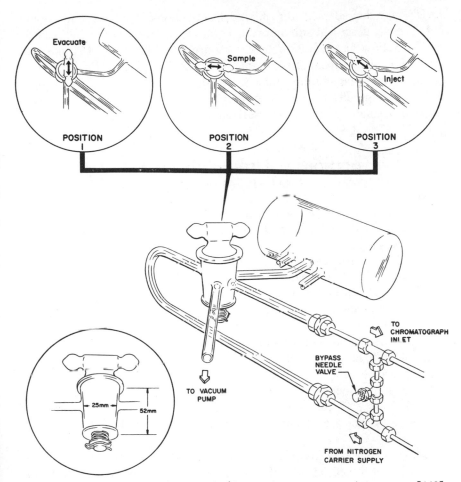

Fig. 7.21. Gas-chromatographic sampler for static gaseous reaction system [142]. (Courtesy *Anal. Chem.*)

rotated clockwise through 60 degrees to position 2, the vapor in the side arm and reaction vessel expanding into the bore. Another 60-degree clockwise rotation puts the stopcock in position 3, and the sample is injected as a "plug" into a parallel-flowing stream of carrier gas leading directly to the gas chromatograph. Rapp and co-workers [143] have reviewed the procedures for studying the aroma constituents of grape musts, wines, and spirits.

Weeden [143a] has fabricated an apparatus for accurately measuring the volume of gas evolved during the fermentation of yeasted dough. This device may be adapted for microdeterminations. Galbacs and Csanyi [144] have described an automated gas-measuring device that can be used to follow gas evolution or gas absorption. Liburne [145] has constructed a micro vacuum-fusion apparatus capable of collecting 2 mm³ of gas at normal temperature and pressure from metal samples. Absorption and scattering cells have been described for obtaining ultraviolet, infrared, and Raman spectra of pressurized liquids [146, 147] and corrosive gases [148].

VII. METHODS FOR THE "ENRICHMENT" OF AN ANALYTICAL SAMPLE

When the amount of analytical sample collected in one operation is insufficient for the subsequent mode of characterization or determination, it is necessary to accumulate the collected species. Such a process has been called "enrichment" or "concentration." At present, preparative chromatography (see Chapter 4) is frequently utilized to this end. For instance, Stedman and co-workers [9] have developed a method for concentrating polynuclear aromatic hydrocarbons in cigarette-smoke condensate as follows: The neutral fraction of cigarette-smoke condensate is separated by chromatography or activated silicic acid and eluted successively with light petroleum, 25% benzene in light petroleum, benzene, ether, and methanol. The residues from the first three eluates are individually partitioned between cyclohexane and dimethyl sulfoxide. As a result, more than 97% of the benzopyrene occurs in one fraction, and the concentration is increased from less than 1 ppm (in smoke condensate) to 105 ppm.

Adsorption is a convenient technique for enrichment. Kubeczka [91] has proposed a method whereby compounds separated by gas chromatography are retained by adsorption on short, packed columns and subsequently concentrated by repeated collection of the material adsorbed. In addition, with the technique described and using the head-space technique (see Section VI.B) but without cooling traps, it is possible to obtain odorous compounds from fungus cultures in sufficient quantity for gaschromatographic examination.

A variation of the adsorption method has been employed by Kaiser [149] for the enrichment of several analytical species from the atmosphere on a single temperature-gradient tube. The schematic diagram is shown in Fig. 7.22. The assembly comprises two concentric tubes. The inner tube (2- to 4-mm. i.d. made of glass or stainless steel, is packed with sorbing material of 30- to 40-mesh size in lengths from a few centimeters to 30 cm, depending on the use of the system. The outer tube, of steel or

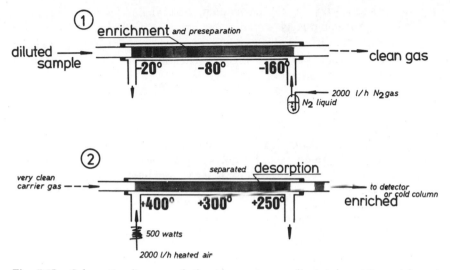

Fig. 7.22. Schematic diagram of the temperature-gradient tube: (1) enrichment, and (2) elution steps [149]. (Courtesy *Anal. Chem.*)

brass, acts as a cooling or heating device; its i.d. is 2 to 3 mm greater than the o.d. of the inner tube. At the enrichment step, the temperature gradient is produced by countercurrent flow of cold nitrogen, bubbled through liquid nitrogen. About 4 liters of air sample is pumped through the gradient tube; the amount is measured carefully by volume, temperature, and pressure, as these data are the basis for calibration-free quantitative data processing. The temperature-gradient tube is now sealed. At the elution step, the tube is heated under gradient with flowing air, and carrier gas at 4-liter/hr flow rate is passed through the reopened collection tube to conduct the analytical species to a specific detector or to the inlet of a gas chromatograph, temperature programmed from cryogenic temperatures. This procedure has been used to determine ethylene and other hydrocarbons up to C_{15} in air samples. It is worthy of note that ethylene, a very active hormone for certain plants [150], kills some flowers within 24 hr when its concentration in air reaches $10^{-7}\%$ by volume, while ethylene emission from a chemical factory can produce levels higher than $10^{-5}\%$ by weight in an area 15 miles away. Without the aid of enrichment, it is impossible to determine ethylene at such low levels.

Lehmann and Hahn [151] have developed a method for the enrichment of water-soluble food dyes in aqueous or ethanolic solution by adsorbing the dyes on polyamide powder and subsequently eluting them under mild conditions. A zone-freezing technique has been proposed by Kurdyumov and co-workers [152] for the enrichment of acyl halides and acid anhydrides at the 10^{-7}-mole level. A p-dioxane solution of the sample is placed in an ampul, which is passed through a system of annular heating (35°C) and cooling (-10°C) units arranged vertically. Thus the enrichment is accomplished by crystallization.

REFERENCES

1. J. A. Ambrose, *Ann. N. Y. Acad. Sci.,* **196**, 295 (1972).
2. Occupational Safety and Health Act, *Fed. Regist.,* **36**, 10503 (1971).
3. Intersociety Committee, *Methods of Air Sampling and Analysis.* Am. Public Health Assoc., Washington, 1972.
4. R. K Skogerboe, D. Dick, and J. Seeley, presented at the Eastern Analytical Symposium, New York, 1971.
5. P. L. Hanst, A. S. Lefohn, and B. W. Gray, Jr., *Appl. Spectrosc.,* **27**, 188 (1973).
6. M. Gruenfold and R. Ginell, *Appl. Spectrosc.,* **24**, 380 (1970).
7. W. L. Chiou and L. D. Smith, *J. Pharm. Sci.,* **59**, 863 (1970).
8. E. W. Wynder and D. Hoffmann, *Adv. Cancer Res.,* **8**, 249 (1964).
9. R. L. Stedman, R. L. Miller, L. Lakritz, and W. J. Chamberlain, *Chem. Ind.,* **1968**, 394.
10. F. G. Hamlyn and J. K. R. Gasser, *Chem. Ind.,* **1970**, 1142.
11. M. Bergstrom-Nielsen, *Dtsch. Lebansm.-Rundsch,* **65**, 163 (1969).
12. B. A. Lodge, *Can. J. Pharm. Sci.,* **5**, 74 (1970).
13. L. Erdey and I. Buzas, *Magy. Kem. Foly.,* **61**, 443 (1955).
14. A. Steyermark, *Quantitative Organic Microanalysis,* 2nd ed., p. 318. Academic, New York, 1961.
15. N. D. Cheronis and T. S. Ma, *Organic Functional Group Analysis,* p. 97. Wiley, New York, 1964.
16. J. M. Corliss, *Chem.-Anal.,* **48**, 16 (1959).
17. M. Potterat and H. Eschmann, *Mitt. Lebensm. Hyg. Bern,* **45**, 329 (1954).
18. R. L. Boswall and D. C. Mackay, *Chem.-Anal.,* **52**, 54 (1963).
19. Sampling Manifold, Model 3025, Millipore Corp., Bedford, Mass.
20. R. D. Spencer, *Anal. Chem.,* **44**, 882 (1972).
21. R. Pietsch, *Mikrochim. Acta,* **1955**, 859.
22. A. N. Fletcher, *Anal. Chem.,* **29**, 1387 (1957).
23. S. Gallinger, *G.I.T. Fachz. Lab.,* **13**, 1179 (1969).
24. S. Samuels, *Microchem. J.,* **6**, 107 (1962).
25. W. C. McCrone, R. G. Draftz, and J. G. Delly, *The Particle Atlas.* Ann Arbor Science Pub., Ann Arbor, Mich., 1967.
26. J. Y. Hwang, *Anal. Chem.,* **44**, (14), A21 (1972).
27. L. J. Pardue, R. E. Enrione, R. J. Thompson, and B. A. Bonfield, *Anal. Chem.,* **45**, 527 (1973).
28. J. P. Hatousek and K. G. Brodie, *Anal. Chem.,* **45**, 1606 (1973).
29. J. A. McHugh and J. F. Stevens, *Anal. Chem.,* **44**, 2187 (1972).

30. C. Golden and E. Sawicki, presented in A.C.S. National Meeting, Dallas, 1973.
31. R. K. Patterson, *Anal. Chem.,* **45**, 505 (1973).
31a. A. Sugimae, *Appl. Spectrosc.,* **28**, 458 (1974).
32. J. L. P. Wyndham, *Med. Sci. Law,* **3**, (3), 141 (1963).
33. R. Solbakken, *Lab. Pract.,* **18**, 851 (1969).
34. A. E. Sherwood, *Lab. Pract.,* **18**, 23 (1969).
35. L. E. Makovsky, *Appl. Spectrosc.,* **27**, 43 (1973).
36. B. Brachfeld, *Appl. Spectrosc.,* **27**, 289 (1973).
37. U.S.A.E.C., British Patent 1,124,098 (1955).
38. R. W. Hannah and J. L. Dwyer, *Anal. Chem.,* **36**, 2341 (1964).
39. In-Line Filter Unit, Metro Scientific, Inc., Farmingdale, N. Y.
40. Zenith Microconcentrator, Biomed Instruments, Inc., Chicago, Ill.
41. C. Feldman and J. Y. Ellenburg, *Anal. Chem.,* **29**, 1559 (1957).
42. G. J. Hudson and R. J. Ison, *Lab. Pract.,* **18**, 1188 (1969).
43. L. Riccobini, N. Fochi, E. Ferrari, and G. Tagliavini, *Chim. Ind.,* **38**, 11 (1956).
44. R. J. Casciato, *Anal. Chem.,* **41**, No. 13, A99 (1969).
45. M. A. Millett, W. E. Moore, and J. F. Seaman, *Anal. Chem.,* **36**, 491 (1964).
46. C. J. Clemett, *Anal. Chem.,* **43**, 490 (1971).
47. T. S. Ma and A. A. Benedetti-Pichler, *Anal. Chem.,* **25**, 999 (1953).
48. Brinkmann Spot Collector, Brinkmann Instruments, Inc., Westbury, N.Y.
49. Brinkmann Vacuum Zone Collector, Brinkmann Instruments, Inc., Westbury, N. Y.
50. A Sapse, A. Fono, and T. S. Ma, *Mikrochim. Acta,* **1965**, 1091.
51. J. Polesuk, and T. S. Ma, *Mikrochim. Acta,* **1972**, 313.
52. A. Martinek, *Mikrochim. Acta,* **1972**, 229.
53. M. J. Rix, B. R. Webster, and I. C. Wright, *Chem. Ind.,* **1969**, 452.
54. S. Sandroni and H. Schlitt, *J. Chromatogr.,* **52**, 169 (1970).
55. H. Halpaap, *Chemikerztgr.,* **89**, 835 (1965).
56. S. Asen, *J. Chromatogr.,* **18**, 602 (1965).
57. S. M. V. Hunt, *Lab. Pract.,* **16**, 601 (1967).
58. R. A. Masaracchia and A. M. Gawienowski, *Steroids,* **11**, 717 (1968).
59. F. A Vandenheuvel, *J. Lab. Clin. Med.,* **69.**, 343 (1967).
60. E. C. Kryiakides and J. A. Balint, *J. Lipid Res.,* **9**, 142 (1968).
61. B. Lewis, *Biochem. Biophys. Acta,* **20**, 417 (1956).
62. W. S. Reith, *Nature,* **179**, 580 (1957).
63. S. Lakshminarayana and K. Sebastian, *Lab. Pract.,* **19**, 815 (1970); **17**, 214 (1968).
64. W. E. Ballinger and E. P. Maness, *Hortic. Sci.,* **4**, 11 (1969).
65. J. E. van Lier and L. L. Smith, *J. Chromatogr.,* **36**, 7 (1968).
66. C. S. Du Plessis, *S. Afr. J. Afric. Sci.,* **4**, 189 (1968).
67. B. A. Cooke, *Anal. Biochem.,* **32**, 198 (1969).
68. R. Hardy and J. N. Keay, *J. Chromat.,* **17**, 177 (1965).
69. J. L. Bloomer and W. R. Eder, *J. Gas Chromatogr.,* **6**, 448 (1968).

70. A. S. Curry, J. F. Read, C. Brown, and R. W. Kenkins, *J. Chromatogr.,* **38,** 200 (1968).

71. D. E. Willis, *Anal. Chem.,* **40,** 1597 (1968).

72. E. J. Parish, *Chem.-Anal.,* **56,** 70 (1967).

73. E. Shibata and S. Saito, *J. Chem. Soc. Jap.,* **80,** 604 (1959).

74. K. Yoshimura, *Jap. Anal.,* **11,** 397 (1962).

75. Y. Shigetomi, *Anal. Chem.,* **45,** 411 (1973).

76, N. W. Jacobson and J. Miller, *Chem. Ind.,* **1957,** 621.

77. J. P. Bitz, *Rev. Sci. Instrum.,* **41,** 1366 (1970).

78. M. I. Vass, *Rev. Roum. Chim.,* **14,** 809 (1969).

79. A. F. Hyde and R. A. Redford, *J. Sci. Instrum.* Ser. 2, **1,** 871, (1968).

79a. R. S. Timsit and A. K. C. Kiang, *Rev. Sci. Instrum.,* **44,** 770 (1973).

80. Fraction Collector MiniRac, LKB Instruments, Inc., Rockville, Md.

81. B. Večerek, *Chem. Listy,* **57,** 337 (1963).

82. K. Popper and F. Nury, *Chem.-Anal.,* **53,** 119 (1964).

83. R. J. Soraco and D. P. Borris, *Anal. Biochem.,* **30,** 158 (1969).

84. R. Elsdon-Dew, *Res. Dev.,* Sept. 1964, p. 48.

84a. J. F. K. Huber, A. M. van Urk-Schoen, and G. B. Sieswerda, *Z. Anal. Chem.,* **264,** 257 (1973).

85. R. E. Lovins, S. R. Ellis, G. D. Tolbut, and C. R. McKinney, *Anal. Chem.* **45,** 1553 (1973).

86. F. Armitage, *J. Chromatogr. Sci.,* **7,** 190 (1968).

86a. G. G. S. Dutton and K. B. Gibney, *J. Chromatogr.,* **72,** 179 (1972).

87. J. H. Jones and C. D. Ritchie, *J. Assoc. Off. Agric. Chem.,* **41,** 753 (1958).

88. T. Tsuda and D. Ishii, *J. Chromatogr.,* **47,** 469 (1970).

89. K. Beyermann, A. Kessler, and P. W. Ungerer, *Z. Anal. Chem.,* **251,** 289 (1970).

90. T. H. Parliament, *Anal. Chem.,* **45,** 1792 (1973).

91. K. H. Kubeczka, *J. Chromatogr.,* **31,** 319 (1967).

92. C. W. Dixon, C. T. Malone, and G. R. Umbreit, *J. Chromatogr.,* **35,** 475 1968).

93. L. J. Papa and A. Varon, *J. Gas Chromatogr.,* **6,** 185 (1968).

94. M. G. Moshonas and P. E. Shaw, *Appl. Spectrosc.,* 25, 101 (1971).

95. M. B. Morin, *Methodes Phys. Anal.* **1967,** 157.

96. A. B. Littlewood, *Chromatographia,* **1968,** 37.

97. D. Ursu and I. Mastan, *Stud. Cerc. Chim.,* **16,** 715 (1968).

98. M. Vandewalle and M. Verzele, *J. Gas Chromatogr.,* **8,** 72 (1968).

99. C. G. Hammar and R. Hessling, *Anal. Chem.,* **43,** 298 (1971).

100. W. Henderson and G. Steel, *Anal. Chem.,* **44,** 2302 (1972).

100a. C. F. Simpson, *CRC Crit. Rev. Anal. Chem.,* **3,** 1 (1972).

101. G. Nota, G. Marino, and A. Malorni, *Chem. Ind.,* **1970,** 1294.

102. K. Grob and H. Jaeggi, *Anal. Chem.,* **45,** 1788 (1973).

103. S. P. Markey, *Anal. Chem.,* **42,** 306 (1970).

104. F. Bruner, P. Ciccioli, and S. Zelli, *Anal. Chem.,* **45,** 1002 (1973).

105. H. J. Dutton, presented at A.C.S. National Meeting, Chicago, 1973.

106. P. G. Simmonds, G. P. Shulman, and C. H. Stembridge, *J. Chromatogr. Sci.*, **7**, 36 (1969).

106a. W. E. Crosmer, N. C. Thomas, P. H. S. Tsang, and R. J. Duckett, *Rev. Sci. Instrum.*, **44**, 837 (1973).

107. A. B. Littlewood, *J. Gas Chromatogr.*, **6**, 65 (1968).

108. B. A. Bierl, M. Beroza, and J. M., *Gas Chromatogr.*, **6**, 280 (1968).

109. A. S. Curry, J. F. Read, C. Brown, and R. W. Jenkins, *J. Chromatogr.*, **38**, 200 (1968).

110. J. Bus. and T. J. Liefkins, *Z. Anal. Chem.*, **250**, 294 (1970).

110a. C. F. Hammer, P. C. Rankin, and D. E. Korte, private communication (1973).

111. J. Michod, C. Platsoukas, and J. Frei, *Z. Klin. Chem. Klin. Biochem.*, **7**, 455 (1969).

112. T. S. West and X. K. Williams, *Anal. Chim. Acta*, **45**, 27 (1969).

113. H. L. Kahn, G. E. Peterson, and J. E. Schallid, *At. Absorp. News.*, **7**, 35 (1968).

114. H. T. Delves, *Analyst*, **95**, 431 (1970).

114a. W. F. Fitzgerold, W. B. Lyons, and C. D. Hunt, *Anal. Chem.*, **46**, 1882 (1974).

115. R. J. Soracco and D. P. Borris, *Anal. Biochem.*, **30**, 159 (1969).

116. H. Noll., *Anal. Biochem.*, **27**, 130 (1969).

117. *Standard Methods for the Examination of Water and Waste-Water*, published jointly by the American Public Health Association, the American Water Works Association, and the Water Pollution Control Federation, 13th ed., Washington, 1971.

118. Hg/As/Se Kit, F & J Scientific, Monroe, Conn.

118a. R. Midget, H. Kator, C. Oppenheimer, J. L. Laseter, and E. J. Ledet, *Anal. Chem.*, 46, 1154 (1974).

119. Microchemical Specialties Co., Berkeley, Calif.

120. A. F. Wartburg, J. B. Pate, and J. P. Lodge, Jr., *Environ., Sci. Technol.*, **3**, 767 (1969).

121. J. P. Lodge, Jr., J. B. Pate, B. E. Ammons, and G. A. Swanson, *Air Pollut. Control Assoc. J.*, **16**, 197 (1966).

122. C. Huygen, *Air Pollut. Control Assoc. J.*, **20**, 675 (1970).

123. A. F. Wartburg, H. D. Axelrod, R. J. Jack, M. D. LaHue, and J. P. Lodge, Jr., *Anal. Chem.*, **45**, 423 (1973).

124. F. W. Williams and M. E. Umstead, *Anal. Chem.*, **40**, 2232 (1968).

125. A. A. Biasotti and L. W. Bradford, *J. Forensic Sci.*, **9**, 65 (1969).

126. L. S. Frankel, P. R. Medsen, R. R. Siebert, and K. L. Wallisch, *Anal. Chem.*, **44**, 2401 (1972).

127. D. C. Weber and E. J. Spanier, *Anal. Chem.*, **42**, 546 (1970).

128. W. R. Dudden, *Lab. Pract.*, **4**, 71 (1955).

129. C. Granger, M. Demurc, and M. Bruder, British Patent, 1,198,888 (1968).

130. V. Kalab and J. Linkava, *Chem. Listy*, **52**, 156 (1958).

131. A. K. Seeler and R. W. Cahill, *J. Chromatogr. Sci.*, **7**, 158 (1969).

132. R. Bassette and G. Ward, *Michochem. J.*, **14**, 471 (1969).

132a. V. Spiehler, *Mod. Packag.*, **45**, (3), 46 (1972).
133. L. F. Cullen and G. J. Papariello, *J. Pharm. Sci.*, **59**, 94 (1970).
134. G. Baerwald and G. Miglio, *Mschr. Brau.*, **23**, 288 (1970).
135. E. Krueger and L. Neumann, *Mschr. Brau,* **23**, 269 (1970).
136. R. E. Hurst, *Chem. Ind.,* **1970**, 90.
137. H. G. Maier, *J. Chromatogr.*, **50**, 329 (1970).
137a. B. J. Luberoff, *Chem. Ind.*, **1972**, 714.
137b. E. W. Martin, *Chem. Ind.* **1972**, 664.
138. V. A. Zvonow, H. E. Stewart, and E. S. Starkman, *Rev. Sci. Instrum.*, **39**, 1820 (1968).
139. M. Kraus, *Chem. Listy,* **49**, 929 (1955).
140. E. T. Redley, *J. Chem. Educ.*, **37**, 360 (1960).
141. R. C. Tindle, *Anal. Biochem.*, **26**, 477 (1969).
142. E. R. Allen, *Anal. Chem.*, **38**, 527 (1966).
143. A. Rapp, W. Hövermann, U. Jecht, H. Franck, and H. Ullemeyer, *Chem. Ztg.,* **97**, 29 (1973).
143a. D. G. Weeden, *Chem. Ind.*, **1964**, 1839.
144. M. Z. Galbacs and L. J. Sasnyi, *Anal. Chem.*, **45**, 1784 (1973).
145. M. T. Lilburne, *Talanta,* **14**, 1029 (1967).
146. J. Corset, P. V. Huong, and J. Lascombe, *Spectrochem. Acta,* **24**, 1385 (1968).
147. R. Cavagnat, J. J. Martin, and G. Turrell, *Appl. Spectrosc.*, **23**, 172 (1969).
148. J. C. Cornut and P. V. Huong, *Appl. Spectrosc.*, **27**, 55 (1973).
149. R. E. Kaiser, *Anal. Chem.*, **45**, 965 (1973).
150. F. B. Abeles, *Ann. Rev. Plant Physiol.*, **23**, 257 (1972).
151. G. Lehmann and H. G. Hahn, *Z. Anal. Chem.*, **238**, 445 (1968).
152. G. M. Kurdyumov, O. V. Ivanov, G. V. Galochkina, E. S. Malinina, and V. M. Dziomko, *Zh. Anal. Khim.*, **24**, 1595 (1969).

8

PURIFICATION OF ANALYTICAL SAMPLES

I. GENERAL REMARKS CONCERNING REQUISITE PURITY

After undergoing the processing discussed in the previous chapters, the sample may still need one more step, known as "purification," before it is subjected to identification and characterization (qualitative analysis) or determination (quantitative analysis). The requisite degree of "purity" depends on the nature of the material as well as the purpose of the analysis. Whereas "purification" frequently involves the removal of solvent, there are other facets of this step to be considered. The following remarks are given in order to clarify the situations.

1. If the objective is either to establish the identity or to measure the quantity of a known compound, a "pure" (i.e., suitable) sample means one free from interfering substances that may vitiate the analytical results. Other foreign matter need not be removed. For instance, a sample for the gas-chromatographic analysis of pesticide residue in wheat can be obtained by shaking 1 to 5 g of material with two ball bearings in a centrifuge tube, freezing out fat from acetone solution, and performing the final cleanup on a small Florisil column [1]. Moats and Kotuba [2] have shown that a high elution rate of 250 ml/min does not adversely affect results. Goenechea [3] has proposed the purification of barbiturate sample in sedative mixtures by means of preparative thin-layer chromatography, the chloroform-acetone eluate being subsequently analyzed by infrared spectroscopy. Grady and co-workers [3a] have reported on drug-purity profiles; the purity is determined by phase solubility, differential scanning colorimetry, or thin-layer, gas, or high-pressure liquid chromatography.

2. When the substance to be analyzed is a new compound (obtained by synthesis or isolated from natural product), the objective is usually to characterize the molecule and confirm its proposed structure or formula. In this case, the tolerance on impurities present in the analytical sample is dependent on the precision of the analytical data that is required in order to unequivocally distinguish related possibilities. For instance, a 95%-pure sample of a new metal chloride can be analyzed for chlorine and the formula ascertained to be either MCl_2 or MCl_3. On the other hand,

an organic compound containing C,H, and O may require a sample of much higher purity in order to establish its empirical formula.

3. If the analysis is for extremely pure materials such as spectroscopic standards and semiconductors, special treatment of the sample is necessary. The reader is referred to the work of the U.S. Bureau of Standards [4], and a recent analytical symposium [5].

4. If the material submitted for analysis is itself a mixture (e.g., air, soil, seawater, etc.), the sample should be treated as such. It should not be modified; i.e., no "purification" should be performed on the sample collected.

In microscale manipulations, one should always be aware of the danger of contamination. This is especially important during the step of final "purification". When a chemical reagent is used in the process, the purest reagent should be employed, and the reagent should be stored in small separate containers [6]. For example, when the Girard-T reagent method is utilized to concentrate carbonyl compounds for gas-chromatographic analysis, Gadbois and co-workers [7] have found that large extraneous peaks interfere with interpretation of the scan unless the impurities are removed by purifying the reagents.

II. CONCENTRATION OF SOLUTION AND RECOVERY OF SOLVENT

A. TREATMENT OF VERY DILUTE SOLUTIONS

1. Apparatus

The rotary evaporator (see Fig. 8.1a) available commercially is a convenient laboratory apparatus for concentrating dilute solutions. Radin [8] has improved the performance of a rotary vacuum evaporator by substituting an O-ring taper joint for the usual ground-glass joint, and running the condensate flask to the water aspirator. Runeckles [9] has constructed a multiple-unit, all-glass apparatus, which is better than the commercial evaporators having metallic parts that are subject to corrosion and may cause sample contamination. Ganz [9a] has described an adapter through which a stream of inert gas can be conducted; it is inserted between the rotary evaporator flask fitting and the flask containing the solution to be concentrated, as illustrated in Fig. 8.1b.

Pettit and Bruschweiler [10] have described an inexpensive apparatus suitable for concentrating large quantities of alcoholic and aqueous-alcoholic solutions. As shown in Fig. 8.2, the outlet P is connected to the vacuum pump, and solution is introduced through A to the reservoir V.

When V is about half full, steam is sent into the evaporator E (cross section CE). Vapor and liquid then leave E and pass into V, the vapor continuing into the condenser and collection flask F. At intervals, the concentrated solution is drained off through the outlet B.

Zeineh and co-workers [10a] have described the construction of a single-hollow-fiber microconcentrator. The assembly comprises the concentrator, which consists of a simple hollow fiber (see Chapter 6, Section II.C.3) of microtubular semipermeable membrane with an outlet and inlet; a test-tube cap with three separate implanted tubular steel outlets; a mini-clamp; connecting tubing; and a stand.

2. Operation and Precautions

During the course of concentration, precautions should be taken to prevent the loss of analytical sample or the decomposition of the material. Petuely and Meixner [11] have described a device that permits the continuous concentration without loss, by vacuum distillation, of large volume of liquid containing only small quantities of the analytical sample; concentration to about 5 ml can be achieved at the rate of 250 ml/hr. To avoid severe losses of pesticides when such solutions are concentrated, Burke and co-workers [12] have recommended a micro Snyder column (Fig. 8 3) and a Kuderna-Danish collection tube for the rapid concentration of 10 ml of solution to 0.1–0.3 ml on a steam bath.

Beroza and co-workers [12a] have concentrated solutions in test tubes the lower ends of which are situated below the heating zone of the evaporator block. In this manner, losses of pesticides resulting from inadvertent evaporation of the solutions to dryness are avoided.

An apparatus has been described by Ducellier [13] for concentrating 1 to 2 liters of liquid to one-tenth of its volume without overheating thermolabile substances in solution. The heating surface is so arranged that the liquid is continuously circulated and spreads in a thin layer on the walls of the concentrator. Begeman [14] has constructed a simple evaporator for heat-sensitive compounds by using a windshield-wiper motor to provide an oscillatory rotational force (Fig. 8.4). This device increases the available liquid surface from which volatile molecules can escape, increases the flow of heat to the evaporating surface, and provides for operation at a reduced pressure to facilitate the evaporation of solvents at relatively low temperature. It is explosion-proof because it operates from the laboratory vacuum system, the upper end of the metal tube is connected through a Dry Ice trap to the vacuum pump via elastic tubing.

Short and Good [15] have described a cover glass to fit Erlenmeyer flasks that enables very rapid evaporation to be carried out. Gold [16] has

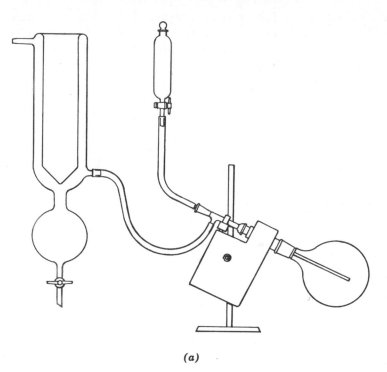

(a)

Fig. 8.1. (*a*) Rotary evaporator. (Courtesy Michrochemical Specialties Co.)

designed an air-stream evaporator for multiple samples. The apparatus is constructed from simple components and can be adjusted to accommodate vessels of various types. It is suitable for concentrating low-boiling liquids, but is not efficient with aqueous solutions. Martin [16a] has described a device for dry-gas-sweep evaporation suitable for the recovery of solvents from high-speed column chromatography fractions.

A two-stage microevaporator has been described by Clarke and Hawkins [17]. The first serves to reduce a volume of 1 ml to 0.05 ml, while the second serves to reduce a volume of 0.05 ml to a few microliters. It works best with chloroform or methanol, although it may be used with less volatile solvents such as ethanol or water. Kazyak and Hrast [18] have designed a solvent evaporator that eliminates solute losses from the bumping or boiling over of very volatile solvents. Heat is applied to the surface of the solvent rather to the bottom of the heating vessel. The apparatus can be adjusted to regulate the evaporation rate, and can be preset to shut off when the solvent has been reduced to a specific volume; e.g., quantitative recovery is virtually complete with 2 μg of solute in 50

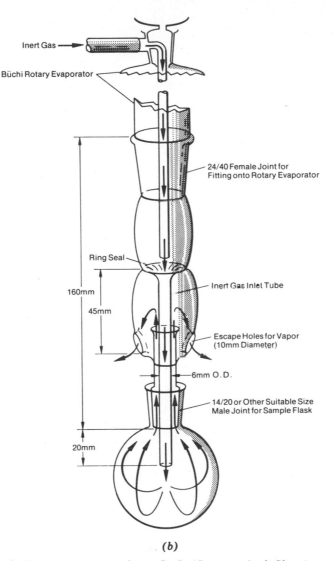

Inert Gas

Büchi Rotary Evaporator

24/40 Female Joint for
Fitting onto Rotary Evaporator

Ring Seal

Inert Gas Inlet Tube

160mm

45mm

Escape Holes for Vapor
(10mm Diameter)

6mm O.D.

14/20 or Other Suitable Size
Male Joint for Sample Flask

20mm

(b)

Fig. 8.1.   (b) Rotary evaporator adapter [9a]. (Courtesy *Anal. Chem.*)

ml of chloroform. Chalmers and Watts [18a] have achieved the quantitative dehydration of aqueous solutions of aliphatic acids by freeze-drying (see Chapter 3, Section III.E).

Devices have been reported in which the solvent can be recovered in a pure state. For example, Naugolnykh [19] has described an apparatus

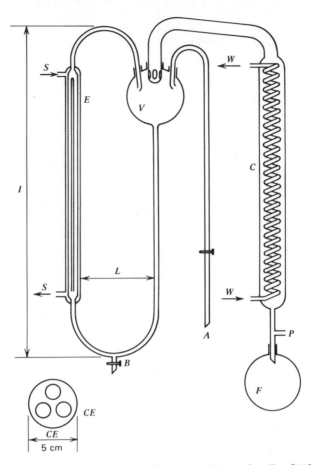

Fig. 8.2. Solvent distillation apparatus of Pettit and Bruschweiler [10]. (Courtesy *Chem. Ind.*)

that comprises a receiving flask and a reflux condenser with an extra socket, into which fits a tube having several joints connecting to flasks from which solvent is to be removed under reduced pressure. It is employed for concentrating steroid solutions, and the solvent recovered is pure enough for immediate reuse.

Ettel [20] has described an evaporating device for the analysis of high-purity liquids in which the solution is evaporated from a dish supported within an enclosed, externally cooled vessel by radiation from an infrared lamp mounted above the assembly. Zief and Barnard [20a] have recommended a single chamber for evaporation, as illustrated in Fig. 8.5.

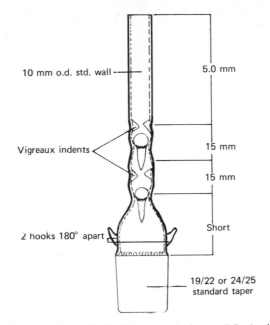
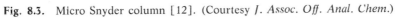

10 mm o.d. std. wall

5.0 mm

Vigreaux indents

15 mm

15 mm

2 hooks 180° apart

Short

19/22 or 24/25
standard taper

Fig. 8.3. Micro Snyder column [12]. (Courtesy *J. Assoc. Off. Anal. Chem.*)

Fig. 8.4. Evaporator of Begeman [14]. (Courtesy *J. Chem. Educ.*)

Fig. 8.5

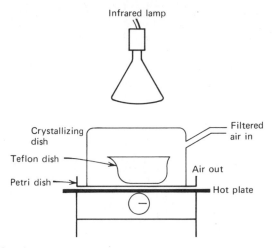

Fig. 8.5.   Chamber for evaporation (after Thiers [20a]). (Courtesy Chemtech.)

B. EVAPORATION OF SMALL VOLUMES OF CONCENTRATED SOLUTIONS

The evaporation of small volumes of concentrated solutions usually presents complications. Bumping and creeping may result in significant loss of the analytical sample. Transferring the material from one container into another vessel may lead to contamination. A simple device that can circumvent these difficulties is shown in Fig. 8.6. The test tube containing the solution is fitted, through a rubber ring cut from a tubing or stopper, into the suction attachment, which is a glass tube with side arm. When suction is applied, clean air is drawn in and carries the solvent vapors to the aspirator [21].

Clements [22] has described an interesting technique for evaporating 3 to 10 ml of solution as follows. The solution is contained in a vial closed with a neoprene stopper carrying a glass tube. The tube is connected to the vacuum line by a length of Tygon tubing (6-mm o.d.), by means of which the vial is suspended from a clamp in a 250-ml plastic beaker, the latter being placed on a magnetic stirrer and filled with water at a suitable temperature. When the stirrer is started and the vacuum applied, the vial is swirled at a frequency depending on the length of tubing between the clamp and the vial. By this means, 5 ml of light petroleum in a 7.5-ml vial can be evaporated in 2 min at 30°C.

Bell [23] has constructed an apparatus for solvent removal at low temperatures. The solution is trickled down a sloping lagged tube, up which warm air can flow. Indentations in the tube cause the air flow to be

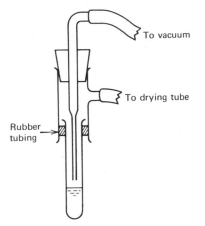

To vacuum

To drying tube

Rubber
tubing

Fig. 8.6. Evaporation of small volume of liquid [21]. (Courtesy *Microchem. J.*)

directed onto the "lakes" of the solution, thus increasing the rate of evaporation. Perold [24] has described a simple apparatus that permits the dust-free evaporation of 1 to 4 ml of solution without creeping. Pierce and Perrine [25] have put forth a method for rapidly evaporating solutions in test tubes while protecting the sample from contamination.

Brisset [26] has used a distillation assembly for the evaporation and recovery of the solvent, as shown in Fig. 8.7. The solution is delivered from the separatory funnel at a controlled rate into the test tube, where the solvent boils and distils via the side arm through a condenser and is collected in a small flask with a second outlet carrying a tube to remove uncondensed vapor.

When the residue is in the form of waxy solid, Smith [27] has recommended a method using a rotary evaporator (see Section II.A) to deposit a film of the solid on the inside of a flask. For decrepitating materials, Riemer [28] has suggested a silica crucible heated electrically by means of a spirally wound tantalum tape. The material to be evaporated (e.g., an organic semiconducting substance) is placed in the crucible and covered with a silica-wool plug. The crucible can be used in any position, including that for downward evaporation.

Wharton and Bazinet [29] have described a concentrating flask for the retention of nanogram amounts of the analytical sample. The device consists of a pear-shaped glass flask tapering to a closed capillary tube of about 2-mm outside diameter. Repeated refluxing or washing down of the material adherent to the wall of the flask is essential to secure its total concentration in the capillary tube. After the sample has been concentrated to dryness in the tube by appropriate means, the tube is cut off and transferred to an analytical instrument such as a mass spectrometer. Glass

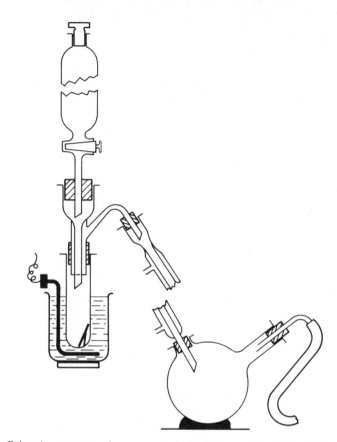

Fig. 8.7. Brisset's apparatus for removal and recovery of solvent [26]. (Courtesy *Ann. Falsif. Fraudes.*)

tubes for sample concentration down to 1 μl in the tip are commercially available [30].

Apparatus for the simultaneous evaporation of mutliple samples have been reported. Carel [30a] has described vacuum-oven manifolding to facilitate solvent evaporation from multiple containers. Komarck [31] has constructed a device consisting of six needle valves, which control streams of nitrogen through nozzle-directing components for the simultaneous evaporation of six small samples. Another design has been published by Albert-Recht [32]. The manifold consists of a main tube, connected with a water pump, having six side tubes that are connected to the sample tubes by plastic tubing and adapters. An apparatus fitted with 24 rotating vials for rapid simultaneous evaporation is available from an equipment manufacturer [33]. Pepper [33a] has developed a dryer for use with multiple fractions of aqueous, organic, or corrosive solvents with recovery in very small volumes down to the microliter level. It utilizes the principle of lyophilization (Chapter 3, Section III.E). Drying is effected by spinning the sample solution in an evacuated bell jar. No difficulties are experienced in handling HCl, methanol, chloroform, pyridine, or acetic anhydride solutions.

Swift [34] has described a simple hanging-drop technique for rapidly drying a series of liquid drops on a glass plate. With the glass plate mounted over a dry work chamber, so that a draft of air can pass under it, one or more rows of droplets of the solution can be laid out. Then, going back to the first one before the solvent has completely disappeared, it can be replenished to its original condition, and so on, until after several replenishments the liquid becomes saturated and solids begin to crystallize. The result, when enough has been accumulated, is a row of separate piles of material, each one of which can be worked with separately.

III. REMOVAL OF THE LAST TRACES OF SOLVENT

The removal of traces of solvent before the sample is subjected to analysis is usually known as "drying." A number of drying devices have been described in Chapter 2, Section VII. It should be noted that the procedure and apparatus best suited for drying the analytical sample is dependent on the nature of the sample as well as the boiling point of the solvent to be expelled. For instance, the removal of traces of acetone (boiling point 56°C) is a simple matter, while expelling dimethylformamide (boiling point 155°C) may be laborious. Contrary to the common laboratory practice, small samples should not be routinely dried under vacuum in a "drying pistol" (Abderhalden apparatus with water in the boiling flask), because many organic compounds have high vapor pressures, resulting in significant loss of the material.

Elevated temperatures are in general required to remove the residual solvent. Drying in open air can be conveniently performed in the apparatus of Maurmeyer and Ma [35] (see Fig. 2.44), while the apparatus of Schenck and Ma [36] (see Fig. 2.45) is recommended for drying under reduced pressure. These two devices permit easy control of the temperature variation to accommodate the nature of the analytical sample.

Modifications of the Abderhalden apparatus have been proposed [37–39] in order to improve its operation. The design suggested by Smith [39] is shown in Fig. 8.8; openings at both ends of the horizontal tube facilitate the delivery of the drying agent and sample.

Skinner [40] has proposed the use of simple vacuum desiccators (17×4.6 cm) to hold sintered glass crucibles or sample bottles. By fitting

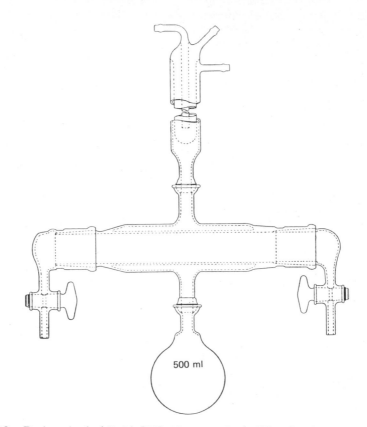

Fig. 8.8. Drying pistol of Smith [39]. (Courtesy *Anal. Chim. Acta.*)

five such desiccators into holes bored in a wooden block (see Fig. 2.46),
five separate drying agents (phosphorus pentoxide, concentrated sulfuric
acid, potassium hydroxide, calcium sulfate, and paraffin) are immediately
available for the removal of solvents that may adhere to the analytical
sample. Wadsworth [41] has described a heated vacuum desiccator in
which the porcelain plate is replaced by a glass heating plate supported
so as to minimize heat transfer to the desiccator. Two holes are drilled
through the wall for the electrical leads. Temperature control is through
the applied voltage. Gardiner [42] has constructed a vacuum-drying oven
of large heat capacity and uniform temperature distribution. It has a
shallow, hollow, but thick metal base holding six sample dishes with a
minimum of free space, and is designed to operate efficiently from 50 to
90°C. Requirements specified for vacuum ovens suitable for drying substances in microchemical vessels have been published by the British
Standards Institution [43].

A drying apparatus for powders has been described by Solomons and Schlosser [44]. It is operated by alternate evacuation and flushing with dry gas. The tendency of the gas to blow the powder into the vacuum manifold is reduced by fitting over the exit of the container a metal bellows, to which is fixed a filter. When the gas pressure is high, partial collapse of the bellows bring the filter into position, but as the pressure falls, the filter is automatically withdrawn and so allows more effective pumping.

IV. REMOVAL OF GASES FROM LIQUIDS

Gases can be removed from liquids by the following techniques: (1) simple boiling; (2) evacuation with or without heating; (3) the evacuation of the frozen material; (4) the displacement of the undesirable gas by another gas. Needless to say, the sample size and the purpose of degassing the liquid govern the technique to be employed. Boiling is a convenient procedure provided that the liquid does not decompose on heating and the loss by volatilization is minimal. The displacement of absorbed gases in a liquid can be carried out by bubbling a gas through it. Thus, the assembly shown in Fig. 8.9 is used for the removal of air trapped in mineral oils [45]. A stream of carbon dioxide passes through the sintered glass disc at the bottom of the sample container holding 5 to 10 ml of oil, and the gas mixture is led into the microazotometer, where carbon dioxide is absorbed by potassium hydroxide, leaving the air to be measured. When microbubbles are observed rising in the microazotometer, the process of displacement is completed.

For the preparation of gas-free liquids, Battino and co-workers [46–48] have critically evaluated the methods and recommended the apparatus shown in Fig. 8.10. It is used for degassing 500 ml of liquid, and the vessel is a 2-liter heavy-wall flask with tubulation. The size of the flask should be reduced when smaller amounts of liquid are to be processed. The stopcocks are quick-opening needle valves with Teflon stems. The O-ring joint provides a grease-free vacuum-tight closure for introducing the stirring bar and the liquid sample. The condenser serves to minimize the loss of sample to the liquid nitrogen trap. Two different procedures are followed, depending on whether the liquid is volatile or nonvolatile. For a volatile liquid, with stopcocks S_1, S_2, and S_3 closed, the system is pumped down to its base pressure (5 to 10 mtorr). Stopcock S_4 is closed, and S_3 opened for 2 to 3 sec. After waiting 1 min for the vapor to freeze out, the pressure is read, and stopcock S_4 is opened to pump down the trap section. When the base pressure is reached, the procedure of expanding from the filter flask to the trap section is repeated. When the pressure

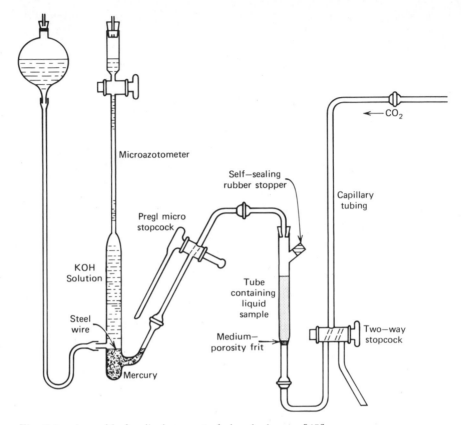

Fig. 8.9. Assembly for displacement of absorbed gases [45].

reaches the base pressure of the system for two successive trials, the liquid is considered degassed. For a nonvolatile liquid, the entire system (S_3, S_4 open) is pumped on for 2 min, S_4 is closed, and the pressure is read after 1 min. The procedure is repeated until the base pressure is attained.

Wiesenfeld and Japar [49] have described a device to degas 100 ml of liquid using a procedure based on the freeze-thaw method. As shown in Fig. 8.11, A and B are Teflon vacuum stopcocks; C is a 1-liter in-line trap containing liquid nitrogen; D is an O-ring seal; E is the vessel that holds the liquid to be degassed. With the trap C empty and B closed, A is opened and the space between the two stopcocks is evacuated. While this is being done, the liquid in E is being frozen. With the material frozen, B is opened and the whole system is allowed to pump down to a suitable pressure. Now C is filled with liquid nitrogen. With A and B remaining open, the material in E is slowly allowed to warm up to room temperature. As the

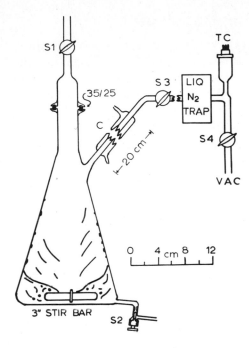

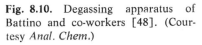

Fig. 8.10. Degassing apparatus of Battino and co-workers [48]. (Courtesy *Anal. Chem.*)

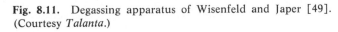

Fig. 8.11. Degassing apparatus of Wisenfeld and Japer [49]. (Courtesy *Talanta.*)

liquid warms up, enough dissolved gas is given off to raise the pressure to about 10^{-3} torr. When the liquid is warm, enough vapor begins to condense on the bottom of C, while almost all of the dissolved gases passes through the pumps. Eventually, a sufficient amount of liquid condenses on the trap so that it starts dripping down to the vessel E. An equilibrium will be set up between the condensation on C and the flow back to E. At 30-min intervals, warm air is applied to the section between C and B as well as the vessel E in order to speed up the circulation between C and E. When the cycle is finished, A is closed, and the liquid in E is frozen as the coolant in C is removed. B is then closed, and the liquid can be stored in E until it is to be used. A similar system has been proposed by Rondeau [50] for degassing individual small samples. Bell and co-workers [51] have described an apparatus to degas 40 ml of liquid by applying vacuum, as in sublimation. Novodoff and Hoyer [52] have fabricated a plastic vessel that simultaneously degasses surfactant solutions and breaks any foam that is generated. It provides for the simultaneous flow of gas through the solution for degassing and above it for breaking the foam. Tanaka and co-workers [52a] have described a simple method for the removal of oxygen from electron-spin-resonance samples in 4-mm-i.d. tubes, using an O-ring coupling.

V. REMOVAL OF TRACE SOLID IMPURITIES

A. FROM SOLID MATERIALS

If a solid sample can be melted and resolidified without any danger of decomposition, the best method for removing traces of solid impurities from it is zone refining [53, 54], wherein the impurities are concentrated in the portion that solidifies last. This technique (see Chapter 3, Section V) can be applied to the purification of metals and of organic and inorganic compounds [55, 56]. A simple automatic apparatus for zone-refining organic substances has been described by Knypl and Zielenski [57]. The tube containing the sample is supported on a base that floats on water in a reservoir. The details are given in Chapter 3, Fig. 3.39. Commercial apparatus [58] for zone refining is available, in which the process of alternate melting and freezing is handled automatically by the heater wires and cooling rings mounted on the movable carriage. This machine can accommodate tubes ranging in size from 4- to 20-mm o.d., and the travel speed can be varied from 2.5 to 60 mm/hr.

The extent of purification by zone refining is affected by a number of factors, such as the size of the melting zone, the speed and direction of zone travel, the temperature gradient at the solid-liquid interface, and the number of passes. Understandably, the efficiency of the process is also dependent on (1) the substance to be purified, (2) the impurities to be removed, and (3) the amount of impurities present. For instance, Jones and McDuffie [59] have studied the distribution coefficients in the removal of Fe(III) and Mg(II) from urethane and from naphthalene. For urethane, the distribution constant k for Mg(II) is 0.31 at 2 ppm, rising to 0.39 at 80 ppm. On the other hand, the k value for Fe(III) in urethane is 0.13 at 5 ppm, rising to 0.27 at 18 ppm, whereas that in naphthalene is 0.13 at 21 ppm. Incorporating complexing agents in the system can facilitate the removal of some metal contaminants [60]. Thus, the addition of 0.1% 8-quinolinal to urethane lowers the k values of both Fe(III) and Mg(II) by about 30%.

Another method of removing solid impurities is microsublimation (see Chapter 3, Section III). This technique is feasible when the pure compound and the impurities differ widely in vapor pressure. Martinck [61] has utilized this method to purify 50-mg samples on a Kofler heating stage.

Recently Gibbins [62] has described a new purification method dependent on vapor-phase extraction. The primary objective is the purification of a nonvolatile compound that is the least soluble component of a mixture when dissolved in a solvent. The compound itself possesses appreciable solubility and cannot be freed of the contaminants by filtration and washing. The method is based on two principles: The solvent used to extract the more soluble component is saturated in these components; the driving force for the spontaneous addition of the solvent to the mixture is determined by the difference in vapor pressure of the pure solvent and the solution formed. The apparatus is shown in Fig. 8.12. The advantage of this technique is that the optimum separation condition

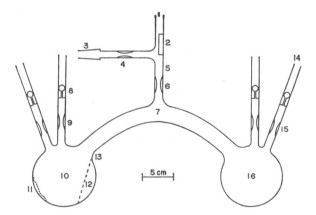

Fig. 8.12. Vapor-phase extractor [62]. (Courtesy *Anal. Chem.*)

may be determined without going beyond a point of no return. Furthermore, as the system is closed, there is no possibility of product loss due to manipulation or reactivity with air, water, or the like.

B. FROM LIQUID SAMPLES

A liquid sample contaminated with trace amounts of solid material should be purified prior to analysis if the suspension interferes with the subsequent method of measurement. A convenient procedure is to centrifuge the sample and use the supernatant liquid for analysis. Alternatively, the sample can be filtered and the filtrate recovered. A sintered glass disc, filter paper, or Millipore filter [63] may be employed. When working with small volumes of liquid, the amount of sample retained on the filter medium should be considered. Gerarde [64] has described a technique applicable to 1–3 ml of sample. The equipment consists of a Unopette reservoir (see Chapter 2, Fig. 2.41), a perforated polyethylene cap, filter paper (13-mm diameter), and a 2-oz rubber bulb fitted with a needle adapter. The liquid to be filtered is placed in the Unopette reservoir, and the polyethylene cap containing a filter-paper disc is fitted over the neck of the reservoir. This assembly is then inverted, and the needle is pushed through the wall of the reservoir; pressure is applied by squeezing the bulb and is maintained until filtration is complete.

There are occasions when fine particles suspended in the liquid sample do not influence the results of the analysis, but are preferably removed because they may interfere with certain operations. For instance, the fine particles may accumulate at the interface between two immiscible liquids, causing the formation of a very persistent emulsion. The presence of microorganisms in the liquid sample may cause deterioration of the material. Appropriate sterilization eliminates this difficulty. One way to sterilize the liquid sample is by ultrafiltration. The use of 0.22-μm Millipore filters in combination with a particular filtration device [65] serves this purpose.

REFERENCES

1. O. W. Grussendorf, A. J. McGinnis, and J. Solomon, *J.A.O.A.C.*, **53**, 1048 (1970).
2. W. A. Moats and A. W. Kotula, *J.A.O.A.C.*, **49**, 973 (1966).
3. C. Goenechea, *Z. Anal. Chem.*, **225**, 30 (1966).
3a. L. T. Grady, S. E. Hays, R. H. King, H. R. Klein, W. J. Mader, D. K. Wyatt, and R. O. Zimmer, *J. Pharm. Sci.*, **62**, 456 (1973).
4. D. H. Freeman and W. L. Zielinski, Jr. (Eds.), NBS Technical Note

549, *Separation and Purification Section: Summary of Activities (1971)*. Nat. Bur. of Stand., Washington.

5. Analytical Problems in Solid State Research and Electronics, 24th Annual Summer Symposium in Analytical Chemistry, Gaithersburg, Md., June 16–18, 1971.

6. T. S. Ma and R. F. Sweeney, *Mikrochim. Acta,* **1956**, 198.

7. D. F. Gadbois, J. M. Mendelsohn, and J. Ronsivalli, *Anal. Chem.,* **37**, 1776 (1965).

8. N. S. Radin, *Chem.-Anal.,* **55**, 117 (1966).

9. V. Runeckles, *Chem.-Anal.* **50**, 23 (1961).

9a. C. R. Ganz, *Anal. Chem.* **45**, 1567 (1973).

10. G. R. Pettit and F. Bruschweiler, *Chem. Ind.,* **1970**, 1018.

10a. R. A. Zeineh, B. J. Fiorella, E. P. Nijm, and G. Dunea, *Anal. Chem.,* **46**, 477 (1974).

11. F. Petuely and N. Meixner, *Mikrochim. Acta,* **1957**, 613.

12. J. A. Burke, P. A. Mills, and D. C. Bostwick, *J.A.O.A.C.,* **49**, 999 (1966).

12a. M. Beroza, M. C. Bowen, and B. A. Bierl, *Anal. Chem.,* **44**, 2411 (1972).

13. G. Ducellier, *Chim. Anal.,* **36**, 181 (1954).

14. C. R. Begeman, *J. Chem. Educ.,* **40**, 427 (1963).

15. F. R. Short and G. Good, *Anal. Chem.,* **28**, 1504 (1956).

16. N. I. Gold, *J. Chromatogr.,* **15**, 431 (1964).

16a. T. T. Martin, *Lab. Pract.* **21**, 347 (1972).

17. E. G. C. Clarke and A. E. Hawkins, *J. Pharm. Pharmacol,* **12**, 509 (1960).

18. L. Kazyak and E. E. Hrast, *J. Gas Chromatogr.,* **6**, 567 (1968).

18a. R. A. Chalmers nad R. W. E. Watts, *Analyst,* **97**, 224 (1972).

19. E. Z. Naugolnylk, *Lab. Pract.,* **19**, 613 (1970).

20. V. Ettel, *Chem. Listy,* **60**, 1237 (1966).

20a. M. Zief and A. J. Barnard, Jr., *Chemtech,* **1973**, 441.

21. R. Roper and T. S. Ma, *Microchem. J.,* **1**, 248 (1957).

22. R. L. Clements, *Chem.-Anal.,* **48**, 20 (1959).

23. E. A. Bell, *Chem. Ind.,* **1953**, 741.

24. G. W. Perold, *Mikrochim. Acta,* **1959**, 251.

25. C. E. Pierce and T. D. Perrine, *Anal. Chem.,* **30**, 2069 (1958).

26. R. Brisset, *Ann. Falsif. Fraudes,* **50**, 333 (1957).

27. I. C. P. Smith, *Chem. Ind.,* **1964**, 54.

28. W. Riemer, *Rev. Sci. Instrum.,* **40**, 1642 (1969).

29. D. R. A. Wharton and M. L. Bazinet, *Anal. Chem.,* **43**, 623 (1971).

30. Labatory Research Co., Los Angeles, Calif.

30a. A. B. Carel, *Lab. Pract.,* **20**, 581 (1971).

31. R. J. Komarck, *J. Lipid Res.,* **8**, 287 (1967).

32. F. Albert-Recht, *Clin. Chim. Acta,* **8**, 976 (1963).

33. R.H.O. Scientific, Inc., Commack, N. Y.

33a. D. S. Pepper, *Lab. Pract.,* **22**, 538 (1973).

34. F. R. Swift, *Microchem. J.,* **7**, 136 (1963).

35. R. K. Maurmeyer and T. S. Ma, *Mikrochim. Acta,* **1957**, 563.
36. R. T. E. Schenck and T. S. Ma, *Mikrochimie,* **40**, 236 (1953).
37. F.S.F. Mehr, *Lab. Pract.,* **15**, 671 (1966).
38. Corning Glass Works, Corning, N.Y.
39. G. F. Smith, *Anal. Chim. Acta,* **17**, 192 (1967); personal communication, 1970.
40. C. G. Skinner, *Anal. Chem.,* **28**, 924 (1956).
41. F. T. Wadsworth, *Anal. Chem.,* **28**, 287 (1956).
42. S. D. Garniner, *Analyst,* **78**, 709 (1953).
43. British Standards Institution, B.S. 1428. Part G 2: 1957.
44. S. Solomons and G. Schlosser, *Rev. Sci. Instrum.,* **33**, 1287 (1962).
45. T. S. Ma and H. M. Halter, unpublished work.
46. R. Battino and F. D. Evans, *Anal. Chem.,* **38**, 1627 (1966).
47. R. Battino, F. D. Evans, and M. Bogan, *Anal. Chim. Acta,* **43**, 518 (1968).
48. R. Battino, M. Banzhof, M. Bogan, and E. Wilhelm, *Anal. Chem.,* **43**, 806 (1971).
49. J. R. Wisenfeld and S. M. Japar, *Talanta,* **16**, 619 (1969).
50. R. E. Rondeau, *J. Chem. Educ.,* **44**, 530 (1967).
51. T. N. Bell, E. L. Cussler, K. R. Harris, C. N. Pepela, and P. J. Dunlop, *J. Phys. Chem.,* **72**, 4693 (1968).
52. J. Novodoff and H. W. Hoyer, *Anal. Chem.,* **44**, 202 (1972).
52a. K. Tanaka, R. P. Ouirk, G. D. Blyholder; and D. D. Johnson, *Appl. Spectros,* **26**; 642 (1972).
53. W. G. Pfann, *Zone Refining,* 2nd ed., Wiley, New York, 1966.
54. E.F. G. Herrington, *Zone Melting of Organic Compounds.* Wiley, New York, 1963.
55. G. Scocciati and A. Vaschetti, *Riv. Ital. Sostanze Grasse,* **41**, 7 (1964).
56. W. R. Wilcox, R. Friedenberg, and N. Beck, *Chem. Rev.,* **64**, 187 (1964).
57. E. T. Knypl and K. Zielenski, *J. Chem. Educ.,* **40**, 352 (1963).
58. Zone Refiner, Fisher Scientific Co., Pittsburgh, Pa.
59. L. N. Jones and B. McDuffie, *Anal. Chem.,* **41**, 65 (1969).
60. G. J. Sloan, *Mol. Cryst.,* **1**, 161 (1966).
61. A. Martinek, *Mikrochim. Acta,* **1971**, 877.
62. S. G. Gibbins, *Sep. Purif. Methods,* **1**, 237 (1972); *Anal. Chem.,* **43**, 1349 (1971).
63. Millipore Corporation, Bedford, Mass.
64. H. W. Gerarde, *Microchem. J.,* **7**, 321 (1963).
65. See *Bibliography—Millipore.* Millipore Corp., Bedford, Mass., 1973.

TRANSFER AND STORAGE OF
ANALYTICAL SAMPLES

I. GENERAL

In this chapter we discuss the techniques and precautions for keeping the analytical sample or transferring it from the storage vessel to the site of the analytical operation. The latter may involve the measurement of a physical property such as spectral absorbance, or a chemical reaction such as nonaqueous titration. For convenience, we also mention in the following sections some microscale vessels that are specially designed for holding certain samples for particular operations such as calorimetry and mass spectrometry. The chief concern is that the material should not undergo any alteration during storage and there should not be any contamination or loss of material while the sample is being transferred.

II. STORAGE OF THE ANALYTICAL SAMPLE

A. STORAGE OF THE SAMPLE AS RECEIVED

If the material as received might be called upon to serve as a reference at a later date, it should be stored under the same conditions under which it is used. Thus, when there is a discrepancy or a contest of the analytical results by different workers, the same material can be resubmitted to analysis as a check. In the laboratories of government regulatory agencies such as the Food and Drug Administration, a common practice is to keep a sample of the inspected material in the original container. For other purposes, however, it is more convenient to transfer a small amount of the material into uniform-size containers [1], so that a large number of samples can be stored in a little space. When commercial sample storage sets [2] are employed, the proper vessel should be used. For example, benzene and acetone solutions should not be stored in vials made of plastics soluble in these liquids. The data on the chemical resistance of plastics given in Table 9.1 should be consulted. Coyne and Collins [2a] have reported the loss of mercury from water during storage in polyethylene bottles (see also

TABLE 9.1. Chemical Resistance of Common Plastics[a]

Plastics	Temp. limit (°C)	Acids (organic)	Alcohols	Aldehydes	Amines	Dimethylsulfoxide	Esters	Ethers	Glycols	Hydrocarbons (aliphatic)	Hydrocarbons (aromatic)	Hydrocarbons (halogenated)	Ketones
Polyethylene conventional	80	E	E	G	G	E	E	G	E	G	G	G	G
Polyethylene linear	120	E	E	G	G	E	E	G	E	G	G	G	G
Polyallomer	130	E	E	G	G	E	E	G	E	G	G	G	G
Polypropylene	155	E	E	G	G	E	E	G	E	G	G	G	G
Teflon	205	E	E	E	E	E	E	E	E	E	E	E	E
Polycarbonate	155	G	G	F	N	N	N	F	G	F	N	N	N
Polystrene conventional	70	G	G	N	G	N	N	F	G	N	N	N	N
Styrene acrylonitrile	95	E	G	F	G	N	N	N	G	E	N	N	N
Polyvinyl chloride	70	G	G	F	N	N	F	F	F	F	N	N	N

[a] E: Excellent; long exposures at room temperature have no effect (up to 1 yr).
G: Good; short exposures at room temperature cause no damage (up to 24 hr).
F: Fair; short exposures at room temperature cause little or no damage under unstressed conditions.
N: Not recommended; short exposures may cause permanent damage.

Chapter 1, Section IV.C). Losses up to 100% in certain instances have been observed in water containing 0.05 mg/liter of mercury.

A dry and cool environment is generally desirable for the storage area. The reader is referred to the discussion of dry boxes in Section III.B. Occasionally, however, moisture is purposely introduced in order to maintain the prescribed conditions. Cross [3] has described a humidity-control system for a small chamber. Wet and dry air streams are mixed in controlled proportions to achieve any desired relative humidity in the range from 5 to 99% inside a 30-liter space at 30°C. The selected value is automatically maintained constant to better than ±1% over a period of several days by a system of monitoring and electronic controls.

Liquid materials should be kept in small sealed ampuls. Even when a fairly large amount of liquid is to be stored, the sample is preferably transferred into a number of 10-ml vials. This is particularly important in the case of volatile and air-sensitive compounds. Gaseous substances by necessity are stored in closed containers, which may be under pressure. A method has been described by Lloyd [4] for the detection of leaks of a specific gas in a system: One face of a membrane that is selectively permeable to the gas is exposed to the sample while the other face of the membrane is exposed to the vacuating system of a mass spectrometer.

Attention should be given to the storage of samples in sealed vials. The techniques for delivering the sample into the vessel and for sealing it are both important. Improper delivery may lead to contamination of the sample by the pyrolytic decomposition products formed on the hot glass. Therefore it is essential that the sample does not touch the neck of the vial (the part that will be heated and sealed). For sealing, an appropriate burner or torch should be used to give a sharp hot flame, so that the time required to soften the glass is short and the heating is localized in a narrow area. Sometimes it is advantageous to use a vial with constriction at the neck, as shown in Fig. 9.1, so that the sealing can be quickly completed at the constriction. The sealing should not produce a heavy glass bead or a thin fiberlike tip. When it is necessary to seal the sample in an inert atmosphere, either of the following two techniques can be used: (1) Displace the air in the vial with nitrogen before introducing the sample; then inject nitrogen by means of a long hypodermic needle for several minutes after filling in the sample; finally remove the needle and seal the constriction of the vial immediately. (2) Connect the vial containing the sample to a vacuum-line manifold (similar to Fig. 9.8 in Section III.B.2); evacuate and flush with nitrogen; then seal the constriction and disconnect the vial from the vacuum line.

Fig. 9.1. Vial for storing sample (before sealing).

B. STORAGE OF THE SAMPLE AFTER PRETREATMENT

If the material requires processing before it can be subjected to analytical reaction or measurement, one must transfer the sample (after processing) to a suitable vessel. Ideally, the sample can be transferred to the vessel where the analysis will be performed (e.g., the absorption cell of the spectrophotometer), but this is generally not the case. The nature and condition of the vessel holding the sample after pretreatment may affect the analytical results. For instance, glass vessels that have been washed by heating in nitric acid or dichromate cleaning solution are liable to contain traces of metal ions, while apparatus cleaned with detergents may contain organic residues (see Chapter 1, Section V). Crawly [5] has recommended a reagent for cleaning glassware that consists of 5% hydrofluoric acid, 33% nitric acid, 2% Teepol, and 60% water. Since it works by removing a thin layer of glass or silica, it should not be used on graduated glassware. Jakab [6] has described an efficient method for removing viscous polymeric substance from nuclear-magnetic-resonance sample tubes. Methanol is added to fill the tube. A copper loop is inserted and the mixture agitated by an up-and-down motion of the loop to precipitate the polymer, which adheres to the loop and is removed.

If the original material subjected to analysis is a complex mixture, the analytical sample obtained after pretreatment will be one fraction or several fractions. The fractionation (e.g., chromatography) is usually performed by the analyst, but the final step of analysis may not be carried out immediately. The analyst should have some knowledge of the nature of the substance in the fraction to be stored. It may still be a mixture, such as certain extracts of pharmaceutical preparations or insecticide residues. Deterioration of the compounds and change of their proportion on storage should be prevented. The storage of separated compounds on the thin-layer chromatographic plate deserves particular attention. Since the amount of material is in the microgram range, the molecules are spread out and come into close contact with air and moisture. If the compound is susceptible to oxidation or hydrolysis, the reaction takes place much more readily (because the adsorbent may act as catalyst) than when it is iso-lated and kept in a capillary tube. It is advisable to make a photographic record of the plate immediately after completion of the separation of the compounds [7], and then again periodically until the compounds are removed from the chromatogram for other operations such as the measure-ment of absorbance. Knight [8] has recommended the use of polyvinyl chloride film for chromatograms with a tendency to fade.

When the sample to be stored is a single compound (e.g., one of the multitude of new organic compounds being synthesized in the research laboratories), the pretreatment of the analytical sample is in general not performed by the analyst. The treatment may involve recrystallization or fractional distillation. While the sample is transferred into the storage container, contamination by extraneous matter such as filterpaper fibers and sintered-glass chips should be prevented.

III. TRANSFER OF SOLID SAMPLE

A. DEVICES FOR TRANSFERRING SOLIDS

1. For General Purposes

The weighing tube shown in Fig. 9.2a is a convenient device for trans-ferring solid samples. Also called the charging tube, it can be made from 10-mm glass tubing (by drawing it out to form the taper) and 2-mm glass rod. The open end is tapered so that the mouth measures about 7 mm in outside diameter, while the bottom is about 5 mm. The bottom should be flat or round but not pointed, in order to facilitate the cleaning or removal of the solid by means of a microspatula. The length of the handle is selected to match the reaction vessel. Thus a weighing tube with a 120-mm handle is used for transferring the sample into a micro Kjeldahl flask, whereas a tube with a 35-mm handle is used for transfer into a 50-ml Erlenmeyer flask.

Microboats made of platinum, quartz, or porcelain are commercially available [9] (Fig. 9.2b). They should be handled with forceps having flat tips. If the microboat and the solid sample are to remain in the reaction vessel during the analysis, the boat is weighed, and the sample then added

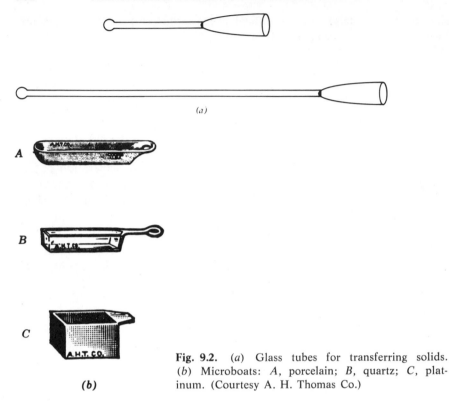

Fig. 9.2. (a) Glass tubes for transferring solids. (b) Microboats: A, porcelain; B, quartz; C, platinum. (Courtesy A. H. Thomas Co.)

and reweighed. On the other hand, if it is not feasible to keep the microboat in the reaction vessel, the total weight of sample and boat is first determined. The microboat is moved to the reaction vessel and turned upside down to empty is contents. After reverting to the upright position, the microboat is reweighed. The difference between the two weighings gives the exact amount of the solid sample that has been transferred into the reaction vessel. In order to avoid cleaning the microboats, Cahn and Cadman [10] have advocated the use of disposable sample-weighing pans (see Chapter 2, Fig. 2.8). Disposable microboats, capsules, and pans made of aluminum, silver, or tin have appeared on the market [10a].

2. For Specific Operations

Sometimes a very simple device suffices in transferring or keeping samples for microchemical operations. For example, Barrett, Hoyer, and Santoro [11] have employed ordinary capillary tubes to perform thermal analysis using 2 to 4 mg of solids and obtained results comparable to those obtained by elaborate equipment. Denney [12] has described a simple apparatus for the addition of small quantities of solids to hot reaction mixtures. As shown in Fig. 9.3, it consists of a hollow stopcock into which part of a test tube has been fitted. So long as the container is not filled beyond half of its capacity, there is little likelihood of the tap binding, and the danger of losing refluxing solvent from open side arms is alleviated.

Since spectroscopic measurements are inherently very sensitive, the problem in microchemical analysis employing spectrophotometric methods involves primarily the transfer and confinement of the minute amounts of sample. The devices mentioned below serve as typical examples. Klein and Ulbert [13] have described a method of preparing micropellets that can be carried out with a normal die. For infrared spectroscopy, the metal disc holder is used to hold the micropellet in the beam of infrared radiation. Good quality spectra have been obtained with the 1.5-mm pellet holder from 1 to 4 μg of solid using a refracting beam condenser with KBr optics. Modification of this method permits the withdrawal of the pellet from the metal holder in cases where the metal may influence the measurement. Brannon [14] has compared ten procedures of sample

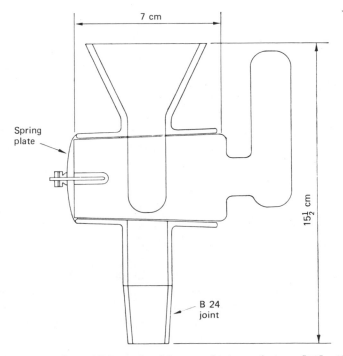

Fig. 9.3. Apparatus for addition of solids to refluxing solutions [12]. (Courtesy *Chem. Ind.*)

preparation for infrared analysis using 15-μg samples and reported that the quality of the spectrum is enhanced by concentrating the sample in the center of the beam. Gilby and co-workers [15] have studied the thin-film and condensed-beam sampling techniques for internal-reflection spectroscopy. Jones [16] has investigated aerosols generated from metal surfaces. Wharton and Bazinet [17] have described a concentrating flask for introduction of micro samples to a mass spectrometer via the solid-insertion probe. Ellen [18] has described an apparatus for producing a graphite pellet containing a 50-μg sample for emission spectroscopy, and an improved version is commercially available [19]. Also available is an argon-arc vacuum furnace for semi-automatically melting and shaping 12 metal samples in 15 min [20]. Elio [20a] has described two devices to transfer 10-mg samples to the tip of the cathode. Walter and Monaci [21] have constructed a rotating-disc sample holder for sparking flat-metal disc samples. Morrision and Talmi [22] have used an induction-heated insulated graphite crucible to vaporize the sample for atomic absorption or emission spectroscopy. Similar technique has been described by Christopher [22a]. Nicky and Rice [23] have described a micro sample support for X-ray fluorescence spectroscopy and noted that sample-handling technique is extremely important in achieving reproducible quantitative results while working with small quantities of material.

Patai and co-workers [23a] have constructed an apparatus for simultaneous differential thermal analysis and thermogravimetric analysis using 20-mg samples. McCrone [23b] and Inman [23c] have discussed the techniques for preparing fine particles for electron-microprobe analysis. Evans and co-workers [23d] have described the mounting of single crystals on glass fibers for the X-ray analysis of lunar samples. Devices for the transfer of solids into a gas chromatograph are described in Chapter 4, Figs. 4.33 and 4.34.

B. TECHNIQUES FOR HANDLING AIR-
AND MOISTURE-SENSITIVE SAMPLES

1. The "Dry Box"

The term "dry box" has been used in the chemical laboratory since World War II to indicate a working chamber with controlled atmosphere. The recent upsurge of studies on oxygen-sensitive organometallic compounds and extremely hygroscopic biochemical materials have enhanced the need of such devices. Ma and Sweeney [24, 25] have described a box suitable for general microchemical operations, including transferring samples, weighing, and running reactions in the milligram region. The box is fitted with service lines and has an extension to house a 10-ml buret. Pack and Libowitz [26]) have described a box to which a stereomicroscope is fitted. A variety of commercial glove boxes are available. They are constructed of stainless steel and marketed for specific purposes.

Many simple controlled-atmosphere devices have been published. Naff [27] has proposed a transfer jar that permits reagents from three bottles to be transferred to a receiver in a dry or inert-gas atmosphere. Gordon and Johannesen [28] have constructed an apparatus that incorporates a plastic bell jar of 16-in. diameter and 14-in. height. The base plate has holes for gloves, for evacuating the chamber, and for admitting inert gases. Similar boxes have been described by Nuti [28a] and Safarik [28b]. Fiebig and co-workers have utilized a glass funnel of 150-mm diameter for dry sampling. As illustrated in Fig. 9.4, operations are conducted through an opening (16×50 mm) slot cut in the side of the funnel, through which inert gas at positive pressure escapes continuously. Similarly, a plastic bag can be mounted over a frame that is purged with a current of inert gas, as suggested by Franklin and Voltz [30]. A more sophisticated apparatus (Fig. 9.5) has been designed by Brueser [31]. It comprises a 10-liter flask with three necks, two being fitted with gloves, and all three being connected through two-way taps to a vacuum pump or source of inert gas. The flask is fitted with an interior base plate. This

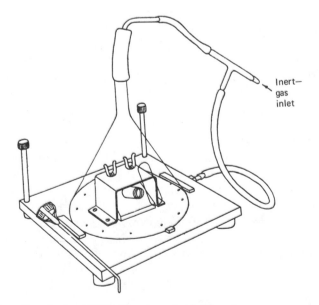

Inert—
gas
inlet

Fig. 9.4. Sampling dry box [29]. (Courtesy *Anal. Chem.*)

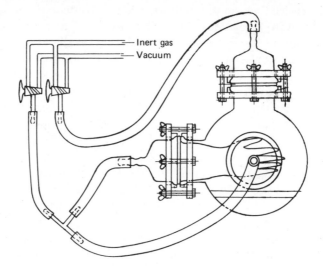

Fig. 9.5. Glove box of Brueser [31]. (Courtesy *Chem. Tech. Berl.*)

apparatus is recommended for use in the preparation of solid samples of air-sensitive organometallic compounds for spectroscopy.

Naturally, the atmosphere to be controlled within the "dry box" depends on the microscale manipulation to be conducted therein. For instance, when the experiment is for the nonaqueous titration of very weak acids [25], carbon dioxide is to be eliminated, but not necessarily oxygen and moisture. In general, a circulating system is employed for atmospheric control in preference to a static system, because more efficient use is made of gas-purification reagents. Several drying agents have been used with good success, but suitable agents for oxygen removal have been more difficult to find [32]. Eubanks and Abbott [33] have described a simple, relatively inexpensive system to remove moisture and oxygen, provide sensitive pressure control, permit the in-line monitoring of moisture and oxygen concentration, and permit the regeneration of the purification agents without interrupting the box operation. The oxygen concentration is kept below 50 ppm, and water vapor below 25 ppm. Dryburgh [34] has designed a system to be used in a commercially available glove box. Air is pumped out of the box, chilled, passed through a column of molecular-sieve 5A drying agent, warmed, and returned to the box. Its resulting moisture content is about 0.1 ppm. For the purpose of providing pure nitrogen atmosphere, Gibb [35] has recommended a high-voltage sodium arc as purifier, and for helium or argon atmosphere free from nitrogen, a lithium arc.

2. Other Devices

Kirsten [36] has described a technique for transferring 0.1 mg of extremely hygroscopic samples for drying and weighing, as illustrated in Fig. 9.6. The solid is placed in a quartz microboat with interchangeable ground joints. The joints are improved by a few turns of individual grinding with carborundum powder, and tightened manually. The weight correction for a weighing time of 10 min at 50% relative humidity and 25°C is 0.5 μg. Redman [37] has proposed a weighing bottle for transferring hygroscopic materials after drying, eliminating the use of desiccators.

Gibbins [38] has described a weighing-transfer apparatus for handling air- and water sensitive solids in high-vacuum systems. As shown in Fig. 9.7, the apparatus comprises three sections: the original sample vessel A, the solvent-sample intermediate transfer vessel B, and the sample receiver C, a weighed fragile bulb. The solid is transferred from A to B by dissolution. The solvent is removed by distillation, and the sample is then dry

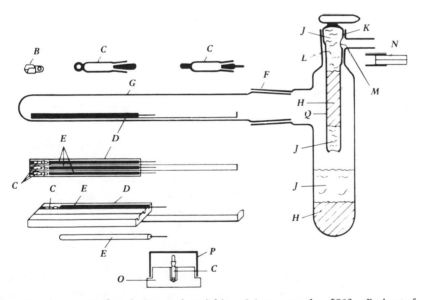

Fig. 9.6. Apparatus for drying and weighing 0.1-mg samples [36]: B, boat for weighing of ordinary solids; C, quartz vessels with interchangeable standard joints, weight about 150 mg; D, metal holder for tubes and rods; E, brass rods that hold tubes and push in stoppers for closing (end surface of rods slightly concave to prevent them from slipping on stoppers); F, ground joint; G, tube of drying pistol; H, magnesium perchlorate; J, glass wool; K, stopcock; L, score in stopcock; M, opening in stopcock; N, fine capillary; O, metal stand for weighing-tubes; P, glass cover for sample tube; Q, drying tube. (Courtesy *Mikrochem. Acta.*)

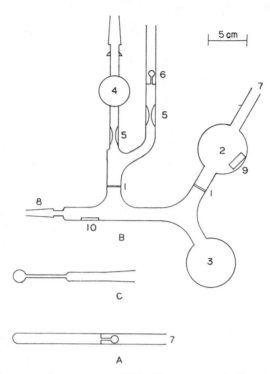

Fig. 9.7. Weighing-transfer apparatus [38]. (Courtesy *Anal. Chem.*)

transferred to *C*. The apparatus is operated in a vertical plane and rotated about a horizontal axis (perpendicular to the apparatus plane) to effect various operations. Stopcocks that contribute contaminating grease are eliminated. A sintered disc prevents glass-chip contamination. The same author has described a technique to subdivide air- and water-sensitive solids for analytical purposes [39]. The apparatus is shown in Fig. 9.8. The sample bulb *A* is sealed onto the sample divider *B* at 2. The apparatus is connected to a high-vacuum line at 3. Additional solvent, if required, is condensed in the bulb 4 (100 ml) by a cold bath. After sealing off the apparatus at the constriction 5 and breaking the fragile bulb 1 with a Teflon-coated bar magnet, an appropriate portion of the sample is transferred by solution and/or slurry manipulations to the fragile-bulb sample tube 6. The constricted portion 8 of the tube is washed free of the solute by condensating the solvent on the walls with (for example) Dry Ice. After 6 and 4 are cooled to a temperature at which the solvent has negligible vapor pressure, 6 is sealed off. The process is then repeated for each of the remaining tubes. Furr and Archibald [39a] have described a

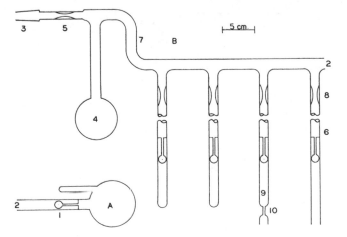

Fig. 9.8. Sample-subdivision apparatus [39]. (Courtesy *Anal. Chem.*)

method for sampling micro amounts of soil from plant roots grown under controlled conditions in the laboratory.

Long and O'Brien [40] have constructed a jointless apparatus for the preparation of saturated solutions from air-sensitive solutes and solvents whose vapors attack stopcock greases. The equipment (Fig. 9.9) comprises a heavy-walled (3-mm) mixing tube 1 connected through the side arm 2 and sintered-glass filter 3 to the receiving ampul 4. The side arm 5 is connected to the vacuum line, while the capillary tube 6 provides a pressure link between the two sides of the filter. The ampuls 7,8 with thin bottoms, containing the solute and solvent, respectively, are introduced into the mixing tube, whose top is then closed with a "Subaseal" 9. After successive evacuation and filling with inert atmosphere, the side arm 5 is sealed off, and the ampuls are broken by impact, solute first. The apparatus is then immersed in the thermostat. After being shaken for the requisite time, the apparatus, still in the bath, is tipped so that the solution runs down the side arm 2 through the filter 3 and into the ampul 4. The ampul is then sealed off at the constriction 10. The insertion of a syringe needle through the "Subaseal" relieves any pressure during sealing off.

In order to overcome the difficulties encountered in working with extremely air-sensitive organoactinides, Eggerman [41] has proposed a simple system for transferring the compounds from Schlenk tubes [42] into capillary tubes for mass-spectral analysis. The design (Fig. 9.10) consists of a glass tube with standard taper joint, a machined Teflon stopper, and a capillary tube that fits tightly to maintain a 1-torr vacuum. The

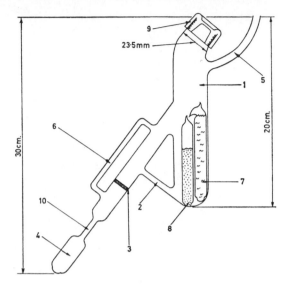

Fig. 9.9. Apparatus for the preparation of saturated solutions from air-sensitive solutes and solvents [40]. (Courtesy *Chem. Ind.*)

Schlenk tube containing the solid sample is placed under a slightly positive nitrogen pressure. The Schlenk stopper is removed and replaced with the Teflon stopper shown in the figure. The transfer system is evacuated and backfilled with nitrogen. After purging, the system is maintained under vacuum. The sample-transfer device is then inverted to transfer the sample into the capillary tube, which is subsequently flamed off and sealed. The sealed capillary is placed in the direct inlet probe of the mass spectrometer. The capillary is easily fractured as the probe is inserted into the source block.

As mentioned above, breaking a thin glass bulb inside a closed apparatus can be carried out by means of an external magnet. This is a common technique for moving small vessels in the closed system. The vessel may be fitted with a sheet-iron collar coated with an epoxy resin, as suggested by Markowitz and Boryta [43]. If iron does not react with the reagents, a thin wire is placed in the vessel, which may be as small as a capillary tube [44].

C. TECHNIQUES FOR TRANSFERRING POWDER MATERIALS

1. Difficulties in the Transfer of Powder Mixtures

The direct transfer of powder substances for analytical purposes often encounters difficulties, especially when the material is a mixture. For

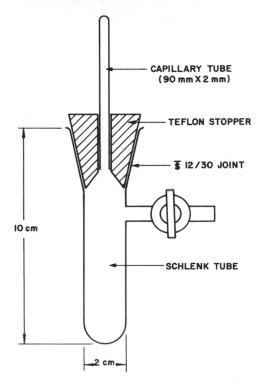

CAPILLARY TUBE
(90 mm X 2 mm)

TEFLON STOPPER

$ 12/30 JOINT

SCHLENK TUBE

10 cm

2 cm

Fig. 9.10. Sample-transfer system from Schlenk tube to capillary tube [41]. (Courtesy *Appl. Spectrosc.*)

instance, the use of a turntable for distributing pulverized samples (see Chapter 1, Section IV.C) becomes impractical at the milligram level. Solbakken [45] has proposed a device for sampling solid powders that are undergoing equilibrium measurements or gas-solid reactions at high temperatures and vacua, without interrupting the process. The apparatus is depicted in Fig. 9.11. The powder is supported on the sintered-glass disc in the reaction tube. Through the sinter is a hole about 4 mm in diameter to hold a loosely fitting glass rod that has a piece of iron imbedded in its lower end. For sampling, a magnet is moved outside the reaction tube. Shaking and lifting the rod makes some of the powder fall through the hole and down into the capillary, where it is collected.

2. Transfer of Solids via Solution or Sublimation

Wherever applicable, the recommended procedure for transferring powder mixtures is to dissolve a known quantity of the material in a suitable

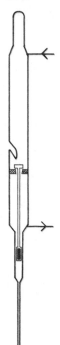

Fig. 9.11. Transferring powder sample into capillary [45]. (Courtesy *Lab. Pract.*)

solvent and take an aliquot of the solution for analysis. (The solvent may be subsequently removed by evaporation, if necessary.) Thus, this technique can be used in the determination of nitrogen in fertilizer by the Kjeldahl method; duplicate analyses using aliquots give more reliable results than multiple determinations made by delivering several solid samples into the micro Kjeldahl digestion flasks [46]. In the preparation of samples for laser microprobe analysis, Rosen [47] has advocated dissolving the material and applying it as droplets to a gel of unfixed spectrographic emulsion. Such a technique gives a homogeneous dispersion of sample, and the gel is ready for analysis upon drying.

The dissolution of the sample is the technique of choice for transferring solids into the gas chromatograph. The device described by Yannone [48], shown in Fig. 9.12, is constructed from a 1-ml disposable syringe. The sample holder consists of a spiral formed by flattening the end of a 32-gauge surgical wire and twisting the wire 150 to 250 times. The other end of the wire is attached to the rubber tip of the plunger. A gastight seal is obtained by inserting a silicone-rubber diaphragm in the hub of the syringe needle through which the wire passes. The sample is deposited on the spiral by evaporation from solution, and the spiral is withdrawn into

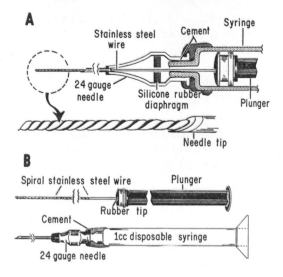

(A) Cross section of needle and adjacent syringe with the twisted wire insert connection to the plunger and its passage through the silicone rubber septum in the hub of the needle. Location of cement between syringe tip and inner needle hub is shown as well as the spiral in the wire.

(B) Withdrawn plunger and attached twisted wire; 1 cc plastic syringe, 24 gauge needle and location of external cement.

Fig. 9.12. Device for injecting solids into gas chromatograph [48]. (Courtesy *J. Gas Chromatogr.*)

the needle; this is inserted through the injection port, and the wire is then pushed forward so that the spiral enters the heated injection zone. An automated injection system for solid samples has been designed by Jenkins and Hunt [49] as follows. Between 35 and 100 μl of a solution containing about 1 μg of the sample in a volatile solvent is injected into a glass sample holder, which is then placed in a magazine taking up to 36 such holders. A gradient heater evaporates the solvent before the magazine is transferred to the injector of the chromatograph, where each holder is automatically transported in turn to the hot injection site. Walker [50] placed the solid sample in a tin alloy capillary tube of 0.7-mm bore, which is then crimped at both ends. A turntable-and-piston arrangement carried the capsule into a pressure-tight zone at 270 to 300°C. As the alloy melts, the sample is swept by the carrier gas into the chromatograph. Levin and

Ikeda [51] have employed a glass tube of 6-mm outside diameter to carry the solid sample, which is held in place with a pad of glass wool. The tube is then inserted in the injection port of the chromatograph.

Sublimation (see Chapter 3, Section III) is another technique that can be utilized to transfer solids. In fact, this is the usual means of transfer in·vacuum-line manipulations. Stahl [52] has applied this principle to transfer solids onto a thin-layer chromatographic plate. The apparatus is shown in Fig. 9.13. The powder mixture D is placed in the glass cartridge B, the end of which is then sealed by a silicone membrane A. The charged cartridge is pushed into the heating block C, adjusted to a given temperature. The tip of the capillary tube projects from the furnace and points to the starting point on the plate F. The thin-layer chromatographic plate is positioned 1 mm from the tip and can be easily removed. It should be noted that not all of the mixture is transferred to one spot on the plate by using this device, since fractionation may occur.

3. Transfer of Powder Material from a Thin-layer Plate

It is apparent that the sample on the thin-layer chromatographic plate is in the form of very fine particles resting on the adsorbent. Transfer of the sample is required (1) when the compound is to be further analyzed, or (2) when a chemical reaction has been performed on the plate and the product is to be recovered. The latter case is exemplified by organic synthesis in the microgram range [7, 53, 54]. Usually the compound to be transferred can be removed (accompanied by the inert adsorbent) from the thin-layer plate by means of a microspatula if it is colored or fluoresces under ultraviolet illumination. The transfer should be carried out as soon as practicable, since many organic compounds are susceptible to oxidation,

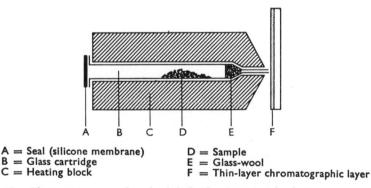

A   B   C   D   E   F

A = Seal (silicone membrane)   D = Sample
B = Glass cartridge   E = Glass-wool
C = Heating block   F = Thin-layer chromatographic layer

Fig. 9.13.   Thermomicro transfer of solids [52]. (Courtesy *Analyst.*)

hydrolysis, photolysis, or the like, and may react with the chemicals present in the environment.

If the compound sublimes without decomposition, and is not affected by the adsorbent upon heating, a convenient way to transfer it is to place the thin-layer plate on a heating stage (Chapter 2, Section IV.B, Fig. 2.17) and cover the spot of the compound with the Kofler vacuum-sublimation chamber (Chapter 3, Section III.C). As depicted in Fig. 9.14, the upper wall on which the sublimate condenses is cooled by suitable means, such as solid carbon dioxide.

Clemett [55] has proposed a device for transferring thin-layer chromatographic fractions for spectroscopic examination. The apparatus (Fig. 9.15) consists of a disposable Pasteur pipet, the tip of which has been turned in slightly and plugged with glass-fiber sheet. The wide end is fitted with a rubber stopper carrying a capillary tube bent as shown. Suction is applied at the plugged end of the pipet, and the mouth of the capillary applied to the spot on the chromatographic plate to be collected. The powder is consolidated in the narrow part of the pipet, forming a miniature column, which allows the sample to be eluted in a very small volume of solvent. The device works extremely well for collecting granular adsorbents such as silica and alumina, but is less successful with fibrous cellulose adsorbents. The eluate from the Pasteur pipet may be used directly for spectroscopic examination. For mass spectrometry, the eluate (5-μg sample) is dropped slowly onto the probe tip, and the solvent allowed to evaporate. For infrared spectrometry, the eluate (20-μg sample) is dropped onto a little KBr powder, which is pressed into a microdisc after the evaporation of the solvent. For nuclear-magnetic-resonance spectrometry the eluate should not be transferred directly into a microcell, as it is generally contaminated with water and the solvents used to develop

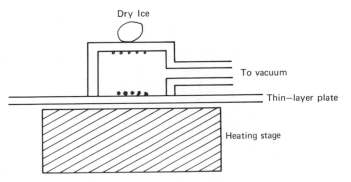

Fig. 9.14.   Transferring solids from thin-layer plate by sublimation.

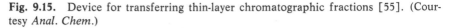

Fig. 9.15. Device for transferring thin-layer chromatographic fractions [55]. (Courtesy *Anal. Chem.*)

the chromatogram. Hence the eluate is transferred into a second Pasteur pipet, the tip of which is sealed. Bridging the neck of this pipet with a drop of solvent prior to adding the eluate confines the latter to the wide portion of the pipet, which facilitates the evaporation of the solvent. After evaporation, the tip is broken off and the material remaining (100 to 200 μg, collected from several chromatograms) is washed down into a microcell with the solvent of choice.

A direct transfer technique for preparing micropellets has been described by Rice [56]. As illustrated in Fig. 9.16, the spot on the thin-layer chromatogram is first outlined in a "teardrop" shape. The TLC support around the spot is removed, and the glass adjacent to the point of the teardrop is

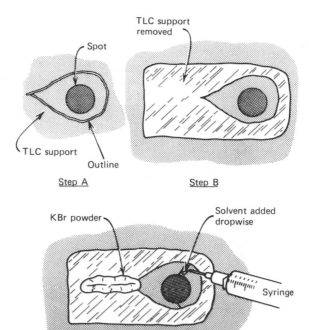

Fig. 9.16. Direct-transfer techniques for preparing micropellets [56]. (Courtesy *Anal. Chem.*)

cleaned with cheesecloth moistened with a volatile solvent. Approximately 15 to 20 mg of KBr is added to the clean area by using a microspatula. The powdered KBr is formed into a line 20 mm wide by 60 mm long, such that one end of the line is in physical contact with the point of the teardrop. A suitable solvent, selected so that the compound will be eluted at or near the solvent front, is added dropwise to the blunt end of the teardrop from a syringe. The point of addition of the solvent is moved so that solvent is continuously added approximately 50 mm behind the sample. As the sample moves across the spot, it will become concentrated at the point of the teardrop. As the solvent front and sample reach the end of the KBr powder, the point of contact of KBr and TLC support should be broken with the flat end of a microspatula to prevent flowback of the solvent onto the TLC support. The solvent is allowed to evaporate, and one-fourth to one-third of the KBr (5 to 7 mg) from the far end of the pile is transferred to a microdie, and a micropellet is pressed. A KBr solid triangle (commercially available) can be used for the same purpose.

4. Transfer of Powder Using the Plunger and Capillery Tube

The plunger and capillary tube are used to transfer powder material into a reaction vessel whose opening is too narrow for the weighing tube or microboat (Fig. 9.2) to pass through. The technique is illustrated in Fig. 9.17. The capillary tube, of convenient length and 1- to 2-mm bore, is fitted with a plunger as shown. The powder material is placed in the

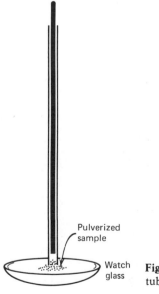

Pulverized
sample

Watch
glass

Fig. 9.17. Transfer of powder material with capillary tube and plunger.

mortar or on a watch glass. On pressing the capillary tip onto the sample (the plunger being raised) a section of the capillary tube is packed with the powder. The capillary tube is removed, and its outside wall is wiped with a camel's-hair brush. It is then inserted into the reaction vessel until its tip nearly reaches the bottom of the vessel. The powder is pushed out of the capillary tube by means of the plunger, and the capillary tube together with the plunger is carefully withdrawn. In this way, the powder does not fall onto the walls of the reaction vessel.

5. Transfer of Particles from Crevices

A device has been described by Schuessler [57] for transferring particles from crevices. The apparatus (Fig. 9.18) works on a soft vacuum and collects the sample on a Millipore filter. A glass capillary is used to pinpoint the sampling area. A short piece of rubber tubing connects the glass capillary to the Millipore-filter housing; this increases the maneuverability of the tool. This device is of general use, since it also can collect sample from any area. It has the advantage of not requiring a carrier to transfer the particles. When a solvent is used as a carrier to transfer solids for infrared spectroscopy, subsequent evaporation of the solvent may cause water to condense on the microsample that is not readily desorbed afterwards, and the resulting infrared spectrum will give questionable absorption bands. Likewise, when collodion is used as the carrier for emission spectroscopy, excess carrier at times will cause the sample to jump away during the analysis, and the material will be lost.

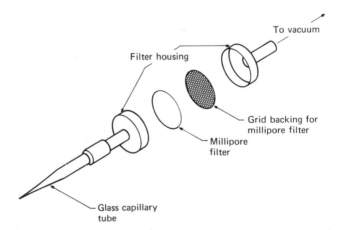

To vacuum

Filter housing

Grid backing for millipore filter

Millipore filter

Glass capillary tube

Fig. 9.18. Tool for transferring solids from crevices [57]. (Courtesy *Appl. Spectrosc.*) *Spectrosc.*)

D. TRANSFER OF IRREPLACEABLE AND RARE SPECIMENS

An important consideration in transferring solids from irreplaceable and rare specimens is to avoid damaging the article that is to be preserved. Examples of this category includes materials submitted for crime detection, historical artifacts, and precious *objets d'art*. The device mentioned in the above section (Fig. 9.18) may be employed for transferring the analytical sample. The hard glass tip may be used to scrape minute particles free, or a tiny area of the specimen may be scratched with the corner of a razor blade. If the dissolution technique is applicable, the following procedure described by Szwarc and co-workers [58] may be adapted; it is for the direct transfer of ink dyes from a document to chromatographic paper. Place successively in a sintered-glass crucible a piece of polyethylene film with a hole in its center, a piece of chromatographic paper, and a piece of the document, with the written side downwards and touching the chromatographic paper. Trace the inscription with a capillary tube containing a solvent. The ink dyes thus dissolved are transferred by gentle suction to the chromatographic paper. Another method of transfer is by means of electrography, which is applicable to metals and some anions [59].

Adams and Tong [60] have used a laser-microscope system (Fig. 9.19) to transfer microsamples from refractory materials. The laser beam volatilizes a chosen area 5 to 200 μm in diameter and 5 to 100 μm deep. The volatilized material is then condensed on a transparent cover plate held 0.2 to 0.5 mm above the specimen surface. Samples in the order of 10 ng transferred from a specimen of irradiated ceramic nuclear fuel have been analyzed by four different methods; alpha spectroscopy, gamma spectroscopy, neutron activation, and mass spectrometry.

IV. TRANSFER OF LIQUID SAMPLES

A. TOOLS FOR THE TRANSFER OF LIQUIDS

1. Weighing Bottles and Capillaries

Weighing bottles for transferring 1 to 100 mg of liquid are shown in Fig. 9.20. Note that the stopper has a hole, so that a hook can go through it to pull the bottle open after the latter has been placed inside the reaction vessel.

Capillaries [61] for holding milligram quantities of liquid are best constructed in the manner illustrated in Fig. 9.21. A glass capillary *a* of 1- to 3-mm bore is heated over a sharp flame to form a constriction *b*. Then one section is heated to seal *c* and drawn out to form the air chamber and handle *e*. Finally the end is drawn out to form the liquid chamber and the

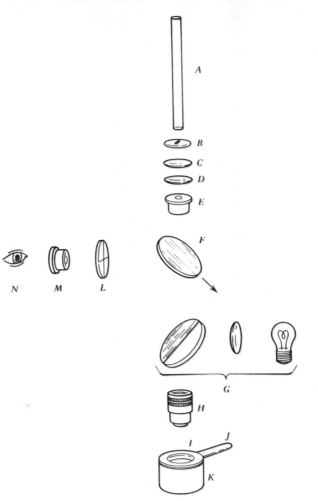

Fig. 9.19. Optical diagram of laser-microscope system for transfer of refractory materials [59]: *A*, ruby laser crystal; *B*, ceramic aperture; *C*, positioning lens; *D*, compensating lens; *E*, eyepiece; *F*, movable mirror; *G*, vertical illuminator; *H*, objective lens; *I*, cover glass; *J*, cover glass support; *K*, specimen in plastic mount; *L*, cross-hair reticle; *M*, eyepiece; *N*, observer. (Courtesy *Anal. Chem.*)

tip of 0.3-mm bore. For transferring the sample into the capillary, the liquid chamber is warmed over a tiny flame, and immediately afterwards the tip is immersed in the liquid, as depicted in Fig. 9.22*a*. The sample enters the capillary owing to the reduction of gas volume inside by cooling. (For very low-boiling liquids, instead of warming the capillary, the liquid chamber is chilled by placing a piece of Dry Ice over it.) When the desired

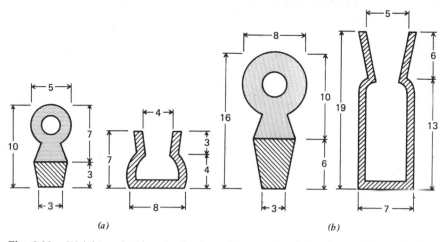

Fig. 9.20. Weighing bottles (scale in millimeters): (*a*) micro; (*b*) semimicro. (Courtesy *Microchem. J.*)

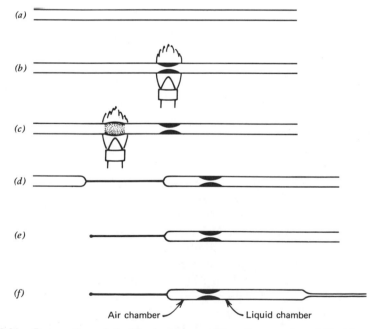

Air chamber — Liquid chamber

Fig. 9.21. Preparation of double-chamber capillary for transfer of liquids.

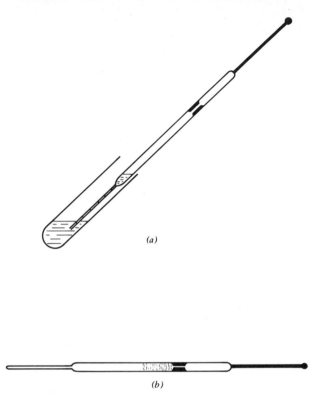

Fig. 9.22. Transferring liquid into capillary.

amount of liquid has entered, the capillary is turned upright, and the sample is made to fall down by gentle tapping until it rests just above the constriction. After sealing (Fig. 9.22*b*), the liquid can be kept in the capillary indefinitely without loss. When the sample is ready for analysis, the tip is cut off, and the capillary is placed at the reaction site. The air in the air chamber expands upon heating and pushes the sample out of the capillary. Bieling [62] has proposed a similar capillary tube that has two small bulbs blown in it. Knoblock and Mudrova [63] have described a capillary tube with a single bulb for preparing solutions by transferring the sample and reagents into the bulb, where they are mixed. Haller and Trutnovsky [64] have suggested a miniature soldering bit to replace the flame for closing the capillary tips. D'Aoust [65] has sealed glass pipets with caps to be opened subsequently in a chamber under pressure.

Unsealed capillaries have been proposed for the transfer of liquids for quantitative analysis. For example, Mitsui and Furuki [66] have described an open quartz capillary in which a platinum wire placed in the center holds the liquid sample. This devise is not suitable for very volatile liquids, nor for amounts over 10 mg.

A capillary tube is generally used for transferring liquids onto the thin-layer chromatographic plate. Abraham and Sasty [67] have designed an apparatus that consists of a small reservoir from which the liquid can drop through a capillary tube onto the plate to produce a uniform straight band. Luis and Sá [68] have constructed a simple attachment to control the delivery of liquid from capillaries as follows (see Fig. 9.23a). Draw out a tube of 10-mm bore to 5 mm (b). Draw out again to 3-mm bore (c); blow a bulb (d) and cut as shown. Cut the other end (a) at 15 to 20 mm from the taper. Obtain an aluminum nut and a screw 2 to 3 mm in diameter and 50 to 70 mm long, the latter being waterproofed by dipping in silicone. Place the nut and screw in position, fill the space a with epoxy resin, and let the resin set for 24 hr to form a screw thread. For use, the bulb d is connected to the capillary by means of a piece of rubber tubing.

Pasternak [69] has undertaken a statistical study of the precision and accuracy of the transfer of liquids using capillaries operated by micrometer or screw, as well as ones calibrated for definite volumes with hairlines. It is reported that 5 μl seems to be the lower limit for routine precise manipulation, the precision being about 2% at a confidence level of 0.95.

2. Syringes and Pipets

The syringe has become a standard tool in the chemical laboratory on account of the popularity of gas chromatography. By now there are instrument manufacturers that specialize in syringes [70–74]; we have come a long way since the analyst had to reshape the needle of a medical syringe for microchemical operations. Discussion of the advantages and limitations of syringes for transferring liquids now seems to be superfluous. Suffice it to say that precision syringes are not inexpensive and require care in use because of their fragility. To meet certain demands, specific types of syringes are also available on the market, such as a syringe for the repetitive dispensing of preset quantities of liquids [72]; for the instantaneous delivery of 2-, 3-, or 4-μl amounts [73]; or with interchangeable plungers of different gauge sizes [74]. Silverman and Gordon [74a] have described a device (Fig. 9.23b) to deliver liquid sample from a syringe without discharging any vapor above the liquid.

Lowenthal and Page [75] have employed a polypropene syringe supported in a metal strip to deliver liquid drops, which can have volumes from 1 μl upwards, and the weight of liquid transferred can be accurately determined on a microbalance. Bacon [76] has constructed a micrometer

(a)

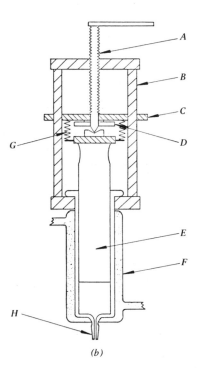

(b)

Fig. 9.23. (*a*) Simple attachment to capillary for transfer of liquids [68]. (Courtesy *Mikrochim. Acta.*) (*b*) Screw-controlled shrinking bottle for delivering liquid sample from syringe [74a]: *A*, screw; *B*, brass frame; *C*, guide for screw; *D*, indicator wheel, 100 marks per rotation (indicator pointer not shown); *E*, 50-ml glass syringe; *F*, glass water jacket; *G*, retaining springs; *H*, Luer tip. (Courtesy *Anal. Chem.*)

syringe to measure accurately the volume of liquid transferred to the thin-layer chromatographic plate. Dirlick and Byck [77] have described a pressure-tight connection between the syringe needle and plastic tubing by screwing onto the end of the tubing of helical spring, the length of which should be at least twice the diameter of the tubing. Bekemeier [77a] has described a sampling method in the microliter region by the use of disposable capillary tubes and a micrometer-syringe.

Gordon and Malacinski [78] have constructed a large-capacity stainless-steel syringe control mounted on a stand to facilitate multiple injections. For the transfer of liquids into a capillary-column (70×0.6 mm) gas chromatograph, Cramers and van Kassel [79] have recommended that one dilutes the sample with a nonvolatile solvent such as silicone oil, injects the whole sample in the syringe, and cools to about $-50°C$ a 10-mm length of the column near to its connection to the precolumn by blowing directly on to the column a jet of carbon dioxide for a period of 10 sec before and after injection. Thus the maximum precision of the microliter syringe is adapted to the minute sample size that can be handled in the capillary column. Crabtree [80] has described a technique for directly transferring the fractions from the gas chromatograph onto a thin-layer chromatographic plate by means of a fine-bore microsyringe. Yayanos [81] has described a method of transferring liquids from a reaction mixture in a pressurized chamber without subjecting the mixture to pressure change, by using a syringe mounted inside the pressurized chamber as the reaction vessel. Cegla [82] has described a pipet for measuring a definite amount of liquid to be mixed with a predetermined volume of diluent before dispensing the mixture. Gerades [82a] has utilized the Unopette (see Chapter 2, Fig. 2.41) for the same purpose.

3. Capsules

Gelatin capsules (commercial) are convenient tools for transferring liquids for elemental analysis using the closed-flask combustion method [44] and for organic functional-group analysis [25], provided that the sample does not exhibit significant vapor pressure. For handling liquid samples in the automatic C-H-N analyzer, Culmo and Fyans [83] have used sealed aluminum capsules, while Ellison [84] has described a procedure to transfer the liquid into a tared aluminum capillary tube, which is then sealed by crimping. Another device, which has been patented [85], comprises a sealable sample holder made of low-melting glass that will shatter under internal pressure and is enclosed in a gas-permeable wire screen to restrain the scattering of glass particles. Anderson and Hoover [85a] have described a technique for containing low-boiling liquid in a sealed platinum capsule.

Metal foils suitably shaped have served as vessels for holding liquids for various purposes. For instance, Delves [86] has made crucibles from nickel foil to transfer 10-μl samples for the rapid determination of lead.

4. Microcells and Devices for Specific Analytical Operations

Since spectroscopic and electrical measurements are inherently sensitive methods, they have become favorites in microanalysis. The main task is to design a suitable vessel to transport and/or hold the small sample for such measurements. Numerous devices have appeared in the literature. Following are some examples.

In investigating the absorption spectra of solutions containing radioactive species, the amount of material available is frequently very limited. Furthermore, hazards are minimized by examining as small a sample as practicable; in all cases, the sample must be quantitatively transferred and recovered after the analysis. Carnall and Fields [87] have observed that commercial spectrophotometer cells made by sealing a window to a cell body tend to leak. These investigators have therefore fabricated fused-silica cells from capillary tubing of about 1-mm bore and length of 20 to 50 mm. The volume of liquid required for a 1×50 mm cell is 70 μl and is suitable for spectral examination in the ultraviolet region down to 200 nm.

Kadoum and Anderson [88] have constructed a dual-chamber micro-cuvette by cutting slits in a methacrylate block and gluing on quartz windows with epoxy glue. It holds 200 μl of aqueous solutions for visible and u.v. spectral analysis. A commercial microcell [89] comprises a cell body and a removable insert, both being opaque over the range from 175 nm to 3 μm. When the insert is placed in the body, a cavity of small cross section and comparatively great length remains to contain the sample. Brown and co-workers [89a] have described a technique for collecting oil spills on aluminum foil for spectral analysis.

Rissmann [90] has described a thin liquid infrared cell for handling aqueous solutions. It consists of two zinc sulfide plates. On the outer portions of the bottom plate is a 3-μm thick silver film deposited by slow vacuum evaporation, which serves as a cell spacer. Nickey and Rice [23] and Bonfiglio [91] have suggested improved microcells for X-ray fluorescence measurements. Raman spectroscopy using a laser source can be carried out using microliter amounts of liquid in a capillary tube [91a]. A sealed glass tube of 10-mm diameter and 3-ml capacity has been used by Beattie and Ozin [92] to study Raman spectra of vapors at high temperatures up to 1000°C. Working in the opposite temperature range, Wong, Bertie, and Whalley have constructed a multiple-sample low-temperature optical cell that can be maintained at −73°C by using solid carbon dioxide as coolant and at −173°C by using liquid nitrogen [92a]. Another low-temperature cell has been described by Brown and co-workers [92b].

Other devices applicable to microscale manipulations may be mentioned. Evans and co-workers [93] have fabricated a microcalorimeter especially suited for holding 1 to 2 ml of liquid mixture liberating up to 400 mcal. Brethschneider and Rogers [94] have conducted calorimetry with 2 to 3mmol of sample in a simple glass tube immersed in water inside a Dewar flask. Tyagi [95] has designed a general-purpose cryostat that enables the specimen temperature to be maintained indefinitely anywhere btween 77 and 600°K. Hello [96] has described a capillary direct-current cell suitable for conductometric titrations. Lugg [97] has constructed a diffusion cell for the production of a constant vapor concentration. The apparatus consists of a water-jacketed precision-bore capillary tube, of which the bottom is connected to a glass syringe reservoir and the top to the carrier-gas stream. Wallach [98] has described a simple system for the rapid formation of small-volume density gradients. Savitsky and Siggia [98a] have described a diffusion cell for introducing known quantities of liquid into gases. Davis [99] has proposed a constant-head device to be attached to the Mariot flask. Croll [100] has advocated the use of metal clips to secure glass stopcocks through which water is transported, in order to prevent contamination of the water by organic polymers or lubricating grease.

B. TECHNIQUE FOR THE TRANSFER OF HYGROSCOPIC AND AIR-SENSITIVE LIQUIDS

When the double-chamber capillary is used to transfer liquids (see Fig. 9.22a), as the fine tip reaches below the liquid surface, the sample that enters the capillary is not contaminated. After sealing the capillary, there is no danger of the sample being exposed to the atmosphere. Maguire [101] has proposed to use a single-bulb capillary made of phosphate glass [102]. Since phosphate glass has low melting point and is soluble in acid, the capillary can be sealed in an electric heating coil inside a "dry box" (Section III.B) and will open when immersed in strong acid medium such as the Kjeldahl digestion fluid.

For working with air-sensitive organosilicon compounds, Jenik and Churacek [103] have stored the liquid sample in 10-ml ampuls under nitrogen and transferred the required amount for analysis with a syringe into capsules (15×1 mm) made from polyethylene tubing. Schuele and McNabb [104] have transferred hygroscopic liquid by a bubble of inert gas into a 1-ml syringe, which is then used as a weight buret. Jotham and Vermeulen [105] have described a device comprising a syringe in a housing with a ground-glass joint through which nitrogen gas flows (Fig. 9.24); it is capable of transferring microliter amounts of liquid into a gas chromatograph entirely under an atmosphere of nitrogen. Bevington and co-workers [105a] have described an apparatus for sampling fixed volumes from reaction mixtures under an inert atmosphere.

The transfer of hygroscopic and air-sensitive solutions for titrimetry can be carried out by various devices [25, 106, 107]. A simple apparatus

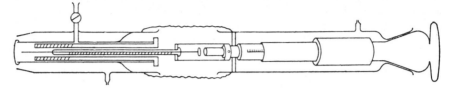

Fig. 9.24. Operation of syringe in nitrogen atmosphere [105]. (Courtesy *Chem. Ind.*)

described by Cleghorn and Lalor [108] is shown in Fig. 9.25. The three-necked flask *A* serves as the titration vessel and contains a magnetic stirrer *F*. The reaction mixture can be raised from *A* into the pH chamber *B* by turning stopcock T_3. Nitrogen continuously enters through T_4 and leaves through T_5. This arrangement is used for pH measurements in the study of the complexation of transition-metal ions such as Co(II).

C. TRANSFER AND EVAPORATION OF SOLUTIONS

The transfer and evaporation of solutions that contain solid solutes have been discussed in Section III.C.2. The procedures described there are applicable to liquids in general. Litt and Adler [109] have designed a manifold that is capable of controlled one-step evaporation of solutions directly into septum-sealed vials for the sampling of micro amounts of solution for gas chromatography or infrared spectrometry. For obtaining infrared spectra of 1 μl of solution, Chen and Gould [110] have described a technique that involves punching an indentation in the matrix (AgCl) in which the sample is evaporated. Brannon [111] has suggested placing the solution in an indentation dug out of a KBr disc where evaporation takes place.

D. TRANSFER OF LIQUID INTO SMALL VOLUMETRIC FLASKS

Since small volumetric flasks have narrow necks, transferring liquids in and out of the flask requires care in order to prevent spilling. Pouring is not advisable. A long delivery tube or disposable pipet is used, so that the liquid can be transferred to near the bottom of the flask. If the liquid has sediment or suspended particles, a simple device illustrated in Fig. 9.26 will transfer the clear sample into the vessel. The apparatus consists of a pipet connected by rubber tubing to a short glass tube with a constriction *C*. A tiny piece of filter paper folded into a roll *P* is inserted as shown. The liquid to be transferred is brought up by suction to the pipet, while

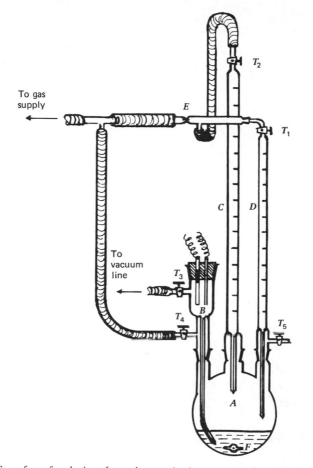

Fig. 9.25. Transfer of solution from burets in inert atmosphere [108]. (Courtesy *Lab. Pract.*)

solid particles are stopped by the filter paper roll. After detaching the rubber connection and glass tube, the clear liquid is transferred from the pipet into the flask.

Kolb [112] has described a procedure for filtering solutions into a volumetric flask by placing the solution inside a sintered-glass funnel. The funnel is fitted with a rubber stopper through which the needle of a syringe is inserted. When the syringe plunger is depressed slowly, the filtrate will pass into the flask. The filtration assembly shown in Chapter 2, Section VI.A (Fig. 2.37) also can be employed, the volumetric flask being placed inside a bell jar or a filtering flask without a bottom. Nemec [113]

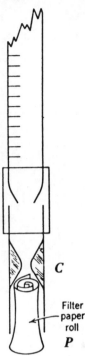

Fig. 9.26. Adapter for removal of suspended particles during transfer of liquids.

has described a filtration adapter suitable for filtering small volumes in an inert atmosphere.

E. DEVICES FOR THE TRANSFER OF A CONSTANT VOLUME OF LIQUID

The versatile micropipet designed by Grunbaum [114, 115] for the transfer of a constant volume of liquid is shown in Fig. 9.27. It is self-adjusting and self-cleaning; it is also a precision pipet, because the same pipet is used for aliquoting blanks, standards, and unknowns [116]. The useful range of this design is between 1 μl and 1 ml. These pipets are commercially available [117]. Another commercial dispenser [118] that delivers a constant volume between 1 and 10 ml can be attached to a motorized drive. O'Hara and May [119] have described a turntable and circuit for measuring equal volumes of a solution into a series of test tubes. Takatsy [120] has constructed an apparatus that, on immersion into a liquid, entrains a known volume of it (0.25 or 0.5 ml) between metal bars and can be transferred to another site with an accuracy of ±1%. Thomas and Owens [121] have described a microburet attached

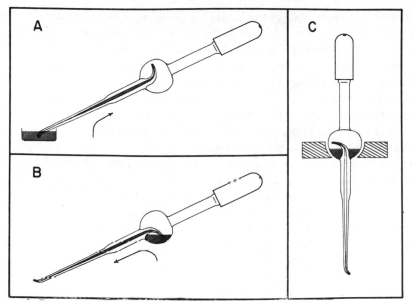

Fig. 9.27. Functional modes of the Grunbaum pipet: (*A*) sampling; (*B*) dispensing; (*C*) reagent storage [116]. (Courtesy *Microchem. J.*)

to a motor-driven gear, which delivers discrete droplets at a uniform rate. Arnold [122] has constructed a microapplicator to deliver drops of five sizes between 0.25 and 5 μl. Varieties of despenser for repetitive applications are commercially available, particularly for use in biochemical experiments.

F. AUTOMATION IN LIQUID TRANSFER

Since automation is the preferred method for the mass production of routine chemical analysis and since automated procedures in general involve handling solutions, a voluminous literature and a large variety of devices related to this subject have appeared. Babjuk and Vecerek [123] published in 1967 a review on the automation of liquid dispensing, which cited 175 references and listed 83 producers. We briefly discuss here the more recent devices and developments.

In general, automated methods of analysis of liquid samples use one of two techniques for measuring the volume of the specimen to be analyzed. One method employs a peristaltic pump to aspirate the liquid [124]; another aspirates the specimen in an autodilutor [125]. In the latter procedure, a plunger retreats to transport the specimen, and a second plunger advances to eject the sample with diluting fluid or reagent. Since neither of these two methods is practical when only small samples are available, Natelson [126] has fabricated an instrument with a micro sample dispenser for liquids contained in capillaries. This apparatus is designed for handling, dilution, and dispensing 25-μl samples of biological fluids in particular. The diluted sample may then be fed directly to an AutoAnalyzer, a flame photometer, or an atomic-absorption spectrometer. Alternatively, the liquid may be split into subsamples for the simultaneous determination of different components. A further application of the dispenser is to the tape system of analysis: the capillary tube containing the liquid sample is brought into contact with an absorbent strip, which is then pressed against, and heated while in contact with, a tape soaked with the appropriate reagent, and the spots produced are measured densitometrically. Up to 100 samples per hour can be processed after dilution, and 300 per hour by the tape. Fig. 9.28 depicts the sampling-cup assembly Fig. 9.29 illustrates the splitting of the liquid with the micro sample dispenser for multicomponent analysis.

Slanina, Frintrop, and Griepink [127] have studied the operation of a tube-insert pump as an automatic microburet and reported standard deviation of 2% for the transfer of 20 μl, 0.6% for 300 μl, 0.3% for 1 ml, and 0.1% for 13 ml. Eckfeldt and Shaffer [127a] have described a semi-automatic precision microburet for volumes up to 1 ml. Various devices for the automatic transfer of liquids have appeared in the patent literature [128–134], and more are expected. In one such pipetting device [128], a relatively large volume (1.5 to 25 ml) of one liquid can be transported to be mixed with a small volume (0.01 to 0.4 ml) of another liquid. Hunt and Cleaver [135] have described a simply constructed apparatus for the aspiration of solutions into a flame photometer. Up to 24 sample tubes are aligned in turn with the transfer probe. This apparatus has been modified to simplify the construction further [136]. Tetlow, Hall, and Fleming [137] have designed a sampling programmer that controls the transfer from a liquid stream or either of two standard solutions. Wilkie, Smith, and Keay [138] have fabricated an instrument capable of dispensing at constant solid:liquid ratio. The assembly consists of a balance that gives an electrical output proportional to the load of the pan, a motor-driven dispensing syringe, and a servo-drive unit, activated by the output from the balance, that controls the syringe that delivers the liquid according to a predetermined ratio. Many improvements of the procedure for transferring liquids in an amino acid analyzer have been reported [139–141]. For instance, Keay and co-workers [139] have shown that movement of the sampling tube through the self-sealing base of a wash vessel into the sample tube beneath by means of a double-acting

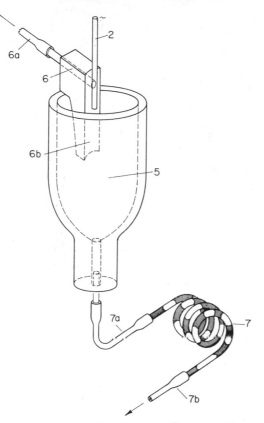

Fig. 9.28. Sampling-cup assembly [126]: 2, capillary; 5, sampling cup, 20 ml; 6, Lucite block drilled and shaped to allow specimen and diluting fluid to drain down through; 6a, lead from peristaltic pump to deliver diluting fluid; 6b, trough extended away from side of cup; 7, mixing coil; 7a, lead from cup to mixing coil; and 7b, lead back to peristaltic pump for processing. (Courtesy *Microchem. J.*)

pneumatic cylinder is much simpler than the use of a solenoid-operated device. Burtis and co-workers have constructed a device for the automated loading of microliter volumes into a miniature fast analyzer [141a].

There are varieties of commercial apparatus for the automatic transfer of samples in paper, thin-layer, gas, and liquid chromatography. Some are easily fabricated. Bailey, Pieters, and Beck [142] have described a simple arrangement for the reproducible semi-automatic transfer of liquids to paper chromatograms. It consists of a microliter pipet clamped in the vertical position with the pipet tip firmly against the surface of the chromatographic paper, which is supported on a perforated metal disc. The pipet is filled and positioned while keeping a finger over the upper orifice.

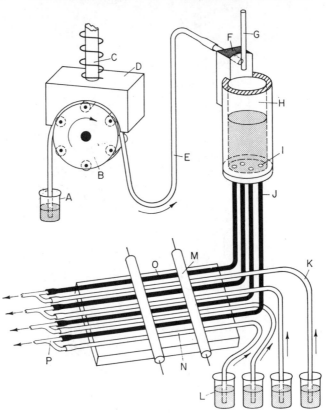

Fig. 9.29. Splitting 25-μl sample with microsample dispenser for multiple component analysis [126]: *A*, diluent; *B*, peristaltic rotor; *C*, spring to hold plate *D* on rotor *B*; *E*, tube leading to cup; *F*, Lucite black-drilled and shaped to cause diluent to flow along wall of cup; *G*, capillary; *H*, diluting cup; *I*, multiple outlets leading to tubes *J* for dividing into aliquots; *K*, tubing leading from different reagents to be added; *L*, reagent containers; *M*, rollers for pumping reagents and aliquots; *N*, tubes containing reagent; *O*, tubes containing aliquots; *P*, connectors to lead mixtures of aliquots and reagents for further processing. (Courtesy Microchem. J.)

Gradual transfer of liquid to paper begins when the finger is released. Stockwell and Sawyer [143] have designed an apparatus for gas chromatography in which the mixing of the liquid with the internal standard and their injection into the gas chromatograph is fully automated. Jentzsch [144] has described a device for repeated injections of liquid onto the column for the purpose of preparative chromatography. Thomson and Eveleigh [145] have constructed an apparatus that transfers several samples, at preselected intervals, onto a single column. Hrdina [146] has patented an unpressurized dosing device in which filled sample containers are successively connected to the suction side of a pump that aspirates the eluent from reservoirs and delivers it and the sample into the closed column. Peterson and co-workers [147] have compared three types of sample introduction valves for liquid chromatography. Two of the valves were of the rotary sliding-chamber type (short and long), while the third was a commercial sample-loop type. The performance of all of the valves was found to be equivalent. For handling radioactive samples, Fortsch and Wade [147a] have described a simple air-controlled device to operate a semi-automatic remote pipetting system.

V. TRANSFER OF GAS SAMPLES

A. GENERAL TOOLS FOR GAS TRANSFER

The literature on gas transfer is sparse when compared with that on the transport of liquids and solids. Standard methods for gas analysis usually specify the manner in which the gas sample should be conducted into the analytical instrument [148]. In general, the gas volume is measured by means of a flowmeter [149, 149a], gas buret [150], or gas cell for spectroscopy [151] when the quantity to be transferred is over 1 ml. The capillary tube [68] and ring chamber [152] are employed for working in the microliter region. Golbacs and Csanyl [153] have described an automated gas-measuring device.

Niedermeyer [154] has described a multipurpose gas-handling apparatus, shown in Fig. 9.30. It consists of a three-necked flask with ground-glass joints. One inner joint terminates with minimum dead volume in a socket holding a silicone seal. The second joint is fitted with a stopcock fabricated to minimize the internal holdup volume. The central inner joint is multifunctional: At its lower end protruding into the flask, it is flared outward slightly so that a rubber balloon may be drawn over it and secured, if necessary, with a rubber band. At the upper end, a stopcock and another fitting holding a silicone seal are attached. The apparatus can be used to transfer gas samples from sources of slightly lowered, atmospheric, and elevated pressures. For transfer from high-pressure regions, one need only flush out the apparatus thoroughly and then seal it off. For transfer from regions of atmospheric or below atmospheric pressure, the balloon is fully inflated to fill the apparatus; then, with the inlet connected to the source to be sampled, the balloon is evacuated to create a vacuum in the apparatus, and the gas is drawn in. Small samples (e.g., for gas-chromatographic analysis) can be taken from the apparatus with a syringe through the side seal without developing significant vacuum. Through inflation of the balloon, more or all of the gas in the apparatus

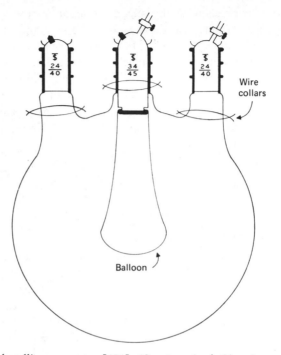

Fig. 9.30. Gas-handling apparatus [154]. (Courtesy *Anal. Chem.*)

can be expelled. Bennett [154a] has described an apparatus for preparing and dispensing gas mixtures.

The transfer of aerosols has received special attention, since generally accepted sampling procedures cannot be used, because the aerosol product contains propellants that are extremely volatile. Tuesley, Sciarra, and Monte-Bori [155] have fabricated a fluorocarbon-plastic apparatus fitted with specially designed valves that allow for transfer of the contents without loss of volatile propellant or active ingredients. Wood and Swann [156] have described a simple sample-introduction device, for use in high vacuum, that incorporates greaseless stopcocks. Hyde and Redford [157] have designed a glass trap, with resistance heating to maintain the temperature between −196 and 0°C, for the controlled release of condensable gases.

B. MANIPULATION OF CLOSED CONTAINERS

Seitz and Emerson [158] have presented a simple method for the accurate determination of the volume of closed containers. It involves measuring the pressure at constant temperature when, in turn, the air (initially at atmospheric pressure) from a container of known volume and from one of unknown volume is allowed to expand into a third container. The volume of closed containers with capacity about 10 ml can be determined to within ±0.2%.

O'Keeffe and Ortman [159] have described a microbottle for dispensing gases in extremely small amounts. The construction and filling of this device are depicted in Fig. 9.31; it consists essentially of a glass envelope and a fluorocarbon insert. The lowest emission rate achieved with propene at room temperature is 4 pg/sec (1 pg = 10^{-12} g). Schachter [160] has designed an apparatus for opening small sealed objects in vacuum. The object is held against a support inside an evacuated vessel by means of a steel point, which operates through a stainless-steel bellows. The sealed container is then punctured by turning a threaded rod, which applies pressure to the steel point.

The head-space gases in hermetically sealed containers are manipulated

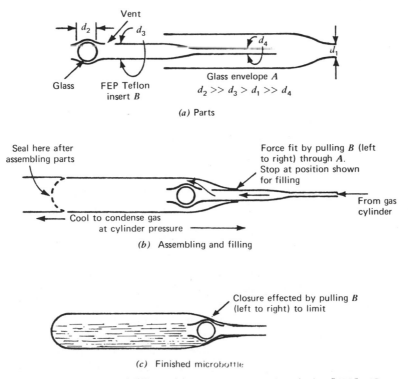

(a) Parts

(b) Assembling and filling

(c) Finished microbottle

Fig. 9.31. Construction and filling of low-range permeation device [159]. (Courtesy *Anal. Chem.*)

in various ways, depending on the nature of the gas and the purpose of the analysis. Hoffmann [161] has constructed an apparatus that includes a gas buret with a total volume of 2 ml for analyzing head-space gases in food packages. Cullen and Papariello [162] have described a procedure for the transfer and analysis of gaseous oxygen in the void-volume head space of ampuls and vials. It involves a specially designed sampling cell and flow system that permit the collection of the head-space gas without atmospheric contamination. Bassette and Ward [163] have evaluated the techniques for direct head-space gas sampling in the qualitative and quantitative analysis of volatile materials in biological fluids; these techniques are suitable for transferring the vapors into gas chromatographs. Zvonow, Stewart, and Starkman [164] have fabricated a sampling valve for transferring gases at high temperature (2700°K) and high pressure (60 atm) from an internal-combustion engine to analytical instruments for the assay of air pollutants.

C. TRANSFER OF ALIQUOTS

A gas-transfer syringe proposed by Rolant [165] is recommended for the storage and delivery of aliquots of milliliter amounts of gaseous mixtures. As illustrated in Fig. 9.32, the apparatus consists essentially of a glass syringe and a short pipet with a two-way stopcock. The handle of the syringe piston is cut off, and about 10 ml (for a 30-ml syringe) of clean, dry mercury is introduced inside the piston A. Molten white paraffin is poured on top of the mercury to form a layer about 10 mm high, which upon solidification holds the mercury in place. The opening of the piston is closed with the stopper C. The nozzle of the syringe is connected, through the Tygon tubing D, to the micropipet E which is constructed as follows. A thinwalled glass tube 50 mm long, with inner diameter not more than 0.5 mm, is soldered onto a vacuum glass two-way stopcock F. A rubber cap with a hole G is pushed onto the tip of the pipet, in order to ensure tight contact with the sample-introduction inlet of a gas-analysis instrument such as the Scholander apparatus [166]. The piston is slightly lubricated with vaseline, and the stopcock is greased with vacuum grease. For sample transfer the syringe is connected by means of its pipet to the gas source and filled by drawing the piston out. The system is flushed several times with gas, using the side outlet of the two-way stopcock. After the final filling of the syringe, the stopcock is closed, and the gas sample may be stored in the syringe for some time. In order to take aliquots for analysis, the gas syringe is held upright. Thus the gas sample is under the pressure of the heavy piston. Stainton [166a] has described

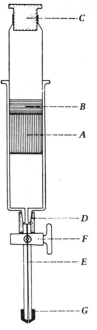

Fig. 9.32. Schematic drawing of the gas-transfer syringe [165]. (Courtesy *Microchem. J.*)

a procedure for the transfer of nanogram quantities of mercury vapor for flameless atomic-absorption spectroscopy.

Weber and Spanier [167] have described an automatic gas-sampling device for the transfer of aliquots from a gas stream at fixed intervals. Allen [168] has designed a quantitative gas-chromatographic sampler for static gaseous-reaction systems, which allows the successive extraction of small aliquots from a reaction zone and their transfer into the chromatograph. Forys and Gawlowski [169] have constructed a sealed device for introducing gaseous samples into a chromatograph. It comprises a manifold tap for operation at reduced or atmospheric pressure, with an additional chamber that is used to evacuate a small chamber at the bottom of the tap seating, thus holding the tap in the seating. For permanent gases and polar vapors a nonpolar grease is used; for non-polar substances such as hydrocarbons a suitable grease is dextran dissolved in diglycerol. Celegin and co-workers [169a] have described a submicroliter sampling device for measuring any gas having low solubility in the sealing fluid.

Magrin [170] has constructed a linear flowmeter for measuring small gas flows. It is applicable to all gases and gaseous mixtures with rates of flow from 1 liter/hr to 1 meter^3/hr. Tindle [171] has described a low-cost variable-proportion gas-metering system for production of mixtures of pure gases. Dual-stage pressure regulators on the gas cylinders are used to minimized pressure fluctuations. Each cylinder is connected to a coil to smooth out the pressure surges, and the temperature of the gas is adjusted to 0°C by passage through a coil immersed in an ice bath. It is then passed through glass capillary tubing to regulate its pressure. The gases are mixed after passage through the capillary tube. Flow rates can be varied from 10 to 100 ml/min with reasonable precision.

REFERENCES

1. T. S. Ma and R. F. Sweeney, *Mikrochim. Acta,* **1956**, 195.
2. Sample Storage Sets, A. Scientific Apparatus Corp., Cleveland, Ohio; Metro Scientific, Inc., Farmingdale, N. Y.
2a. R. V. Coyne ad J. A. Collins, *Anal. Chem.,* **44**, 1093 (1972).
3. N. L. Cross, *J. Sci. Instrum.,* Ser. 2, **1**, 65 (1968).
4. J. T. Lloyd, British Patent 1,121,853 (1964).
5. R. H. A. Crawley, *Chem. Ind.,* **1953**, 1205.
6. A. Jakab, *Anal. Chem.,* **43**, 489 (1971).
7. A. Fono, A. M. Sapse, and T. S. Ma, *Mikrochim. Acta,* **1965**, 1098.
8. C. S. Knight, *Nature,* **199**, 1288 (1963).
9. Arthur H. Thomas Co., Philadelphia, Pa.
10. L. Cahn and W. J. Cadman, *Anal. Chem.,* **29**, 49A (1957).
10a. H. Reinhardt, Basel, Switz.
11. E. J. Barrett, H. W. Hoyer, and A. V. Santoro, *Mikrochim. Acta,* **1970**, 1121.
12. R. C. Denney, *Chem. Ind.,* **1968**, 684.
13. W. J.de Klein and K. Ulbert, *Anal. Chem.,* **41**, 682 (1969).
14. W. L. Brannon, *J. Assoc. Off. Anal. Chem.,* **53**, 599 (1970).
15. A. C. Gilby, J. Cassels, and P. A. Wilks, Jr., *Appl. Spectrosc.,* **24**, 539 (1970).
16. J. L. Jones, Soc. Appl. Spectrosc., New York Meeting, October, 1971.
17. D. R. A. Wharton and M. L. Bazinet, *Anal. Chem.,* **43**, 623 (1971).
18. D. M. Ellen, *J. Forensic Sci. Soc.,* **5**, 196 (1965).
19. Research and Industrial Instrument Co., London, England.
20. *Metallurgia,* **76**, 226 (1967).
20a. M. Elio, *Analusis,* **1**, 374 (1972).
21. P. E. Walters and T. Monaci, *Appl. Spectrosc.,* **25**, 678 (1971).
22. G. H. Morrison and Y. Talmi, *Anal. Chem.,* **42**, 809 (1970).
22a. A. J. Christopher, *Microchem. J.,* **17**, 470 (1972).
23. D. A. Nickey and J. O. Rice, *Appl. Spectrosc.,* **25**, 383 (1971).
23a. S. Patai, Y. Halpern, L. Esterman, and M. Weinstein, *Isr. J. Chem.,* **6**, 445 (1968); **7**, 691 (1969).
23b. W. C. McCrone, *Rev. Sci. Instrum.,* **42**, 1094 (1971).
23c. C. Inman, *Anal. Chem. Acta,* **60**, 486 (1972).

23d. H. T. Evans, Jr. and R. P. Christian, *Appl. Spectrosc.,* **26,** 313 (1972).

24. T. S. Ma and R. F. Sweeney, *Mikrochim. Acta,* **1956,** 198.

25. N. D. Cheronis and T. S. Ma, *Organic Functional Group Analysis.* Wiley, New York, 1964.

26. J. G. Pack and G. F. Libowitz, *Rev. Sci. Instrum.,* **40,** 414 (1969).

27. M. B. Naff, *Chem.-Anal.,* **50,** 54 (1961).

28. C. L. Gordon and R. B. Johannesen, *J. Res. Natl. Bur. Stand.,* **A, 67,** 269 (1963).

28a. V. Nuti, *Farmaco, Ed. Sci.,* **28,** 418 (1973).

28b. L. Safarik and W. R. Heumann, *Chem. Listy,* **66,** 423 (1972).

29. E. C. Fiebig, E. L. Spencer, and R. N. McCoy, *Anal. Chem.,* **29,** 861 (1957).

30. A. J. Franklin and S. E. Voltz, *Anal. Chem.,* **27,** 865 (1955).

31. W. Brueser, *Chem. Tech. Berl.,* **22,** 556 (1970).

32. C. J. Barton, in H. B. Jonassen and A. Weissberger (Eds.), *Technique of Inorganic Chemistry,* Vol. III, p. 29. Wilcy-Interscience, New York, 1963.

33. I. D. Eubanks and F. J. Abbott, *Anal. Chem.,* **41,** 1708 (1969).

34. P. M. Dryburgh, *J. Sci. Instrum.,* **44,** 658 (1967).

35. T. R. P. Gibb, Jr., *Anal. Chem.,* **29,** 584 (1957).

36. W. J. Kirsten, *Mikrochim. Acta,* **1966,** 105.

37. H. N. Redman, *Analyst,* **92,** 584 (1967).

38. S. G. Gibbins, *Anal. Chem.,* **43,** 295 (1971).

39. S. G. Gibbins, *Anal. Chem.,* **43,** 621 (1971).

39a. E. Farr and W. A. Archibald, *Lab. Pract.,* **21,** 349 (1972).

40. A. M. Long and O. O'Brien, *Chem. Ind.,* **1968,** 1764.

41. W. F. Eggerman, *Appl. Spectrosc.,* **26,** 112 (1972).

42. D. F. Shriver, *The Manipulation of Air-Sensitive Compounds,* p. 141. McGraw-Hill, New York, 1969.

43. M. M. Markowitz and D. A. Baryta, *Chem.-Anal.,* **52,** 21 (1963).

44. T. S. Ma, "Quantitative Microchemical Analysis," in F. J. Welcher (Ed.), *Standard Methods of Chemical Analysis,* 6th ed., Vol. 2, p. 452. van Nostrand, Princeton, 1963.

45. A. Solbakken, *Lab. Pract.,* **18,** 80 (1969).

46. T. S. Ma and G. Zuazaga, *Ind. Eng. Chem., Anal. Ed.,* **14,** 280 (1942).

47. R. C. Rosen, *Appl. Spectrosc.,* **19,** 97 (1965).

48. M. E. Yannone, *J. Gas Chromatogr.,* **6,** 465 (1968).

49. A. Jenkins and R. J. Hunt, *Column,* **2,** 2 (1968).

50. J. Q. Walker, *Hydrocarbon Process,* **46,** 122 (1967).

51. R. J. Levins and R. M. Ikeda, *J. Gas Chromatogr.,* **6,** 331 (1968).

52. E. Stahl, *Analyst,* **94,** 723 (1969).

53. T. S. Ma et al., *Mikrochim. Acta,* **1965,** 1098; **1967,** 960; **1968,** 436; **1969,** 352; **1970,** 677; **1971,** 267, 662; **1972,** 313.

54. J. Polesuk and T. S. Ma, *J. Chromatogr.,* **57,** 315 (1971).

55. C. J. Clemett, *Anal. Chem.,* **43,** 490 (1971).

56. D. D. Rice, *Anal. Chem.,* **39,** 1906 (1967).

57. P. W. H. Schuessler, *Appl. Spectrosc.*, **25**, 501 (1971).
58. A. Szwarc, J. Purzycki, and M. Owoc, *Chem. Anal.*, **10**, 753 (1965).
59. H. W. Hermance and H. V. Wadlow, "Electrography and Electro-spot Testing," in W. G. Berl (Ed.), *Physical Methods in Chemical Analysis*, Vol. 2, p. 156. Academic, New York, 1951.
60. M. D. Adams and S. C. Tong, *Anal. Chem.*, **40**, 1762 (1968).
61. T. S. Ma and K. Eder, *J. Chin. Chem. Soc.*, **15**, 112 (1947).
62. H. Bieling, *Z. Anal. Chem.*, **203**, 345 (1964).
63. V. Knobloch and B. Murdrova, *Mikrochim. Acta*, **1970**, 235.
64. P. Haller and H. Trutnovsky, *Mikrochim. Acta*, **1970**, 588.
65. B. G. D'Aoust, *Anal. Biochem.*, **29**, 533 (1969).
66. T. Mitsui and C. Furuki, *Mikrochim. Acta*, **1960**, 170.
67. K. O. Abraham and V. L. Sasty, *Lab. Pract.*, **19**, 1038 (1970).
68. P. Luis and A. Sá, *Mikrochim. Acta*, **1965**, 627.
69. A. Pasternak, *Chem. Anal.*, **13**, 593 (1968).
70. Hamilton Co., Whittier, Calif.
71. Precision Sampling Corp., Baton Rouge, La.
72. Bolab Inc., Derry, N.H.
73. Kensco, Emeryville, Calif.
74. Wahl-Henius Institute, Inc., Chicago, Ill.
74a. R. A. Silverman and G. Fordon, *Anal. Chem.*, **46**, 178 (1974).
75. G. C. Lowenthal and V. Page, *Anal. Chem.*, 42, 815 (1970).
76. M. F. Bacon, *Chem. Ind.*, **1965**, 1692.
77. P. Dirlick and R. Byck, *Rev. Sci. Instrum.*, **33**, 1472 (1962).
77a. H. Bekemeier, *Zentralbl. Pharm.*, **112**, 367 1973.
78. G. Gordon and G. M. Malacinski, *Microchem. J.*, **15**, 686 (1970).
79. C. A. Cramers and M. M. van Kassel, *J. Gas Chromatogr.*, **6**, 577 (1968).
80. A. N. Crabtree, *Lab. Pract.*, **15**, 311 (1966).
81. A. A. Yayanos, *Rev. Sci. Instrum.*, **40**, 961 (1969).
82. J. H. Cegla, *G.I.T. Fachz. Lab.*, **16**, 1425 (1972).
82a. H. W. Gerarde, *Lab. Manage.*, Mar., 1969; *Microchem. J. Symp. Ser.*, **2**, 1009. (1962).
83. R. Culmo and R. Fyans, *Mikrochim. Acta*, **1968**, 816.
84. M. Ellison, *Analyst*, **93**, 264 (1968).
85. Perkins-Elmer Corp., British Patent 1,180,668 (1967).
85a. B. C. Anderson and C. L. Hoover, *Rev. Sci. Instrum*, **44**, 644 (1973).
86. H. T. Delves, *Analyst*, **95**, (1970).
87. W. T. Carnoll and P. R. Fields, *Appl. Spectrosc.*, **25**, 503 (1971).
88. A. M. Kadoum and C. C. Anderson, *J. Agric. Food Chem.*, **18**, 322 (1970).
89. Beckman Instruments, Inc., British Patent 1,038,109 (1964).
89a. C. W. Brown, P. F. Lynch, and M. Ahmadjian, *Anal. Chem.*, **46**, 183
90. E. F. Rissmann, *Anal. Chem.*, **44**, 644 (1972).
91. S. Bonfiglio, *Appl. Spectrosc.*, **26**, 113 (1972).
91a. B. J. Bulkin, K. Dill, and J. J. Dannenberg, *Anal. Chem.*, **43**, 974 (1971).
92. I. R. Beattie and G. A. Ozin, *Spex*, **14**, (4), 1 (1969).

92a. P. T. T. Wong, J. E. Bertie, and E. Whalley, *Rev. Sci. Instrum.*, **41**, 283 (1970).

92b. C. W. Brown, A. G. Hopkin, and F. P. Daly, *App. Spectrosc.*, **28**, 194 (1970).

93. W. J. Evans, E. J. McCourtney, and W. B. Carney, *Anal. Chem.*, **40**, 262 (1968).

94. E. Bretschneider and D. W. Rogers, *Mikrochim. Acta*, **1970**, 484.

95. R. C. Tyagi, *J. Sci. Instrum.*, Ser. 2, **2**, 995 (1969).

96. O. Hello, *Anal. Chem.*, **44**, 646 (1972).

97. G. A. Lugg, *Anal. Chem.*, **41**, 1911 (1969).

98. D. F. H. Wallach, *Anal. Chem.*, **37**, 138 (1970).

98a. A. C. Savitsky and S. Siggia, *Anal. Chem.*, **44**, 1712 (1972).

99. J. G. Davis, *J. Chromatogr.*, **40**, 169 (1969).

100. B. T. Croll, *Chem. Ind.*, **1970**, 1295.

101. D. A. Maguire, *Chem. Ind.*, **1963**, 1655.

102. W. Zimmermann, *Mikrochemie*, **31**, 15 (1944).

103. J. Jenik and J. Churacek, *Chem. Listy*, **54**, 966 (1960).

104. W. J. Schuele and W. M. McNabb, *Chem.-Anal.*, **46**, 101 (1957).

105. R. W. Jotham and P. J. Vermeulen, *Chem. Ind.*, **1968**, 1764.

105a. J. C. Bevington, B. J. Hunt, and S. Sevcik, *Anal. Chim. Acta*, **52**, 149 (1970).

106. T. S. Ma and J. V. Earley, *Mikrochim. Acta*, **1959**, 129.

107. R. K. Maurmeyer, M. Margosis, and T. S. Ma, *Mikrochim. Acta*, **1959**, 177.

108. H. P. Cleghorn and G. C. Lalor, *Lab. Pract.* **19**, 394 (1970).

109. G. J. Litt and N. Adler, *Anal. Chem.*, **38**, 1096 (1966).

110. J. T. Chen and J. H. Gould, *Appl. Spectrosc.*, **22**, 5 (1968).

111. W. L. Brannon, *Mikrochim. Acta*, **1970**, 327.

112. J. J. Kolb, *Chem.-Anal.*, **50**, 23 (1961).

113. J. Nemec, *Chem. Listy*, **62**, 593 (1968).

114. B. W. Grunbaum, *Microchem. J.*, **9**, 46 (1965).

115. B. W. Grunbaum and P. L. Kirk, *Anal. Chem.*, **27**, 333 (1955).

116. B. W. Grunbaum, *Microchem. J.*, **15**, 680 (1970).

117. Labindustries, Berkeley, Calif.

118. Microchemical Specialties Co., Berkeley, Calif.

119. K. A. O'Hara and C. P. May, *Lab. Pract.*, **18**, 549 (1969).

120. G. Takatsy, British Patent 1,135,847 (1966).

121. W. U. Thomas and E. J. Owens, *Am. Ind. Hyg. Assoc. J.*, **25**, 405 (1964).

122. A. J. Arnold, *Lab. Pract.*, **16**, 56 (1967).

123. J. Babjuk and B. Vecerek, *Chem. Listy*, **61**, 925 (1967).

124. L. T. Skeggs, *Am. J. Clin. Pathol.*, **28**, 316 (1957).

125. D. A. Patient, *Ann. N. Y. Acad. Sci.*, **87**, 830 (1960).

126. S. Natelson, *Microchem. J.*, **13**, 433 (1968); Rohe Scientific Corp., Santa Anata, Calif.

127. J. Slanina, P. C. M. Frintrop, and B. Griepink, *Z. Anal. Chem.*, **246**, 28

127a. E. L. Eckfeldt and E. W. Shaffer, Jr., *Anal. Chem.*, **37**, 1624 (1965).

128. Medincinsk-Kemiska Laboratoriet, British Patent 1,163,281 (1966).
129. Joyce, Loebl and Co., Ltd., British Patent 1,149,383 (1965).
130. Centrala Automationslaboratoriet, British Patent 1,095, 272 (1965).
131. Ceskoslovenska Akademie Ved, British Patent 1,119,198 (1966).
132. Shandon Scientific Industries, Ltd., British Patent 1,119, 197 (1966).
133. Evans Electroselenium Ltd., British Patent 1,169,614 (1967).
134. Warner-Lambert Pharmaceutical Co., British Patent 1,121,412 (1967).
135. J. Hunt and J. L. Cleaver, *Lab. Pract.,* **17,** 471 (1968).
136. J. Hunt, personal communication, 1970.
137. J. A. Tatlow, G. Hall, and A. J. Fleming, *Lab. Pract.,* **17,** 62 (1968).
138. A. Wilkie, P. Smith, and J. Keay, *Lab. Pract.,* **19,** 103 (1970).
139. J. Keay, P. Menage, and A. Wilkie, *Lab. Pract.,* **18,** 59 (1968).
140. C. G. Bird and J. A. Owen, *Clin. Chim. Acta,* **24,** 305 (1969).
141. J. J. T. Gerding and K. A. Peters, *J. Chromatogr.,* **43,** 256 (1969).
141a. C. A. Burtis, W. F. Johnson, and J. B. Overton, *Anal. Chem.,* **46,** 786 (1972); *Clin. Chem.,* **18,** 753 (1972).
142. R. E. Bailey, H. P. Pieters, and J. H. Beck, *Nature,* **204,** 485 (1964).
143. P. B. Stockwell and R. Sawyer, *Anal. Chem.,* **42,** 1136 (1970).
144. D. Jentzsch, *J. Gas Chromatogr.,* **5,** 226 (1967).
145. A. R. Thomson and J. W. Eveleigh, *Anal. Chem.,* 41, 1073 (1969).
146. J. Hrdina, British Patent 1,118,769 (1965); 1,119,513 (1965).
147. J. I. Peterson, F. Wagner, F. Anderson, and G. M. Thomas, Jr., *Anal. Biochem.,* **32,** 128 (1969).
147a. E. M. Fortsch and M. A. Wade, *Anal. Chem.,* **46,** 2065 (1974).
148. ASTM Standards, D1071–55, Methods for Measurement of Gaseous Fuel Samples, 1969. D1145–53, Methods of Sampling Natural Gas, 1969. Am. Soc. Test. Mater., Philadelphia, Pa.
149. Microflowmeter, Gilmont Instruments, Great Neck, N.Y.
149a. A. B. Waugh and P. W. Wilson, *Anal. Chem.,* **44,** 2118 (1972).
150. A. H. Thomas Co. gas burets.
151. Barnes Engineering Co., Stamford, Conn.
152. G. T. Chang and A. A. Benedetti-Pichler, *Microchem. J.,* **4,** 459 (1960). A. A. Benedetti-Pichler, *Identification of Materials,* p. 159. Springer, Wien, 1964.
153. M. Z. Galbacs and L. J. Csanyi, *Anal. Chem.,* **45,** 1784 (1973).
154. A. O. Niedermeyer, *Anal. Chem.,* **42,** 310 (1970).
154a. W. D. Bennett, *Lab. Pract.,* **20,** 583 (1971).
155. S. P. Tuesley, J. J. Sciarra, and A. J. Monte-Bovi, *J. Pharm. Sci.,* **57,** 488 (1968).
156. D. F. Wood and D. A. Swann, *Analyst,* **95,** 828 (1970).
157. A. F. Hyde and R. A. Redford, *J. Sci. Instrum.,* Ser. 2, 1, 871 (1968).
158. C. A. Seitz and D. E. Emerson, *Anal. Chem.,* **40,** 260 (1968).
159. A. E. O'Keeffe and G. C. Ortman, *Anal. Chem.,* **39,** 1047 (1967).
160. G. P. Schachter, *Anal. Chem.,* **31,** 161 (1959).
161. G. Hoffmann, *Dtsch. Lebensm. Rundsch.,* **61,** 177 (1965).
162. L. F. Cullen and G. J. Papariello, *J. Pharm. Sci.,* **59,** 94 (1970).

163. R. Bassette and G. Ward, *Microchem. J.*, **14**, 471 (1969).
164. V. A. Zvonow, H. E. Stewart, and E. S. Starkman, *Rev. Sci. Instrum.*, **39**, 1820 (1968).
165. F. Rolant, *Microchem. J.*, **16**, 315 (1971).
166. P. F. Scholander, *J. Biol. Chem.*, **167**, 235 (1957).
166a. M. P. Stainton, *Anal. Chem.*, **43**, 625 (1971).
167. D. C. Weber and E. J. Spanier, *Anal. Chem.*, **42**, 546 (1970).
168. E. R. Allen, *Anal. Chem.*, **38**, 527 (1966).
169. M. Forys and J. Gawlowski, *Chem. Anal.*, **15**, 1053 (1970).
169a. M. Celegrin, R. Hanson, and G. Sundstroem, *Scand. J. Clin. Lab. Invest*, **27**, 367 (1971).
170. M. Magrin, *Chim. Analyst.*, **52**, 170 (1970).
171. R. C. Tindle, *Anal. Biochem.*, **26**, 477 (1968).

APPENDIX

A

LIST OF ILLUSTRATIONS

426

SUPPLY HOUSES AND MANUFACTURERS OF MICROCHEMICAL APPARATUS

Following is a partial list of instrument manufacturers and laboratory supply houses in the United States, Japan, and Europe. It should be noted that the apparatus specified are not necessarily the only items of a particular company, nor are they available only from that company.

Company Name	Apparatus
Ace Glass, Inc. 1430 Northwest Blvd. Vineland, N.J. 08360	Glassware
Alltech Associates, Inc. 202 Campus Drive Arlington Heights, Ill. 60004	Mini-vials
American Instrument Co. 8030 George Ave. Silver Spring, Md. 20910	General
Analtech, Inc. 75 Blue Hen Drive Newark, Del. 19711	Thin-layer chromatography
Arion 11 rue Marceau 38000 Grenoble, France	Microcalorimeter
Badisches Glasbearbeitungswerk 698 Wertheim/Main West Germany	Glassware
Baird & Tatlock 18 Great Marlborough St. London W1, England	General
Barnes Engineering Co. 30 Commerce Road Stamford, Conn. 06902	Infrared accessories
Bausch and Lomb Bausch St. Rochester, N.Y. 14602	Microscopes

Beckman Instruments, Inc.   Spectrophotometers
2500 Harbor Road
Fullerton, Cal. 92634

F. Bergmann   General
Berliner Str. 25
Berlin, West Germany

Bio-Rad Laboratories   Liquid chromatograph
32nd and Griffin Ave.
Richmond, Cal. 94804

Boleb Inc.   Thin-layer chromatography
Derry, New Hampshire   accessories

F. Bonet   Glassware
Rua José Domingos Barriero 17
Lisbon, Portugal

Brinkmann Instruments   General, thin-layer chromatography
Cantiague Road
Westbury, N.Y. 11590

Bruker Spectrospin   Polarograph
B.P. 56
67160 Wissebourg, France

Büchi Laboratory-Technique Ltd. Drying oven
CH-9230 Flawil
Switzerland

Cahn Instruments   Electrobalance, thermal analyzer
500 Jefferson St.
Paramount, Cal. 90723

Camag, Inc.   Thin-layer chromatography,
2855 S. 163rd St.   electrophoresis apparatus
New Berlin, Wis. 53151

Carle Instruments, Inc.   Gas chromatograph
1141 E. Ash Ave.
Fullerton, Cal. 92631

Carlo Erba   Automated CHNO apparatus
Via Asinari di Bernezzo 70
Torino, Italy

Central Scientific Co.   General
2600 S. Kostner Ave.
Chicago, Ill. 60623

Chatas Glass Co.   Glassware, drying apparatus
570 Broadlawn Terace
Vineland, N.J. 08360

Chemical Data Systems, Inc.   Gas chromatograph accessories
Oxford, Pa. 19363

Chyo Balance Corp.   Microbalances
376-2, Tsukiyama-cho, Kuze
Minami-ku, Kyoto, Japan

Cogébé-Phywe   Centrifuge
276 Longue rue d'Argille
Antwerp, Belgium

Cole-Parmer Instruments Co.   Dialysis apparatus
7425 N. Oak Park Ave.
Chicago, Ill. 60648

Corning Glass Works   Glassware
Houghton Park
Corning, N.Y. 14830

Custom Glassblowing Service   Glassware
35 Henry Ave.
Baltimore, Md. 21236

D. & E. Glass Ltd.   Glassware
72 Henley St.
Sparkbrook, Birmingham, England

Dohrmann Division,   Trace analysisapparatus for S,N,Cl
Environtech Corp.
3240 Scott Blvd.
Santa Clara, Cal. 95050

Drummond Scientific Co.   Micropipets
500 Parkway
Broomall, Pa. 19008

Du Pont Co. Instrument   Liquid chromatograph, thermal
Products Division   analyzer, biomedical instruments
1007 Market St.
Wilmington, Del. 19898

E. M. Laboratories, Inc.   Prepacked liquid chromatography
500 Executive Blvd.   columns
Elmsford, N.Y. 10523

Engelhard Industries Division 700 Blair Road Carteret, N.J. 07008	Platinum apparatus
Eurolabo 35 rue de Meaux Paris 19^e, France	Electrobalance, infrared spectrophotometer
Faglaviks Glasbruk Faglavik, Sweden	Glassware
Fisher & Porter, Lab-Crest Scientific Division 523 Warminster Road Warminster, Pa. 18974	General
Fisher Scientific Co. 585 Alpha St. Pittsburgh, Pa. 15230	General, gas chromatograph, zone refining apparatus
D. Fritz Gmbh Feidstr. 1 D6238 Hofheim am Taunus West Germany	Glassware, microspinning band column
A. Gallemkamp Christopher St. London EC2, England	General
Gelman Instrument Co. 600 S. Wagner Road Ann Arbor, Mich. 48106	Membrane filter, electrophoresis apparatus
C. Gerhardt Bornheimer Str. 100 Boon, West Germany	General
Glasblazerijen der Schelde 45 rue de Termonde Schelle, Belgium	Glassware
Glass-Col Apparatus Co. 711 Hulman St. Terre Haute, Ind. 47802	Heating mantles and tapes
Glassexport Liberec, Czechoslovakia	Glassware
Grant Instruments Barrington Cambridge, England	Instruments

Greiner Scientific Corp. 22 N. Moore St. New York, N.Y. 10013	General, μl-burets
Griffen & George Ealing Road, Alperton Wembly, Middlesex, England	General
P. Haack Garnisongasse 3 Vienna 9, Austria	Glassware
E. Haage Hauskampstr. 58 433 Mülheim, West Germany	General
Hamilton Co. P.O. Box 17500 Reno, Nev. 89510	Microsyringes
Heat Systems-Ultrasonics Inc. 47 East Mall Plainview, N. Y. 11803	Ultrasonic disintegrater, glassware cleaner
W. C. Heraeus 645 Hanau West Germany	Microcombustion apparatus
Hewlett-Packard Route 41 Avondale, Pa. 19311	Gas chromatograph, GC-MS interface, CHN apparatus
Hitachi Co. Instrument Division 2, Sakuragawa-cho, Shiba-Nishikubo Minako-ku, Tokyo, Japan	Gas chromatograph, liquid chromatograph, CHN apparatus, spectrophotometers, mass spectrometer
Hydrex 24 rue de la Buanderie Brussels, Belgium	General
Invitro Twijnstr. 54 Utrecht, Netherlands	Glassware
Isolab, Inc. Drawer 4350 Akron, Ohio 44321	Thin-layer chromatography accessories
Jouan, S. A. 113 boulevard Saint-Germain Paris 6e, France	General

Kahl Scientific Instrument Corp.   Glassware
P.O. Box 1166
El Cajon, Cal. 92022

E. Keller   General
Veltstr. 102
Basel, Switzerland

Kontes Glass Co.   Glassware
Spruce St.
Vineland, N.J. 08360

Kovo   General
Trida Dukelskych Hrdinu 47
Prague 7, Czechoslovakia

La Pine Scientific Co.   General
6001 S. Knox Ave.
Chicago, Ill. 60629

The Lab Apparatus Co.   Thin-layer chromatography
18901 Cranwood Parkway   accessories
Cleveland, Ohio 44128

Lab Glass, Inc.   Glassware
1172 Northwest Blvd.
Vineland, N.J. 08360

Labindustries   Repipet dispenser, tube rocker
1802 Second St.
Berkeley, Cal. 94710

Laboratory Data Control   Liquid chromatograph
P.O. Box 10235
Interstate Industrial Park
Riviera Beach, Fla. 33404

Laboratory Supplies Co.   Chromatography accessories
Hicksville, N.Y.

Leitz   Microscopes
Jena, East Germany

Mashpriborintorg   General
Moscow G 200
U.S.S.R.

Matthey Bishop, Inc.   Platinum apparatus
Malvern, Pa. 19355

Measuring & Scientific Equipment Ltd. Manor Royal Crawley, Sussex, England	Instruments
Medexport Smolenskaja-Sennaja pl. 32/34 Moscow G 200 U.S.S.R.	General
Metro Scientific, Inc. 2121 Brood Hollow Road Framingdale, N.Y. 11735	General
Mettler Pelikanstr. 19 Zurich 1, Switzerland	Microbalances
Microchemical Specialties Co. Berkeley, Cal.	General
Millipore Corp. Ashby Road Bedford, Mass. 01730	Microfilters
Minex Krakowskie Przedmiescie 79 Warsaw, Poland	Glassware
N. V. Instrumentfabrik Postbus 7 Delft, Netherlands	General
Nippon Kogaku Co. Fuji Building, 3-2-3, Marunouchi Chiyoda-ku, Tokyo, Japan	Microscopes
Oertling Cray Valley Works St. Mary Cray, Orpington Kent, England	Balances
Olympus Optical Co. 3-7, Ogawa-cho, Kanda Chiyoda-ku, Tokyo, Japan	Microscopes
Paris-Labo S.A.R.L. 7 rue du Cardinal-Lemoine Paris 5e, France	General

Perkin-Elmer Corp.
702 Main Ave.
Norwalk, Conn. 06856

Spectrophotometers, gas
chromatograph, liquid
chromatograph, CHN apparatus

Pharmacia Fine Chemicals, Inc.
800 Centential Ave.
Piscataway, N.J. 08854

Column chromatography, gel
filtration apparatus

Pierce Chemical Co.
P.O. Box 117
Rockford, Ill. 61106

Thin-layer chromatography
accessories

Prolabo
12 rue Pelée
75011 Paris, France

General

Quartz & Silice
8 rue d'Anjou
Paris 8ᵉ, France

Glassware

Quartz General Corp.
12240 Exline St.
El Monte, Cal. 91732

Quartz apparatus

Quickfit & Quartz
Stone, Staffs,
England

Glassware

C. Rafel Mares
Valencia No. 333
Barcelona, Spain

General

H. Reeve Angel & Co.
9 Bridewell Place
Clifton, N.J. 07014

Paper chromatography

RHO Scientific, Inc.
P.O. Box 295
Commack, N.Y. 11725

Fraction evaporator

Roucaire
20 av. de L'europe
78140 Velizy, France

General

Sargent-Welch Scientific Co.
7300 N. Linder Ave.
Skokie, Ill. 60076

General

F. Sartorius
3401 Rauschenwasser
Göttingen, West Germany

Microbalances

A. Schwinherr 7070 Schwäbisch Gmünd West Germany	General
Scientific Glass & Instruments, Inc. P.O. Box 18353 Houston, Tex. 77023	Glassware
Scientific Manufacturing Industries 1399-64th St. Emeryville, Cal. 94608	Micropipets
SGA Scientific, Inc. 735 Broad St. Bloomfield, N.J. 07003	Glassware
Shimadzu Co. Kawaramachi-Nijo Nakagyo-ku, Kyoto- Japan	Microbalances, gas chromatograph, liquid chromatograph, CHN apparatus, spectrophotometers
H. Steinbuch Mittersteig 26 Vienna 5, Austria	General
W. Stoffer Kapellenstr. 33 Basel, Switzerland	Glassware
Technicon Industrial Systems 511 Benedict Ave. Tarrytown, N.Y. 10591	Automated analysis equipment
Technoimport 5 Doamuel St. Bucharest, Roumania	General
A. H. Thomas Co. Third and Vine St. P.O. Box 779 Philadelphia, Pa. 19105	General, microcombustion apparatus
Touzart & Matignon 3 rue Amyot 75005 Paris, France	Gas chromatograph
Toyo Roshi Co. 3-7, Nihonbashi-honcho Chiyoda-ku, Tokyo, Japan	Thin-layer chromatography
Universal Glass Co. 89 Queen St. Manchester 11, England	Glassware

Varian Instrument Division 611 Hanson Way Palo Alto, Cal. 94303	Spectrometers, gas chromatograph, liquid chromatograph
Varimax Wilzca 50 Warsaw, Poland	General
VEB Jena Glaswerk 69 Jena East Germany	Glassware
Veiligglas Realengracht 20 Amsterdam, Netherlands	Glassware
Vetreria Lusvardi Viale Toscana 13 Milan, Italy	Glassware
Vidrerlas Masip Calle General Mola, Cornella de Llobregat Barcelona, Spain	Glassware
G. Wagner Schulstr. 18 19 Munich, West Germany	General
Waters Associates, Inc. Maple St. Milford, Mass. 01757	Liquid chromatography
C. J. Wennbergs Vikengatan 11 Karlstad, Sweden	General
Wilkens-Anderson Co. 4525 W. Division St. Chicago, Ill. 60651	General
Wilks Scientific Corp. P.O. Box 449 South Norwalk, Conn. 06856	Infrared accessories
Winkelcentrifug Norrtullsgatan 23 Stockholm 16, Sweden	General
Yanagimoto Co. (Yanaco) 28, Joshungamae-cho, Shimotoba Fushimi-ku, Kyoto, Japan	General, gas chromatograph, liquid chromatograph, CHN apparatus

J. Young & Co.   Glassware
Colville Road, Acton
London W3, England

C. Zeiss   Microscopes
Oberkochen, West Germany

Zentralwerkstatt Göttingen Gmbh   General
Bunsenstr. 10
34 Göttingen, West Germany

INDEX

445